Springer Undergraduate Mathematics Series

T0226006

Agustí Reventós Tarrida

Affine Maps,
Euclidean Motions
and Quadrics

Springer

Agustí Reventós Tarrida
Department of Mathematics
Universitat Autònoma de Barcelona
Cerdanyola del Vallès
08193 Bellaterra, Barcelona, Spain
agusti@mat.uab.cat

ISSN 1615-2085
ISBN 978-0-85729-709-9   e-ISBN 978-0-85729-710-5
DOI 10.1007/978-0-85729-710-5
Springer London Dordrecht Heidelberg New York

British Library Cataloguing in Publication Data
A catalogue record for this book is available from the British Library

Library of Congress Control Number: 2011930403

Mathematics Subject Classification: 11E10, 11E39, 15A21, 15A63, 14R99, 51M05, 51N10, 51N20, 97G70

Cover design: VTeX UAB, Lithuania

Printed on acid-free paper

Springer is part of Springer Science+Business Media (www.springer.com)

Preface

About three hundred years before our era, the great Greek geometer Euclid[1] attempted to compile and logically order the many results of geometry known by that time.

The Greeks were well acquainted with the notion of proof; logical arguments were used to obtain new results from old. However, these old theorems had to be derived from other known theorems, which in turn were to be deduced from more primitive results, and so on. Hence, in order to give a logically coherent exposition of geometry, avoiding an infinite regress, it became necessary to posit a number of *first theorems*, that is, to decide at what point one must begin the chain of reasoning that allows new theorems to be proved from those that have already been established. Results were sought that were considered to be so self-evident that it would not be necessary to prove them.

Euclid began his great masterpiece, *The Elements*, with a list of five postulates that play the role of *first theorems* and which are considered to be so obvious as to be accepted without proof.

Explicitly, these postulates are:

1. A straight line can be drawn from any point to any point.
2. A finite straight line can be produced continuously in a straight line.
3. A circle can be described with any center and distance.
4. All right angles are equal to one another.
5. If a straight line falling on two straight lines makes the interior angles on the same side less than two right angles, the two straight lines, if produced

---

[1] Very little is known about Euclid. We cannot even be certain of his existence. *The Elements* is a monumental achievement, perhaps more than one man could conceivably have produced, so it is possible that Euclid simply directed the work.

indefinitely, meet on that side on which the angles are less than the two right angles.

Prior to this, in order to clarify what was being discussed, Euclid gave twenty-three definitions (a *point* is that which has no parts; a *line* is a breadth-less length, etc.) together with five logical rules or common notions (things which are equal to the same thing are also equal to one another; if equals be added to equals, the wholes are equals, etc.). See, for instance, [13] or [26].

From this, with a level of rigor that was considered exemplary until the 19th century, Euclid recovered all the theorems of elementary geometry. Some more involved topics, such as conic sections, for example, although known in Euclid's time, do not appear in *The Elements*.[2]

In the paper [12], A. Dou makes a remark concerning *The Elements* that is worth quoting here:

> The geometry of *The Elements* is a geometry that today should be called physical geometry, because for Euclid and Aristotle the terms of the propositions in *The Elements* refer precisely to the real natural bodies of the physical world [...] *It is a geometry that tries to study the structure of physical space.*

Euclid's points and straight lines are not physical points and straight lines, but an abstraction of them. When one tries to give a rigorous foundation of geometry, some fundamental questions arise, such as: What is a point? What is a straight line? The definitions given by Euclid are not satisfactory, they must be considered only as descriptions, and it is clear that the definition of a point (that which has no parts) and straight line (a length without width) in *The Elements* do not provide much in the way of rigor.

For more than two thousand years after Euclid many mathematicians tried to prove the fifth postulate. Unable to find a direct proof, they instead tried an indirect approach, hoping to arrive at a contradiction by assuming its negation. Curiously, the assumption that the fifth postulate is false does not lead to a contradiction, but rather opens the door to new and marvelous geometries.

The first to develop this approach, and realize that it does not lead to contradiction, were N. Lobatchevski and J. Bolyai, who independently[3] discovered

---

[2] *The Elements* is composed of thirteen books; books I, III, IV, XI, XII and XIII are concerned with geometry, while books VII, VIII and IX cover arithmetic. The remaining four books are dedicated to algebra. The results of the first four books, the seventh and the ninth, are principally due to the Pythagorean school, the fifth and the sixth are due to Eudoxus (4th century BC), and the tenth and thirteenth are due to Teatetus (368 BC). *The Elements* contain a total of 131 definitions and 465 propositions.

[3] Lobatchevski's first publications on non-Euclidean Geometry appear in the *Kazan Messenger* (1829–1830). The work of Bolyai, known as the *Appendix*, because it

Hyperbolic Geometry (a totally coherent geometry that does not satisfy the fifth postulate).

To establish these new geometries rigorously it was necessary to reformulate Euclid's *Elements*, removing all ambiguities such as that of the definition of a straight line mentioned above. Several works appeared in this direction and the efforts culminated in 1899 with D. Hilbert's *Grundlagen der Geometrie*.

Hilbert's new axiomatization was a synthesis of the work of many mathematicians (including, for instance, M. Pasch). The main novelty of Hilbert's approach with respect to Euclid's work was that, at the outset, he assumed only a pair of sets (here we restrict our discussion to plane geometry) whose elements are *not defined* but that are called respectively *points* and *straight lines*. Among the elements of these sets there are certain *relations*[4] that again are *not defined*, but which satisfy some axioms or properties: incidence, order, continuity and congruence.

The *incidence* axioms describe the fact that through any two points only one straight line can pass. The *order* axioms enable us to talk about line segments, and those of *continuity* permit the construction of the real numbers. The *congruence* axioms assert, essentially, that "given a line segment AB and a straight half-line with origin C, there exists a unique point D on this straight half-line such that the line segment AB is congruent to the line segment CD", together with an analogous assertion for angles.

From this one can reproduce the results of *The Elements*. A simplified axiomatic development of Euclidean Plane Geometry can be found in [24].

This approach immediately suggests some natural questions: What happens if we do not include the order axioms, or the axioms of continuity? Is it essential to identify the points of a straight line with the set of real numbers? Can we "do geometry" identifying, for instance, the points of straight lines with the complex numbers?

We shall see that, in fact, we can do geometry identifying the points of straight lines with the elements of an arbitrary field.

Next we state the axioms of Affine Plane Geometry as they appear in [1].

Recall that we only have a pair of sets whose elements are undefined but which we suggestively call, respectively, *points* and *straight lines*. Let us assume that there is a relation, called *incidence*, among the elements of the first set and the elements of the second. When an element P of the first set is *related*

---

was published as an addendum to his father's book *Tentamen Juventutem Studiosam in Elementa Matheseos Purae Introducendi*, dates from 1831, but the year of publication of the *Tentamen* itself was 1829. Bolyai elaborated his geometry in outline in 1823 and the final work, in German, was completed in 1826.

[4] A *relation* among the elements of two sets is a subset of the Cartesian product of these two sets.

to an element l of the second, we shall write $P \in l$, and we shall say that P *belongs to* l, or that the straight line l *passes through* the point P.

We further assume that this (undefined) relation satisfies the following axioms:

Axiom 1

Given two distinct points P and Q, there exists a unique straight line l such that $P \in l$ and $Q \in l$.

Axiom 2

Given a point P and a straight line l, there exists a unique straight line m such that $P \in m$ and $l \| m$.

The notation $l \| m$ means that either $l = m$ or there is no point P satisfying both $P \in l$ and $P \in m$. One says that l is *parallel* to m.

Axiom 3

There are three non-collinear points.

The next axioms are more easily stated if we first define the concepts of *dilation* and *translation*.

Definition

A map σ from the set of points into itself is called a *dilation* if it has the following property: Let P, Q be distinct points and let l be the straight line they determine. Then the straight line l' determined by $\sigma(P)$ and $\sigma(Q)$ is parallel to l. A dilation without fixed points is called a *translation*.

The bundle of straight lines, each line determined by a point and its image under a given translation, is called the *direction* of the translation.

We shall postulate that there must be 'sufficiently many' dilations and translations. More precisely:

Axiom 4

Given two distinct points P and Q there exists a translation τ such that $\tau(P) = Q$.

Axiom 5

Given three distinct collinear points P, Q and R there exists a dilation σ such that $\sigma(P) = P$ and $\sigma(Q) = R$.

With these last two axioms we can define *coordinates* in the plane in such a way that each point has two coordinates and each straight line is given by a *linear equation*. These coordinates are not necessarily real numbers, but elements of some field k. This field can be constructed from axioms 1, 2, 3 and 4. In fact k is formed by certain morphisms of the group T of translations. Concretely (for details, see [1]) we have

$$k = \{\alpha : T \longrightarrow T : \alpha \text{ satisfies the following two properties}\}$$

1. For all $\tau, \sigma \in T$, $\alpha(\tau \circ \sigma) = \alpha(\tau) \circ \alpha(\sigma)$, i.e. α is a group homomorphism.
2. For all $\tau \in T$, τ and $\alpha(\tau)$ have the same direction.

All axiomatic theories face the problem of *consistency*. A system is consistent if no contradiction can be derived within it. To prove that a theory is consistent, one provides a mathematical model in which all the axioms are satisfied. If this model is consistent, so too is the axiomatic system. The axiomatic system is, at least, as consistent as the model.

This is what we are going to do in this book. We shall construct an algebraic model, based on the concept of a vector space over a field, which, with suitable definitions, will satisfy all the axioms of Affine and Euclidean Geometry.

Organization   In Chapter 1 we introduce the most fundamental concept of these notes: affine space. This is a natural generalization of the concept of vector space but with a clear distinction between points and vectors. This distinction is not often made in vector spaces: it is commonplace, for example, not to distinguish between the point $(1, 2) \in \mathbb{R}^2$ and the vector $v = (1, 2) \in \mathbb{R}^2$. The problem is that $\mathbb{R}^2$ is both a set of points and a vector space.

In the study of vector spaces, the vector subspaces and the relations among them (the Grassmann formulas) play a central role. In the same way, in the study of affine spaces, the affine subspaces and the relations among them (the affine Grassmann formulas) also play an important role.

The simplest figure that we can form with points and straight lines is the triangle. In this chapter we shall meet two important results that refer to triangles and the incidence relation: the theorems of Menelaus and Ceva.

In Exercise 1.5 of this chapter, page 38, we verify Axioms 1, 2 and 3 of Affine Geometry.

In Chapter 2 we introduce a class of maps between affine spaces: affine maps, or *affinities*. The definition is a natural one, indeed affinities are revealed to be simply those maps that take collinear points to collinear points.

We shall also see that there are 'enough' affinities. In fact, in an affine space of dimension n, given two subsets of $n+1$ points, there exists an affinity which maps the points of the first subset onto the points of the second.

In Exercise 2.9 of this chapter, page 88, we verify Axioms 4 and 5 of Affine Geometry.

In Chapters 3 and 4 we answer the natural question of how many affinities there are. To do so we first define an equivalence relation between affinities and study its equivalence classes. In low dimensions the problem is not too hard and is solved explicitly in Chapter 3. However, in arbitrary dimensions the problem is rather involved, since it depends upon the classification of endomorphisms, and in particular on the Jordan normal form of a matrix. In Chapter 4 we provide the full details of this classification, since we have not been able to find it in the literature.

In Chapter 5 we consider affine spaces on which a distance has been defined. Thus we have a model of classical Euclidean Geometry, where, for instance, Pythagoras' Theorem holds.

In Chapter 6 we study distance preserving maps, that is, the *Euclidean motions*. Since there are fewer Euclidean motions than affinities, their classification is simpler. We also introduce a natural equivalence relation among Euclidean motions, similar to that for affinities, and we characterize each equivalence class by a finite sequence of numbers (the coefficients of a polynomial and a *metric invariant*).

In Chapter 7 we study Euclidean motions in dimensions 1, 2 and 3. In dimension three, for instance, there are only three types of Euclidean motion: *helicoidals* (which include rotations, translations and the identity mapping), *glide reflections* (which include mirror symmetries) and *anti-rotations*.

In Chapter 8 we study *quadrics*, two quadrics being considered equivalent if there is an affinity that maps one onto the other. Quadrics are the zeros of quadratic polynomials, and therefore they are the most natural objects to consider after straight lines (the zeros of linear polynomials). From this perspective, there are only three inequivalent quadrics (conics) in the plane: the ellipse, the hyperbola and the parabola. We also give the complete list of quadrics in three dimensions.

In Chapter 9 we study quadrics, this time considering two quadrics to be equivalent if there is a Euclidean motion that maps one onto the other. From this perspective there are infinitely many quadrics (conics) in the plane, since ellipses, parabolas and hyperbolas of different sizes are inequivalent to one another. Nevertheless, we shall give the classification in dimensions two and three, representing the quadrics by a finite sequence of real numbers. For example, in the plane, there are as many ellipses as pairs (a, b) of real numbers, with $0 < a \leq b$ and as many hyperbolas as pairs (a, b) of real numbers, with $0 < a$, $0 < b$.

We give a faithful list of all quadrics in arbitrary dimensions. For this we need to introduce a suitable definition of a *good order* among various real numbers. Most textbooks are not concerned with the faithfulness of this list, that is, that each quadric appears in the list once and only once; for this reason this notion of good order is, as far as we know, new in this context.

Finally we have collected together in the appendices the results from linear algebra that we have used in the text: bilinear maps and their diagonalization, isometries, the classification of isometries, diagonalization of symmetric bilinear maps, the method of completing the squares, orthogonal diagonalization, simultaneous diagonalization (the spectral theorem), the Nullstellensatz, and so on.

There are many interesting books on Affine Geometry, treated in many cases from the viewpoint of projective geometry. In order to complement the presentation given in this book the reader may also wish to consult, for example, [2–5, 11, 14–16, 18, 19, 22, 28, 29, 31–34] or [36].

Barcelona, Spain   Agustí Reventós Tarrida

Acknowledgements

I would like to thank professor Laia Saumell for handing over to me the collection of exercises she used in her courses during the years 2004–2007. I also want to acknowledge the contribution of the many professors who have taught the subject before me, such as Eduard Gallego, Joan Porti, Carmen Safont and Enric Nart, and other colleagues with whom I have discussed the subject, including Ferran Cedó, Joan Girbau, Gregori Guasp, Abdó Roig, Gil Solanes and Maria Sancosmed, who taught the problem course during the year 2007–2008. To all of them, my gratitude. Thanks are also due to Albert Ruíz, who helped me with some TEX details.

I also want to mention the help of those who have listened to me during this time: my students. They are the raison d'être of this book. For instance, some very astute comments of Albert Ferreiro contributed to improvements in the final version of these notes, which are based on the notes of my course taken by Judit Abardia, a very good student at the time, a prestigious mathematician today.

Finally I am grateful for the support of my family, without whom this book could not have be written. Thanks, therefore, to Berta, Oriol, Vane, Sergi and Neus.

Contents

1
Affine Spaces

1.1 Introduction

In this chapter we introduce the concept of an affine space as an algebraic model in which the axioms of Affine Geometry listed in the introduction are all fulfilled. For this we will need to define the terms *straight line*, *plane*, etc. We shall also demonstrate some classical theorems, such as those of Thales and Menelaus.

1.2 Definition of Affine Space

Let k be a field. A review of the definitions of a field and of a vector space over a field, as well as their more elementary properties, can be found in [8].

Definition 1.1

Let E be a k-vector space. An *affine space over E* is a set $\mathbb{A}$ together with a map

$$\mathbb{A} \times E \longrightarrow \mathbb{A}$$
$$P, v \longmapsto P + v$$

such that:

A. Reventós Tarrida, *Affine Maps, Euclidean Motions and Quadrics*,
Springer Undergraduate Mathematics Series,
DOI 10.1007/978-0-85729-710-5_1, © Springer-Verlag London Limited 2011

(1) $P + \vec{0} = P$ for all $P \in \mathbb{A}$, where $\vec{0}$ is the identity element of E;
(2) $P + (v + w) = (P + v) + w$ for all $P \in \mathbb{A}$ and $v, w \in E$; and
(3) given $P, Q \in \mathbb{A}$, there exists a unique $v \in E$ such that $P + v = Q$.

1.2.1 Observations

Observation 1. Note that if $P \in \mathbb{A}$ and $v \in E$, the notation $P + v$ means only
the image of the pair (P, v) via the above map $\mathbb{A} \times E \longrightarrow \mathbb{A}$.
Hence, the four signs "+" appearing in condition 2 have different meanings:
three of them represent the above map, and the other, ordinary vector
addition in the vector space E.

Observation 2. The unique vector determined by the points P and Q is denoted
by $\overrightarrow{PQ}$. Hence, we have the fundamental relation

$$P + \overrightarrow{PQ} = Q.$$

Observation 3. Let G be a group. A map

$$\mathbb{A} \times G \longrightarrow \mathbb{A}$$
$$P, v \longmapsto P + v$$

such that:
(1) $P + e = P$ for all $P \in \mathbb{A}$, where e is the identity element of G; and
(2) $P + (v + w) = (P + v) + w$ for all $P \in \mathbb{A}$ and $v, w \in G$,
is called an *action* of G on $\mathbb{A}$. If, in addition, the following holds:
(3) for all pairs of points $P, Q \in \mathbb{A}$, there exists a unique $v \in G$ such that
$P + v = Q$,
then the action is said to be *simply transitive*.
Hence, we can also define affine space as follows:

Definition 1.2

An affine space is a simply transitive action of the additive group of a k-vector
space E on a set $\mathbb{A}$.

Observation 4. Notice that, since the action of the additive group is transitive,
for all $P \in \mathbb{A}$ the map

$$\varphi_P : \mathbb{A} \longrightarrow E$$
$$Q \longmapsto \overrightarrow{PQ},$$

is a well-defined bijection. The injectivity and surjectivity of φ_P are immediate; see Proposition 1.4.

1.3 Examples

Example 1. The standard example of an affine space is given by $\mathbb{A} = E$, that is, the "points" of this affine space are the elements of the vector space. The action is

$$\mathbb{A} \times E \longrightarrow \mathbb{A}$$
$$P, v \longmapsto P + v,$$

where the sum is ordinary vector addition.

Let us verify that the three conditions of the definition of affine space are satisfied.

(1) $P + \vec{0} = P$. This is obvious, since $\vec{0}$ is the identity element of vector addition in E.

(2) $P + (u + v) = (P + u) + v$. This is precisely the associative property of vector addition in E.

(3) Given the points $u, v \in \mathbb{A} = E$, there exists a unique vector $w \in E$ such that $u + w = v$. It is sufficient to take $w = v - u$. That is, $\overrightarrow{uv} = v - u$.

Thank heavens we do not put arrows over the elements of E, because otherwise we would be forced to write such cumbersome expressions as

$$\overrightarrow{\vec{u}\vec{v}} = \vec{v} - \vec{u}. \quad \text{(Whew!)}$$

This "confusion" between points and vectors can bring about some problems. For instance, if $\mathbb{A} = E = \mathbb{R}^2$, it is not clear a priori if the pair $(1, 2)$ represents a point or a vector.

Example 2. Although only a special case of the above example, the following is of independent interest. As the set of points we take $\mathbb{A} = k^n$, and for the k-vector space we take $E = k^n$. The action is

$$\mathbb{A} \times E \longrightarrow \mathbb{A}$$
$$P, v \longmapsto P + v,$$

where the sum $P + v$ is the componentwise sum. That is, if $P = (p_1, \ldots, p_n)$ and $v = (v_1, \ldots, v_n)$, then

$$P + v = (p_1 + v_1, \ldots, p_n + v_n).$$

Example 3. Let us take $\mathbb{A} = \{(x, y) \in \mathbb{R}^2 : y > 0\}$ and, as the $\mathbb{R}$-vector space, $E = \mathbb{R}^2$. As the action of the vector space on the set we take

$$\mathbb{A} \times E \longrightarrow \mathbb{A}$$
$$(x, y), (u_1, u_2) \longmapsto (x + u_1, e^{u_2} y).$$

Note that, for all $u_2 \in \mathbb{R}$, $e^{u_2} y > 0$, and hence $(x + u_1, e^{u_2} y) \in \mathbb{A}$. It is easy to prove that this action is simply transitive.

Example 4. Let $\mathbb{A} = \{(x_1, \ldots, x_n) \in \mathbb{R}^n : x_1 + \cdots + x_n = 1\}$ be the set of points and let $E = \{(u_1, \ldots, u_n) \in \mathbb{R}^n : u_1 + \cdots + u_n = 0\}$ be the $\mathbb{R}$-vector space. As the action of the vector space on the set, we take

$$\mathbb{A} \times E \longrightarrow \mathbb{A}$$
$$(x_1, \ldots, x_n), (u_1, \ldots, u_n) \longmapsto (x_1 + u_1, \ldots, x_n + u_n).$$

Note that $(x_1 + u_1, \ldots, x_n + u_n) \in \mathbb{A}$.

It is easy to prove that this action is simply transitive.

Example 5.[1] Let us take

$$\mathbb{A} = \{(x, y, z) \in \mathbb{R}^3 : x > 0, y > 0, z > 0, x + y + z = 1\}$$

and as the $\mathbb{R}$-vector space we take $E = \mathbb{R}^2$. As the action of the vector space on the set we take

$$\mathbb{A} \times E \longrightarrow \mathbb{A}$$
$$(x, y, z), (u_1, u_2) \longmapsto \frac{1}{xe^{u_1} + ye^{u_2} + z}(xe^{u_1}, ye^{u_2}, z).$$

Note that

$$\frac{1}{xe^{u_1} + ye^{u_2} + z}(xe^{u_1}, ye^{u_2}, z) \in \mathbb{A}.$$

It is easy to prove that this action is simply transitive.

1.4 The Dimension of an Affine Space

An affine space is formed by three mathematical objects: a set $\mathbb{A}$, a k-vector space E, and an action of E on $\mathbb{A}$. Nevertheless, to simplify the language, we normally speak of *the affine space* $\mathbb{A}$; where it is understood that we are not only referring to the set $\mathbb{A}$.

The *dimension* of an affine space $\mathbb{A}$ is defined to be the dimension of its associated vector space E. We shall write $\dim \mathbb{A} = \dim E$. In this book we only consider finite dimensional affine spaces.

---

[1] This example, which appears in statistics, was suggested to me by Carles Barceló. See *Mathematical foundations of compositional data analysis*, C. Barceló, J.A. Martín, V. Pawlowsky, Preprint UdG, 2000.

Observation 1.3

Note that in the definition of affine space the vector addition of E appears explicitly, for instance, when writing $P + (u + v) = (P + u) + v$. However, scalar multiplication does not appear at all.

The following situation is possible. On a given vector space E, with vector addition "$+$" and scalar multiplication "$\cdot$", there could be a second scalar multiplication "$\bullet$" on E such that $E_1 = (E, +, \bullet)$ is a vector space different from $E_2 = (E, +, \cdot)$. In fact, it may even be the case that E_1 and E_2 do not have the same dimension.

It follows, from Example 1, that $\mathbb{A}_1 = E_1$ and $\mathbb{A}_2 = E_2$ are automatically affine spaces. The set of points and the action is exactly the same in both cases. Nevertheless, they are different as affine spaces. In fact, since $\dim \mathbb{A}_1 = \dim E_1$ and $\dim \mathbb{A}_2 = \dim E_2$, they could even have different dimensions.

Ferran Cedó suggested to me the following example of this phenomenon. Take the field of rational functions $k = \mathbb{R}(x)$ (see, for instance, [8]). Every field is a vector space over itself of dimension 1. Now we define a second scalar multiplication "$\bullet$" by

$$k \times \mathbb{R}(x) \longrightarrow \mathbb{R}(x)$$
$$p(x), q(x) \longmapsto p(x) \bullet q(x) = p(x^2) \cdot q(x).$$

Then $k = \mathbb{R}(x)$ with the usual addition of rational functions and this scalar multiplication is a k-vector space of dimension 2, since 1 and x are linearly independent.

In summary, if we put $k_1 = (\mathbb{R}(x), +, \cdot)$ and $k_2 = (\mathbb{R}(x), +, \bullet)$ we have two k-vector spaces, on the same set and with the same addition, but such that $\dim k_1 = 1$ and $\dim k_2 = 2$. If we consider them as affine spaces we have two affine spaces over the same set and with the same action but of different dimensions.

1.5   First Properties

Proposition 1.4

Let $P, Q, R, S \in \mathbb{A}$ and $u, v \in E$ be arbitrary points and vectors. The following properties are satisfied.
(1) $P + u = P + v$ implies $u = v$.
(2) $P + u = Q + u$ implies $P = Q$.
(3) $\overrightarrow{PQ} = \vec{0}$ if and only if $P = Q$.
(4) $\overrightarrow{PQ} = -\overrightarrow{QP}$.

(5) Given $P \in \mathbb{A}$, $v \in E$, there is a unique $Q \in \mathbb{A}$ such that $\overrightarrow{PQ} = v$.

(6) $\overrightarrow{PQ} + \overrightarrow{QR} = \overrightarrow{PR}$.

(7) $\overrightarrow{PQ} = \overrightarrow{PR}$ implies $Q = R$.

(8) $\overrightarrow{PQ} = \overrightarrow{RS}$ implies $\overrightarrow{PR} = \overrightarrow{QS}$.

Proof

(1) Let $Q = P + u = P + v$. Since there is a unique vector that added to P gives Q, we have $u = v$.

(2) Let us assume $P + u = Q + u$, and subtract u. We obtain $(P + u) - u = (Q + u) - u$. Hence, $P + (u - u) = Q + (u - u)$, that is, $P = Q$.

(3) If $\overrightarrow{PQ} = \vec{0}$, we have $Q = P + \overrightarrow{PQ} = P + \vec{0} = P$. If $P = Q$, we have $P + \overrightarrow{PP} = P$, and by uniqueness we have $\overrightarrow{PP} = \vec{0}$.

(4) Let us compute

$$P + (\overrightarrow{PQ} + \overrightarrow{QP}) = (P + \overrightarrow{PQ}) + \overrightarrow{QP} = Q + \overrightarrow{QP} = P.$$

Hence, $\overrightarrow{PQ} + \overrightarrow{QP} = \vec{0}$.

(5) Given $P \in \mathbb{A}$ and $v \in E$, we take $Q = P + v$, and we have $\overrightarrow{PQ} = v$.

(6) Let us compute

$$P + (\overrightarrow{PQ} + \overrightarrow{QR}) = (P + \overrightarrow{PQ}) + \overrightarrow{QR} = Q + \overrightarrow{QR} = R = P + \overrightarrow{PR}.$$

Hence, $\overrightarrow{PQ} + \overrightarrow{QR} = \overrightarrow{PR}$.

(7) Let us assume $\overrightarrow{PQ} = \overrightarrow{PR}$. Then

$$\overrightarrow{QR} = \overrightarrow{QP} + \overrightarrow{PR} = \overrightarrow{QP} + \overrightarrow{PQ} = \vec{0}.$$

Hence, $Q = R$.

(8) Let us assume $\overrightarrow{PQ} = \overrightarrow{RS}$. Then

$$\overrightarrow{PR} = \overrightarrow{PQ} + \overrightarrow{QR} = \overrightarrow{RS} + \overrightarrow{QR} = \overrightarrow{QR} + \overrightarrow{RS} = \overrightarrow{QS}.$$

$\square$

Property (1) says that we can "cancel" points and Property (2) says that we can "cancel" vectors.

Property (3) means that the vector with origin and end-point P is the zero vector and Property (4) refers to the "direction" of the vectors.

Properties (5) and (7) are, respectively, surjectivity and injectivity of the map φ_p of Observation 4 in Section 1.2.

Property (6), also known as *Chasles' identity*, tells us that vector addition satisfies the "parallelogram law" and Property (8) tells us that "parallels between parallels are equal" (see Figure 1.1).

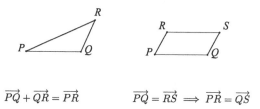

$$\overrightarrow{PQ} + \overrightarrow{QR} = \overrightarrow{PR} \qquad\qquad \overrightarrow{PQ} = \overrightarrow{RS} \implies \overrightarrow{PR} = \overrightarrow{QS}$$

Figure 1.1. Chasles' identity and the parallelogram law

1.6 Linear Varieties

We ask when a subset $\mathbb{B}$ of $\mathbb{A}$ inherits the properties of an affine space and therefore can itself be considered as an affine space.

Definition 1.5

A subset $\mathbb{B}$ of an affine space $\mathbb{A}$ is an *affine subspace* of $\mathbb{A}$, with associated vector space a vector subspace F of E, if
(0) for all $P \in \mathbb{B}$ and for all $v \in F$ one has $P + v \in \mathbb{B}$.
Moreover, the map

$$\mathbb{B} \times F \longrightarrow \mathbb{B}$$
$$P, v \longmapsto P + v$$

satisfies
(1) $P + \vec{0} = P$, for all $P \in \mathbb{B}$, $\vec{0} \in F$;
(2) $P + (v + w) = (P + v) + w$, for all $P \in \mathbb{B}$ and $v, w \in F$; and
(3) for each pair of points $P, Q \in \mathbb{B}$ we have $\overrightarrow{PQ} \in F$.

1.6.1 Observations

Observation 1. Property (0) says that the action of the vector space E over $\mathbb{A}$ restricts to an action of the vector subspace F over $\mathbb{B}$.
Observation 2. It is not necessary to impose properties (1) and (2) since they are immediate consequences of condition (0). Condition (3), however, must

be imposed; see Exercise 1.4. Thus $(\mathbb{B}, F)$ is an affine space. Notice that $\dim \mathbb{B} = \dim F$.

Observation 3. Affine subspaces are also called *linear subvarieties* or *linear varieties*. If they have dimension 1, they are called *affine straight lines* (or simply straight lines); if they have dimension 2, they are called *affine planes* (or simply planes); if they have dimension equal to $\dim \mathbb{A} - 1$, they are called *affine hyperplanes* (or simply hyperplanes).

For each point $P \in \mathbb{A}$ and for each vector subspace F of E, we put

$$P + [F] = \{Q \in \mathbb{A} : Q = P + v, \text{ with } v \in F\}.$$

Proposition 1.6

If $\mathbb{B}$ is an affine subspace of $\mathbb{A}$, with associated vector space F, and $P \in \mathbb{B}$, then

$$\mathbb{B} = P + [F].$$

Proof

$P + [F] \subset \mathbb{B}$ is condition (0) in the definition of affine subspace.

Conversely, to see that $\mathbb{B} \subset P + [F]$, let us take $Q \in \mathbb{B}$ and put $Q = P + \overrightarrow{PQ}$. From Property (3) we have $\overrightarrow{PQ} \in F$. Hence, $Q \in P + [F]$ and this completes the proof.   □

If $\mathbb{B} = P + [F]$ we say that $\mathbb{B}$ is the linear variety through P *directed* by F. We also say that F is the *direction* of $\mathbb{B}$.

Proposition 1.7

Let $\mathbb{B} = P + [F]$ be a linear variety of an affine space $\mathbb{A}$. Then
(1) $Q \in P + [F]$ if and only if $\overrightarrow{PQ} \in F$;
(2) if $Q \in P + [F]$, then $Q + [F] = P + [F]$; and
(3) if $Q, R \in P + [F]$, then $\overrightarrow{QR} \in F$.

Proof

(1) The vector $\overrightarrow{PQ}$ is the unique vector such that $Q = P + \overrightarrow{PQ}$.
(2) $Q + [F] = P + \overrightarrow{PQ} + [F] = P + [F]$.
(3) $\overrightarrow{QR} = \overrightarrow{QP} + \overrightarrow{PR}$, but both $\overrightarrow{QP}$ and $\overrightarrow{PR}$ are vectors of F.   □

1.7   Examples of Straight Lines

Example 1. Let us consider the affine space of Example 1 in Section 1.3, that
is, $\mathbb{A} = E$, and let $F = \langle v \rangle$ be a vector subspace of E of dimension 1. Let
$P \in \mathbb{A}$. Then the set $\mathbb{B} = P + [F] = \{P + \lambda v : \lambda \in k\}$ is the straight line
through P with direction v (see Figure 1.2). The sum $P + \lambda v$ is the sum
of two elements of E.

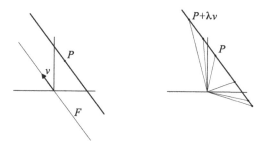

Figure 1.2.   Straight lines

The elements of $\mathbb{B}$ are points of the affine space; but, in this example, they
are also vectors of E, since they are obtained as a sum of elements of E.

Example 2. Let us consider the affine space of Example 2 in Section 1.3, that
is, $\mathbb{A} = k^n$, and let $F = \langle v \rangle$ be a vector subspace of $E = k^n$ of dimension 1.
Let $P \in \mathbb{A}$. Then the set $\mathbb{B} = P + [F] = \{P + \lambda v : \lambda \in k\}$ is the straight line
through P with direction v.

If $P = (p_1, \ldots, p_n)$ and $v = (v_1, \ldots, v_n)$, the points $X = (x_1, \ldots, x_n)$ of $\mathbb{B}$
satisfy

$$x_i = p_i + \lambda v_i, \quad i = 1, \ldots, n.$$

Example 3. Let us consider the affine space of Example 3 in Section 1.3. Figure 1.3 below represents the straight lines through the point $(0, 1)$ with direction vectors $u = (1, 1)$ and $v = (1, 2)$ respectively. In fact, $(0, 1) + t(1, 1) =$
$(0, 1) + (t, t) = (t, e^t)$, and hence the graph of the first straight line is the
graph of the exponential function $y = e^x$. Analogously $(0, 1) + t(1, 2) =$
$(0, 1) + (t, 2t) = (t, e^{2t})$, and hence the graph of the second straight line is
the graph of the exponential function $y = e^{2x}$.

Example 4. Let us consider the affine space of Example 4 in Section 1.3. We
draw the straight line through the point $(1/3, 1/3, 1/3)$ with direction vector $v = (1/4, 1/4, -1/2)$. Recall that the sum of the three components
of the vector v must be zero. The straight line is formed by the points
$(1/3 + t/4, 1/3 + t/4, 1/3 - t/2)$ (see Figure 1.4).

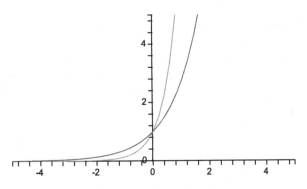

Figure 1.3. Straight lines

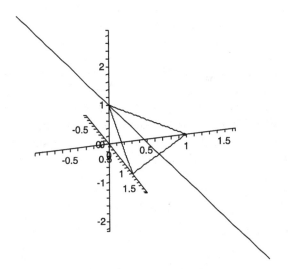

Figure 1.4. Straight lines

Example 5. Let us consider the affine space of Example 5 in Section 1.3. Figure 1.5 represents the straight line passing through the point $(1/3, 1/3, 1/3)$ with direction vector $u = (1, 2)$. In fact,

$$(1/3, 1/3, 1/3) + t(1, 2) = \left(\frac{e^t}{1 + e^t + e^{2t}}, \frac{e^{2t}}{1 + e^t + e^{2t}}, \frac{1}{1 + e^t + e^{2t}} \right).$$

This coincides, up to parametrization, with the intersection of $\mathbb{A}$ with the cone $x^2 - yz = 0$.

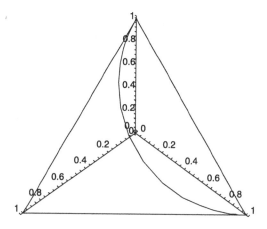

Figure 1.5. Straight lines

1.8 Linear Varieties Generated by Points

Definition 1.8 (The linear variety generated by r points)

Let us consider the points $P_1, \ldots, P_r \in \mathbb{A}$. The linear variety *generated* by these r points, denoted by $\langle P_1, \ldots, P_r \rangle$, is the smallest linear variety containing them.

Proposition 1.9

Let $P_1, \ldots, P_r \in \mathbb{A}$. Then

$$\langle P_1, \ldots, P_r \rangle = P_1 + \langle \overrightarrow{P_1 P_2}, \overrightarrow{P_1 P_3}, \ldots, \overrightarrow{P_1 P_r} \rangle.$$

Proof

Let $L = P_1 + [F]$ be a linear variety containing the points P_i, $i = 1, \ldots, r$. Since $P_i \in L$, we have $\overrightarrow{P_1 P_i} \in F$ and, hence,

$$P_1 + \langle \overrightarrow{P_1 P_2}, \overrightarrow{P_1 P_3}, \ldots, \overrightarrow{P_1 P_k} \rangle \subset L.$$

Thus $P_1 + \langle \overrightarrow{P_1 P_2}, \overrightarrow{P_1 P_3}, \ldots, \overrightarrow{P_1 P_r} \rangle$ contains P_i (since $P_i = P_1 + \overrightarrow{P_1 P_i}$), and it is contained in any other linear variety containing these points. □

Notice that the role played by P_1 in the above proposition can be played by any of the other points P_i.

We observe that $\dim \langle P_1, \ldots, P_r \rangle \leq r - 1$. When $\dim \langle P_1, \ldots, P_r \rangle = r - 1$, that is, when the vectors $\overrightarrow{P_1 P_2}, \ldots, \overrightarrow{P_1 P_r}$ are linearly independent, we say that *the points $P_1, \ldots, P_r$ are affinely independent.* Again, the role played by P_1 in this definition can be played by any of the other points P_i, see Exercise 1.7, page 38.

1.9 The Affine Grassmann Formulas

We first study the intersection of two linear varieties.

Proposition 1.10

Let $P + [F]$ and $Q + [G]$ be two linear varieties of an affine space $\mathbb{A}$. Then $(P + [F]) \cap (Q + [G]) \neq \emptyset$ if and only if $\overrightarrow{PQ} \in F + G$.

Proof

If there exists a point R such that $R \in P + [F]$ and $R \in Q + [G]$, then $\overrightarrow{PQ} = \overrightarrow{PR} + \overrightarrow{RQ} \in F + G$.

Conversely, if $\overrightarrow{PQ} \in F + G$, we can write $\overrightarrow{PQ} = u + v$, with $u \in F$ and $v \in G$. Thus, from $Q = P + \overrightarrow{PQ} = P + u + v$, we deduce $P + u = Q - v \in (P + [F]) \cap (Q + [G])$, and the intersection is non-empty. This completes the proof. $\qquad\square$

Proposition 1.11

Let $P + [F]$ and $Q + [G]$ be two linear varieties of an affine space $\mathbb{A}$. If $R \in (P + [F]) \cap (Q + [G])$, then

$$(P + [F]) \cap (Q + [G]) = R + [F \cap G].$$

Proof

We know, from point 2 of Proposition 1.7, that $P + [F] = R + [F]$ and that $Q + [G] = R + [G]$. Now, the equality

$$(R + [F]) \cap (R + [G]) = R + [F \cap G]$$

is clear. $\qquad\square$

Definition 1.12

The *linear variety sum* of two linear varieties L_1 and L_2 is the smallest linear variety containing them, and is denoted by $L_1 + L_2$.

Proposition 1.13

Let $P + [F]$ and $Q + [G]$ be two linear varieties of an affine space $\mathbb{A}$. Then

$$(P + [F]) + (Q + [G]) = P + [F + G + \langle \overrightarrow{PQ} \rangle].$$

Proof

Let L be a linear variety containing $P + [F]$ and $Q + [G]$. Since $P, Q \in L$, there is a vector subspace H such that

$$L = P + [H] = Q + [H],$$

and $\overrightarrow{PQ} \in H$. In particular $\langle \overrightarrow{PQ} \rangle \subset H$.

But, $P + [F] \subset P + [H]$ implies $F \subset H$, and $Q + [G] \subset Q + [H]$ implies $G \subset H$.

Hence,

$$F + G + \langle \overrightarrow{PQ} \rangle \subset H,$$

and

$$P + [F + G + \langle \overrightarrow{PQ} \rangle] \subset P + [H] = L.$$

That is, $P + [F + G + \langle \overrightarrow{PQ} \rangle]$ is contained in every linear variety containing $P + [F]$ and $Q + [G]$. Moreover, it is clear that these two linear varieties are contained in $P + [F + G + \langle \overrightarrow{PQ} \rangle]$. $\square$

Theorem 1.14 (Grassmann formulas)

Let $L_1 = P + [F]$ and $L_2 = Q + [G]$ be linear varieties of an affine space $\mathbb{A}$.
If $L_1 \cap L_2 \neq \emptyset$, then

$$\dim(L_1 + L_2) = \dim L_1 + \dim L_2 - \dim(L_1 \cap L_2).$$

If $L_1 \cap L_2 = \emptyset$, then

$$\dim(L_1 + L_2) = \dim L_1 + \dim L_2 - \dim(F \cap G) + 1.$$

Proof

Let us first assume $L_1 \cap L_2 \neq \emptyset$. Then, by Proposition 1.10, we have

$$F + G + \langle \overrightarrow{PQ} \rangle = F + G,$$

and hence, by Proposition 1.13, we have

$$L_1 + L_2 = P + [F + G].$$

Thus,

$$\dim(L_1 + L_2) = \dim(F + G)$$
$$= \dim F + \dim G - \dim(F \cap G)$$
$$= \dim L_1 + \dim L_2 - \dim(L_1 \cap L_2),$$

since, by Proposition 1.11, $\dim(L_1 \cap L_2) = \dim(F \cap G)$.

Now, let us assume that $L_1 \cap L_2 = \emptyset$. Then, by Proposition 1.10, we have

$$(F + G) \cap \langle \overrightarrow{PQ} \rangle = \vec{0},$$

and hence, by Proposition 1.13, we have

$$\dim(L_1 + L_2) = \dim(F + G + \langle \overrightarrow{PQ} \rangle)$$
$$= \dim(F + G) + \dim\langle \overrightarrow{PQ} \rangle$$
$$= \dim F + \dim G - \dim(F \cap G) + 1$$
$$= \dim L_1 + \dim L_2 - \dim(F \cap G) + 1,$$

and this completes the proof. Note that in this case we cannot replace $\dim(F \cap G)$ by $\dim(L_1 \cap L_2)$. $\qquad\square$

Definition 1.15

Two linear varieties $L_1 = P_1 + [F_1]$ and $L_1 = P_2 + [F_2]$ are said to be *parallel* when $F_1 \subset F_2$ or $F_2 \subset F_1$.

In particular, if $\dim F_1 = \dim F_2$, the above linear varieties are parallel if and only if $F_1 = F_2$.

Proposition 1.16

If two parallel linear varieties meet, then one of them is contained in the other.

Proof

Let us assume that $L_1 = P + [F]$ and $L_2 = Q + [G]$ are *parallel* (for instance, $G \subset F$) and that they meet. It follows, by Proposition 1.10, that $\overrightarrow{PQ} \in F + G \subset F$. Hence,

$$Q + [G] = P + \overrightarrow{PQ} + [G] \subset P + [F],$$

and this completes the proof.   □

Proposition 1.17 (Parallels between parallels are equal)

Let us suppose that two parallel straight lines r and s are cut by two parallel straight lines r' and s'. Let $A = r' \cap s$, $B = s' \cap s$, $C = r \cap r'$ and $D = r \cap s'$ be the intersection points. Then $\overrightarrow{AB} = \overrightarrow{CD}$ and $\overrightarrow{AC} = \overrightarrow{BD}$.

Proof

By hypothesis we have $\overrightarrow{AB} = \lambda \overrightarrow{CD}$ and $\overrightarrow{BD} = \mu \overrightarrow{AC}$. But the point D can be written as

$$D = A + \overrightarrow{AC} + \overrightarrow{CD},$$

or as

$$D = A + \overrightarrow{AB} + \overrightarrow{BD} = A + \lambda \overrightarrow{CD} + \mu \overrightarrow{AC}.$$

Equating these expressions, and taking into account that the vectors $\overrightarrow{AC}$ and $\overrightarrow{CD}$ are linearly independent, one obtains $\lambda = \mu = 1$. Hence $\overrightarrow{AB} = \overrightarrow{CD}$ and $\overrightarrow{AC} = \overrightarrow{BD}$, see Figure 1.6, and this completes the proof.   □

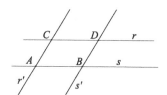

Figure 1.6. Parallels between parallels

Example 1.18

Study the relative position of two planes in an affine space of dimension 4.

Solution

Let $L_1 = P_1 + [F_1]$, $L_2 = P_2 + [F_2]$ be linear varieties of an affine space $\mathbb{A}$, with $\dim \mathbb{A} = 4$ and $\dim L_i = 2$, $i = 1, 2$.

If $L_1 \cap L_2 \neq \emptyset$, we have

$$\dim(L_1 + L_2) = 2 + 2 - \dim(L_1 \cap L_2) = 4 - \dim(F_1 \cap F_2).$$

Hence, we have three possibilities:

1. If $\dim(F_1 \cap F_2) = 0$, they meet in a point.
2. If $\dim(F_1 \cap F_2) = 1$, they meet in a straight line.
3. If $\dim(F_1 \cap F_2) = 2$, they coincide.

If $L_1 \cap L_2 = \emptyset$, we have

$$\dim(L_1 + L_2) = 2 + 2 - \dim(F_1 \cap F_2) + 1 = 5 - \dim(F_1 \cap F_2).$$

Hence, we only have two possibilities:

4. If $\dim(F_1 \cap F_2) = 1$, they cross (they neither cut nor are parallel).
5. If $\dim(F_1 \cap F_2) = 2$, they are parallel ($F_1 = F_2$).

If we take a basis (e_1, e_2, e_3, e_4) of E, we can easily construct examples of the above five cases. For instance:

1. Take $L_1 = P + [F_1]$, $L_2 = P + [F_2]$, with $F_1 = \langle e_1, e_2 \rangle$, $F_2 = \langle e_3, e_4 \rangle$.
2. Take $L_1 = P + [F_1]$, $L_2 = P + [F_2]$, with $F_1 = \langle e_1, e_2 \rangle$, $F_2 = \langle e_2, e_3 \rangle$.
3. Take $L_1 = P + [F_1]$, $L_2 = P + [F_2]$, with $F_1 = F_2$.
4. Take $L_1 = P + [F_1]$, $L_2 = Q + [F_2]$, with $Q = P + e_4$ and $F_1 = \langle e_1, e_2 \rangle$, $F_2 = \langle e_2, e_3 \rangle$.
5. Take $L_1 = P + [F_1]$, $L_2 = Q + [F_2]$, with $Q = P + e_3$ and $F_1 = F_2 = \langle e_1, e_2 \rangle$.

$\square$

1.10 Affine Frames

Definition 1.19

An *affine frame* in an affine space $\mathbb{A}$ is a set $\mathcal{R} = \{P; (e_1, \ldots, e_n)\}$ formed by a point $P \in \mathbb{A}$ and a basis $(e_1, \ldots, e_n)$ of the associated vector space E. The point P is called the *origin* of this affine frame.

When we fix an affine frame, the points $Q \in \mathbb{A}$ have coordinates, called *affine coordinates*, defined in the following way: let us consider the vector $\overrightarrow{PQ}$ given by the origin P of the affine frame and by the point Q. Then, if

$$\overrightarrow{PQ} = q_1 e_1 + \cdots + q_n e_n, \quad q_i \in k,$$

we say that Q has *affine coordinates* $(q_1, \ldots, q_n)$. We also say that $(q_1, \ldots, q_n)$ are the coordinates of Q with respect to $\mathcal{R}$, or in $\mathcal{R}$.

Slightly abusing notation, we write $Q = (q_1, \ldots, q_n)$. Note that for the origin P of $\mathcal{R}$ we have $P = (0, \ldots, 0)$.

If $Q = (q_1, \ldots, q_n)$ and $R = (r_1, \ldots, r_n)$, then

$$\overrightarrow{QR} = \overrightarrow{QP} + \overrightarrow{PR}$$

$$= -(q_1 e_1 + \cdots + q_n e_n) + (r_1 e_1 + \cdots + r_n e_n)$$

$$= (r_1 - q_1)e_1 + \cdots + (r_n - q_n)e_n,$$

that is, the i-th component of the vector $\overrightarrow{QR}$ is equal to the i-th coordinate of the point R minus the i-th coordinate of the point Q.

Equivalently, if $v = v_1 e_1 + \cdots + v_n e_n$, the affine coordinates $(r_1, \ldots, r_n)$ of the point $R = Q + v$ are given by

$$r_i = q_i + v_i, \quad i = 1, \ldots, n.$$

In the particular case of the affine space $\mathbb{A} = k^n$ (Example 2 on page 3) there is a privileged affine frame: $\mathcal{C} = \{P; (e_1, \ldots, e_n)\}$, with $P = (0, \ldots, 0)$ and $e_i = (0, \ldots, 0, 1, 0, \ldots, 0)$ (where the 1 is in the i-th position), $i = 1, \ldots, n$. This affine frame is called the *canonical affine frame* of k^n. It has the great advantage that the components of a point coincide with its coordinates.

1.10.1 Change of Affine Frame

Let us assume given two affine frames

$$\mathcal{R} = \{P; (e_1, \ldots, e_n)\} \quad \text{and} \quad \mathcal{R}' = \{P'; (v_1, \ldots, v_n)\}$$

in the same affine space $\mathbb{A}$. We want to find the relationship between the coordinates $(x_1, \ldots, x_n)$ of a point X with respect to $\mathcal{R}$ and the coordinates $(x'_1, \ldots, x'_n)$ of the same point X with respect to $\mathcal{R}'$ (see Figure 1.7).

To find this relationship we put

$$\overrightarrow{PX} = \sum_{i=1}^{n} x_i e_i,$$

$$\overrightarrow{P'X} = \sum_{j=1}^{n} x'_j v_j,$$

$$\overrightarrow{PP'} = \sum_{i=1}^{n} b_i e_i,$$

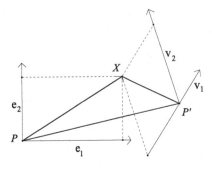

Figure 1.7. Coordinates of X with respect to $\mathcal{R}$ and $\mathcal{R}'$

$$v_j = \sum_{i=1}^{n} a_{ij} e_i.$$

Substituting, one obtains

$$\overrightarrow{P'X} = \sum_{i=1}^{n} x'_j v_j = \sum_{i,j=1}^{n} x'_j a_{ij} e_i, \qquad (1.1)$$

and also

$$\overrightarrow{P'X} = \overrightarrow{P'P} + \overrightarrow{PX} = -\sum_{i=1}^{n} b_i e_i + \sum_{i=1}^{n} x_i e_i. \qquad (1.2)$$

Equating coefficients in (1.1) and (1.2) one obtains

$$x_i = \sum_{j=1}^{n} x'_j a_{ij} + b_i, \quad i = 1, \dots, n.$$

These n equations can be written as

$$\begin{pmatrix} x_1 \\ \vdots \\ x_n \end{pmatrix} = \begin{pmatrix} a_{11} & \cdots & a_{1n} \\ \vdots & \vdots & \vdots \\ a_{n1} & \cdots & a_{nn} \end{pmatrix} \begin{pmatrix} x'_1 \\ \vdots \\ x'_n \end{pmatrix} + \begin{pmatrix} b_1 \\ \vdots \\ b_n \end{pmatrix}.$$

To simplify the notation we shall write

$$\boxed{x = Ax' + b}$$

with

$$x = \begin{pmatrix} x_1 \\ \vdots \\ x_n \end{pmatrix}, \qquad b = \begin{pmatrix} b_1 \\ \vdots \\ b_n \end{pmatrix}, \qquad A = (a_{ij}), \qquad x' = \begin{pmatrix} x'_1 \\ \vdots \\ x'_n \end{pmatrix}.$$

This equality is equivalent to

$$
\begin{pmatrix} x_1 \\ \vdots \\ x_n \\ 1 \end{pmatrix} = \begin{pmatrix} a_{11} & \cdots & a_{1n} & b_1 \\ \vdots & \vdots & \vdots & \vdots \\ a_{n1} & \cdots & a_{nn} & b_n \\ 0 & \cdots & 0 & 1 \end{pmatrix} \begin{pmatrix} x'_1 \\ \vdots \\ x'_n \\ 1 \end{pmatrix}.
$$

This $(n+1) \times (n+1)$ matrix is denoted by $M(\mathcal{R}', \mathcal{R})$ and is called the *matrix of the change of affine frame* or the *matrix of the change of coordinates*. So, we have

$$
\begin{pmatrix} x_1 \\ \vdots \\ x_n \\ 1 \end{pmatrix} = M(\mathcal{R}', \mathcal{R}) \begin{pmatrix} x'_1 \\ \vdots \\ x'_n \\ 1 \end{pmatrix}. \tag{1.3}
$$

We shall write, briefly,

$$
\boxed{ \begin{pmatrix} x \\ 1 \end{pmatrix} = M(\mathcal{R}', \mathcal{R}) \begin{pmatrix} x' \\ 1 \end{pmatrix} = \begin{pmatrix} A & b \\ 0 & 1 \end{pmatrix} \begin{pmatrix} x' \\ 1 \end{pmatrix} }
$$

1.11 Equations of a Linear Variety

Let us fix an affine frame $\mathcal{R} = \{P; (e_1, \ldots, e_n)\}$ in an affine space $\mathbb{A}$, and consider a linear variety $L = Q + [F]$. Let $Q = (q_1, \ldots, q_n)$.

Let us fix a basis $(v_1, \ldots, v_r)$ of F. Put $v_j = \sum_{i=1}^n a_{ij} e_i$, $j = 1, \ldots, r$. It is clear that $X = (x_1, \ldots, x_n) \in L$ if and only if there are scalars λ_j, $j = 1, \ldots, r$ such that

$$
x_i = q_i + \sum_{j=1}^r \lambda_j a_{ij}, \quad i = 1, \ldots, n. \tag{1.4}
$$

Equations (1.4) are called *parametric equations* of the linear variety L.

These equations impose restrictions between the affine coordinates x_i of the points of L. We shall see that they are solutions of a certain linear system. We will use some well-known properties of the rank of a matrix, which can be found, for instance, in [8], page 200.

Proposition 1.20

Let $L = Q + [F]$ be a linear variety of dimension r in an affine space $\mathbb{A}$ and
let $\mathcal{R} = \{P; (e_1, \ldots, e_n)\}$ be an affine frame. Then the coordinates of the points
of L in $\mathcal{R}$ are a solution of a linear system $AX = B$, of $n - r$ equations, n
unknowns and rank $n - r$. Moreover, the components of any vector of F in the
basis $(e_1, \ldots, e_n)$ are a solution of the homogeneous system $AX = 0$.

Proof

Let us assume $F = \langle v_1, \ldots, v_r \rangle$, and let $Q = (q_1, \ldots, q_n)$. A point $X = (x_1, \ldots, x_n)$
belongs to L if and only if the vector $\overrightarrow{QX}$ is a linear combination of the vectors
$v_1, \ldots, v_r$. Equivalently,

$$
\mathrm{rank}
\begin{pmatrix}
v_{11} & \cdots & v_{1r} & x_1 - q_1 \\
\vdots & \vdots & \vdots & \vdots \\
v_{n1} & \cdots & v_{nr} & x_n - q_n
\end{pmatrix}
= r,
$$

since the last column is a linear combination of the first r columns (which are
linearly independent). Note that column j, for $j = 1, \ldots, r$, is formed by the
components of the vector $v_j = (v_{1j}, \ldots, v_{nj})$.

Permuting, if necessary, the rows of this matrix, we may assume that the
$r \times r$ minor,

$$
\delta =
\begin{vmatrix}
v_{11} & \cdots & v_{1r} \\
\vdots & \vdots & \vdots \\
v_{r1} & \cdots & v_{rr}
\end{vmatrix}
$$

is non-zero.

Since all the $(r + 1) \times (r + 1)$ minors of the above matrix must be zero, the
coordinates $(x_1, \ldots, x_n)$ of the points of L satisfy the following linear system
of $n - r$ equations and n unknowns:

$$
\begin{vmatrix}
v_{11} & \cdots & v_{1r} & x_1 - q_1 \\
\vdots & \vdots & \vdots & \vdots \\
v_{r1} & \cdots & v_{rr} & x_r - q_r \\
v_{j1} & \cdots & v_{jr} & x_j - q_j
\end{vmatrix}
= 0, \quad j = r+1, \ldots, n. \tag{1.5}
$$

Note that this system can be written as

$$
AX = B,
$$

where X and B are matrices of 1 column and A is a matrix of $n - r$ rows and n columns. In fact, the matrix A is of the form

$$A = \begin{pmatrix} * & \cdots & * & \delta & 0 & \cdots & 0 \\ * & \cdots & * & 0 & \delta & \cdots & 0 \\ \vdots & \cdots & \vdots & \vdots & \vdots & \vdots & \vdots \\ * & \cdots & * & 0 & 0 & \cdots & \delta \end{pmatrix},$$

and, hence, the system has rank $n - r$ (it has a non-zero $(n - r) \times (n - r)$ minor). This proves the first part of the proposition.

The second part states that

$$A \begin{pmatrix} v_{1j} \\ \vdots \\ v_{nj} \end{pmatrix} = \begin{pmatrix} 0 \\ \vdots \\ 0 \end{pmatrix}, \quad j = 1, \ldots, r,$$

since if the components of the vectors of the basis of F are solutions of the homogeneous system $AX = 0$, then the components of any other vector of F will also be a solution of this system.

Notice that each of the determinants in system (1.5) is a sum of two determinants, those obtained by considering the last column as a sum of two columns. In particular, one of these determinants does not contain any x_i.

Inspired by this observation, we consider the system of $n - r$ equations

$$\begin{vmatrix} v_{11} & \cdots & v_{1r} & x_1 \\ \vdots & \vdots & \vdots & \vdots \\ v_{r1} & \cdots & v_{rr} & x_r \\ v_{j1} & \cdots & v_{jr} & x_j \end{vmatrix} = 0, \quad j = r+1, \ldots, n, \tag{1.6}$$

which can be written as $AX = 0$, where the matrix A is exactly the same matrix A as in system (1.5).

It is now clear that, by substituting in (1.6) the variables $(x_1, \ldots, x_r, x_j)$ respectively by $(v_{1i}, \ldots, v_{ri}, v_{ji})$, for $i = 1, \ldots, r$, and $j = r+1, \ldots, n$, all these determinants are zero, because they have two equal columns. Hence,

$$A \begin{pmatrix} v_{1j} \\ \vdots \\ v_{nj} \end{pmatrix} = \begin{pmatrix} 0 \\ \vdots \\ 0 \end{pmatrix}, \quad j = 1, \ldots, r,$$

and this completes the proof.   □

Observation 1.21

We can arrive at the same result by row-reducing the matrix to row-reduced echelon form

$$\begin{pmatrix} v_{11} & \cdots & v_{1r} & x_1 - q_1 \\ \vdots & \vdots & \vdots & \vdots \\ v_{n1} & \cdots & v_{nr} & x_n - q_n \end{pmatrix},$$

since, if the rank is equal to r, we obtain a matrix of the form

$$\begin{pmatrix} \bullet & * & \cdots & * & * \\ 0 & \bullet & \cdots & * & * \\ \vdots & \vdots & \ddots & \vdots & \vdots \\ 0 & 0 & \cdots & \bullet & * \\ 0 & 0 & \cdots & 0 & \text{Expression } 1 \\ \vdots & \vdots & \vdots & \vdots & \vdots \\ 0 & 0 & \cdots & 0 & \text{Expression } n - r \end{pmatrix},$$

where $\bullet$ denotes any non-zero element (the r pivots), $*$ denotes an arbitrary element, and "*Expression i*" is an expression of the form

$$a_{i1}x_1 + \cdots + a_{in}x_n - b_i, \quad i = 1, \ldots, n - r.$$

Since the rank of this reduced matrix must also be r, we must have

$$a_{11}x_1 + \cdots + a_{1n}x_n = b_1,$$

$$\vdots$$

$$a_{n-r,1}x_1 + \cdots + a_{n-r,n}x_n = b_{n-r}.$$

This is the linear system of $n - r$ equations, n unknowns and rank $n - r$ satisfied by the coordinates $x_1, \ldots, x_n$ of the points of L. The argument proving that this system has rank $n - r$ is the same as that used in the above proof.

This system is equivalent to system (1.6), in the sense that they have the same solutions.

The equations given by the linear system $AX = B$ are known as *Cartesian equations* of the linear variety.

Since the systems $AX = B$ and $CAX = CB$, where C is an invertible matrix, have the same solutions, it is clear that Cartesian equations of a linear variety are not unique.

Example 1.22

Find Cartesian equations for the plane of the affine space $\mathbb{R}^4$ given by $L = P + [F]$, with $P = (1, 0, 1, 0)$ and $F = \langle(1, -1, 0, 0), (0, 0, 1, 1)\rangle$.

Solution

First method. We must have

$$\text{rank} \begin{pmatrix} 1 & 0 & x-1 \\ -1 & 0 & y \\ 0 & 1 & z-1 \\ 0 & 1 & t \end{pmatrix} = 2. \tag{1.7}$$

Since the minor

$$\begin{vmatrix} 1 & 0 \\ -1 & 0 \end{vmatrix}$$

is equal to zero, we must permute the rows of (1.7). For instance, we can put

$$\text{rank} \begin{pmatrix} 1 & 0 & x-1 \\ 0 & 1 & z-1 \\ -1 & 0 & y \\ 0 & 1 & t \end{pmatrix} = 2.$$

Then, the equations of L are

$$\begin{vmatrix} 1 & 0 & x-1 \\ 0 & 1 & z-1 \\ -1 & 0 & y \end{vmatrix} = 0, \qquad \begin{vmatrix} 1 & 0 & x-1 \\ 0 & 1 & z-1 \\ 0 & 1 & t \end{vmatrix} = 0.$$

Equivalently

$$\begin{cases} x + y = 1, \\ z - t = 1. \end{cases}$$

Second method. Row-reducing the matrix to row-reduced echelon form

$$\begin{pmatrix} 1 & 0 & x-1 \\ -1 & 0 & y \\ 0 & 1 & z-1 \\ 0 & 1 & t \end{pmatrix},$$

one obtains

$$\begin{pmatrix} 1 & 0 & x-1 \\ 0 & 1 & z-1 \\ 0 & 0 & t-z+1 \\ 0 & 0 & y+x-1 \end{pmatrix}.$$

Since the rank must be 2, we have

$$\begin{cases} x+y=1, \\ z-t=1. \end{cases}$$

Finally we remark that, for instance,

$$\begin{cases} x+y+2z-2t=3, \\ 3x+3y+4z-4t=7 \end{cases}$$

are also Cartesian equations of L. $\square$

When, in k^n, the equations of a linear variety are given without specifying an affine frame, as was done in the previous example, we implicitly assume that we are considering the canonical affine frame.

Observation 1.23

We have seen that the coordinates of the points of $L = Q + [F]$ are solution of a linear system $AX = B$, and that the components of the vectors of F are solution of the homogeneous system $AX = 0$.

Conversely, given a linear system $AX = B$ and an affine frame, we can interpret the solutions of this system as a linear variety $L = Q + [F]$. It suffices to take

$$F = \left\{ \sum x_i e_i : (x_1, \dots, x_n) \text{ a solution of the system } AX = 0 \right\},$$

where $(e_1, \dots, e_n)$ is the basis of the given affine frame and Q the point $Q = (q_1, \dots, q_n)$, where $(q_1, \dots, q_n)$ is any solution of the system $AX = B$.

1.11.1 Equations of Straight Lines

The equations of a straight line are given by (1.5) with $r = 1$. Hence, we have

$$\begin{vmatrix} v_{11} & x_1 - q_1 \\ v_{j1} & x_j - q_j \end{vmatrix} = 0, \quad j = 2, \dots, n. \tag{1.8}$$

These equations are usually written as

$$\frac{x_1 - q_1}{v_1} = \cdots = \frac{x_n - q_n}{v_n},$$

where $v = (v_1, \ldots, v_n)$ denotes a direction vector of the straight line. We should not worry about division by zero: at least one v_i is non-zero.

1.11.2   Equations of Hyperplanes

The equation of a hyperplane is given by (1.5) with $r = n - 1$. Hence, we have

$$\begin{vmatrix} v_{11} & \cdots & v_{1r} & x_1 - q_1 \\ \vdots & \vdots & \vdots & \vdots \\ v_{n1} & \cdots & v_{nr} & x_n - q_n \end{vmatrix} = 0. \tag{1.9}$$

This is usually written (expanding by the last column) as

$$a_1 x_1 + \cdots + a_n x_n = b,$$

with at least one $a_i \neq 0$. By Proposition 1.20, the equation of the direction of this hyperplane is

$$a_1 x_1 + \cdots + a_n x_n = 0.$$

Since two hyperplanes are parallel if and only if they have the same director subspace, the hyperplanes

$$a_1 x_1 + \cdots + a_n x_n = b,$$
$$a_1' x_1 + \cdots + a_n' x_n = b'$$

are parallel if and only if the equations

$$a_1 x_1 + \cdots + a_n x_n = 0,$$
$$a_1' x_1 + \cdots + a_n' x_n = 0 \tag{1.10}$$

have the same solutions.

Since the space of solutions of each one of these equations has dimension $n - 1$, the space of solutions of system (1.10) also has dimension $n - 1$. Hence, by Rouché-Frobenius' theorem (see [8], page 260), we have

$$\text{rank} \begin{pmatrix} a_1 & \cdots & a_n \\ a_1' & \cdots & a_n' \end{pmatrix} = n - (n - 1) = 1,$$

and therefore one row is a multiple of the other. That is, there exists a $\lambda \in k$ such that

$$a_i' = \lambda a_i, \quad i = 1, \ldots, n.$$

Equivalently, every hyperplane parallel to

$$a_1 x_1 + \cdots + a_n x_n = b$$

has equation

$$a_1 x_1 + \cdots + a_n x_n = b',$$

with $b' \in k$.

Observation 1.24

The above result on parallel hyperplanes can also be obtained as a consequence of the following lemma.

Lemma 1.25

Let $f, g : k^n \longrightarrow k$ be surjective linear maps such that $\ker f = \ker g$. Then there exists a $\lambda \in k$ such that $f = \lambda g$.

Proof

Surjectivity implies $\dim \ker f = \dim \ker g = n - 1$.

Let $(e_1, \ldots, e_{n-1})$ be a basis of $\ker f = \ker g$, and let $(e_1, \ldots, e_{n-1}, e_n)$ be a basis of k^n. Put $f(e_n) = \lambda g(e_n)$. It is clear that $f(e_i) = \lambda g(e_i)$, $i = 1, \ldots, n$. Hence, $f = \lambda g$, and this completes the proof. $\qquad\square$

Corollary 1.26

The hyperplanes

$$a_1 x_1 + \cdots + a_n x_n = b,$$

$$a_1' x_1 + \cdots + a_n' x_n = b'$$

are parallel if and only if there exists a $\lambda \in k$ such that

$$a_i' = \lambda a_i, \quad i = 1, \ldots, n.$$

Proof

Apply the previous lemma with $f(x) = \sum_i a_i x_i$ and $g(x) = \sum_i a'_i x_i$.   □

1.12 Barycenter

Definition 1.27

Let $P_1, \ldots, P_r$ be points of an affine space $\mathbb{A}$. The barycenter G of these r points is the point

$$G = P_1 + \frac{1}{r}(\overrightarrow{P_1 P_2} + \cdots + \overrightarrow{P_1 P_r}).$$

Therefore we need r (i.e. the sum of r times the unit element of the field k) to be invertible in k.

Proposition 1.28

The barycenter of r points $P_1, \ldots, P_r \in \mathbb{A}$ is the unique point G such that

$$\overrightarrow{GP_1} + \cdots + \overrightarrow{GP_r} = \vec{0}.$$

Proof

We have

$$\overrightarrow{GP_1} + \cdots + \overrightarrow{GP_r} = \overrightarrow{GP_1} + (\overrightarrow{GP_1} + \overrightarrow{P_1 P_2}) + \cdots + (\overrightarrow{GP_1} + \overrightarrow{P_1 P_r})$$

$$= r\overrightarrow{GP_1} + (\overrightarrow{P_1 P_2} + \cdots + \overrightarrow{P_1 P_r})$$

$$= r\overrightarrow{GP_1} + r\overrightarrow{P_1 G} = \vec{0}.$$

Hence, the barycenter satisfies the given condition.

Moreover, it is the only point satisfying it. To see this let us assume that G' satisfies

$$\overrightarrow{G'P_1} + \cdots + \overrightarrow{G'P_r} = \vec{0}.$$

Then

$$(\overrightarrow{G'G} + \overrightarrow{GP_1}) + \cdots + (\overrightarrow{G'G} + \overrightarrow{GP_r}) = \vec{0}.$$

Hence, $r\overrightarrow{G'G} = \vec{0}$, and so $G = G'$.   □

It is now clear that the role played by P_1 in the definition of barycenter can be played by any of the points P_i, $i = 1, \ldots, r$. That is, we also have

$$G = P_i + \frac{1}{r}(\overrightarrow{P_i P_1} + \cdots + \overrightarrow{P_i P_r}).$$

The barycenter of two points is called the *midpoint* between them. That is, the midpoint between P_1 and P_2 is the point

$$G = P_1 + \frac{1}{2}\overrightarrow{P_1 P_2}.$$

1.12.1 Computations in Coordinates

Let $\mathcal{R}$ be an affine frame of $\mathbb{A}$, and let us denote by

$$P_i = (x_{i1}, \ldots, x_{in}), \quad i = 1, \ldots, r,$$

$$G = (g_1, \ldots, g_n)$$

the coordinates of the points P_i and G in $\mathcal{R}$. It is easy to see that

$$g_j = \frac{x_{1j} + \cdots + x_{rj}}{r}, \quad j = 1, \ldots, n.$$

1.13 Simple Ratio

Definition 1.29

Let $A, B, C \in \mathbb{A}$ be three distinct collinear points. The *simple ratio* of these three points is the unique scalar $\lambda = (A, B, C) \in k$ such that

$$\overrightarrow{AB} = \lambda \overrightarrow{AC}.$$

Equivalently, the simple ratio of three distinct collinear points is the scalar $(A, B, C) \in k$ such that

$$\boxed{B = A + (ABC)\overrightarrow{AC}}$$

Note that the order of the points is very important.[2] It is clear, for instance, that by permuting the last two points, the simple ratio is inverted. In fact, we have

---

[2] Not all authors take the points in the same order in the definition of simple ratio. We have adopted, as a sign of recognition and respect, the definition used by Puig Adam, [23], volume 2, page 108.

$$(A,B,C) = \lambda,$$

$$(A,C,B) = \frac{1}{\lambda},$$

$$(B,A,C) = \frac{\lambda}{\lambda - 1},$$

$$(B,C,A) = \frac{\lambda - 1}{\lambda},$$

$$(C,A,B) = \frac{1}{1 - \lambda},$$

$$(C,B,A) = 1 - \lambda.$$

1.13.1 Characterization of the Points of a Line Segment $(k = \mathbb{R})$

Let $\mathbb{A}$ be a real affine space, that is, such that its associated vector space E is an $\mathbb{R}$-vector space. Then every pair of points $P, Q \in \mathbb{A}$ determine a line segment $\overline{PQ}$ defined by

$$\overline{PQ} = \{X \in \mathbb{A} : X = P + \lambda\overrightarrow{PQ}, 0 \leq \lambda \leq 1\}.$$

Proposition 1.30

Let $\mathbb{A}$ be a real affine space, and let $X, P, Q \in \mathbb{A}$ be three distinct collinear points. Then

$$X \in \overline{PQ} \quad \text{if and only if} \quad 0 < (P,X,Q) < 1.$$

Proof

Let $X = P + \lambda\overrightarrow{PQ}$. Then $\overrightarrow{PX} = \lambda\overrightarrow{PQ}$, and hence $(P,X,Q) = \lambda$. That is, X belongs to the line segment $\overline{PQ}$ if and only if $0 < (P,X,Q) < 1$.   □

Since the simple ratio depends on the order, we also have the following, for example.

Proposition 1.31

Let $\mathbb{A}$ be a real affine space, and let $X, P, Q \in \mathbb{A}$ be three distinct collinear points. Then

$$X \in \overline{PQ} \quad \text{if and only if} \quad (X, P, Q) < 0.$$

Proof

If $(P, X, Q) = \lambda$, then

$$(X, P, Q) = \frac{\lambda}{\lambda - 1}.$$

$\square$

1.13.2 Characteristic Property of the Barycenter of a Triangle

The barycenter of a triangle[3] is the barycenter of its vertices.

Proposition 1.32

The straight line joining the vertex A of a triangle $\triangle ABC$ with the barycenter G of this triangle meets the opposite side in the midpoint A' of the points B, C. That is,

$$(B, A', C) = \frac{1}{2}.$$

Moreover,

$$(A, G, A') = \frac{2}{3}.$$

Proof

Let us cut the straight line $A + \lambda \overrightarrow{AG}$ with the straight line $B + \mu \overrightarrow{BC}$ (see Figure 1.8). Recall $B = A + \overrightarrow{AB}$. We have

$$A + \lambda \overrightarrow{AG} = B + \mu \overrightarrow{BC}$$
$$= A + \overrightarrow{AB} + \mu \overrightarrow{BC}.$$

---

[3] Also called the *centroid* or *geometric center*.

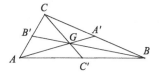

Figure 1.8. Barycenter

Hence,

$$\lambda\overrightarrow{AG} = \overrightarrow{AG} + \overrightarrow{GB} + \mu\overrightarrow{BG} + \mu\overrightarrow{GC}.$$

Since $\overrightarrow{GA} + \overrightarrow{GB} + \overrightarrow{GC} = \vec{0}$, we have

$$(\lambda - 1)\overrightarrow{GA} + (1 - \mu)\overrightarrow{GB} + \mu(-\overrightarrow{GA} - \overrightarrow{GB}) = \vec{0},$$

and hence

$$\lambda - 1 - \mu = 0,$$

$$1 - \mu - \mu = 0,$$

that is, $\lambda = \frac{3}{2}$ and $\mu = \frac{1}{2}$.

Thus, $A' = A + \frac{3}{2}\overrightarrow{AG}$, that is $(A, G, A') = \frac{2}{3}$.

And also $A' = B + \frac{1}{2}\overrightarrow{BC}$, that is $(B, A', C) = \frac{1}{2}$. $\qquad\square$

The straight lines joining each vertex of a triangle with the midpoint of the opposite side are called *medians*. The above proposition implies that the three medians of a triangle meet in a point: the barycenter.

1.13.3 The Real Plane as a Complex Straight Line

Let $\mathbb{A} = \mathbb{R}^2$ be the real affine plane. We have $\dim_{\mathbb{R}} \mathbb{A} = 2$. Let $A = (a_1, a_2)$, $B = (b_1, b_2)$, $C = (c_1, c_2) \in \mathbb{A}$. If these points are not collinear, it makes no sense to talk about their simple ratio. However, we can think of $\mathbb{A}$ as $\mathbb{C}$. Then $\dim_{\mathbb{C}} \mathbb{A} = 1$ (the real plane is the complex straight line), and hence the complex numbers $A = z_1 = a_1 + ia_2$, $B = z_2 = b_1 + ib_2$, $C = z_3 = c_1 + ic_2$ are collinear. So, there exists a $\lambda \in \mathbb{C}$ such that

$$z_2 - z_1 = \lambda(z_3 - z_1)$$

(simply take λ as the ratio of these two complex numbers). Thus,

$$(z_1, z_2, z_3) = \frac{z_2 - z_1}{z_3 - z_1}.$$

As an application we have that three complex numbers z_1, z_2, z_3 are the vertices of an equilateral triangle in $\mathbb{C}$ if and only if

$$(z_1, z_2, z_3) = \frac{1}{2} + i\frac{\sqrt{3}}{2}.$$

1.14 Theorems of Thales, Menelaus and Ceva

Theorem 1.33 (Thales' theorem)

Let us assume that three parallel straight lines r, s, t meet two concurrent straight lines in points A, A' (on r), B, B' (on s) and C, C' (on t). Then $(A, B, C) = (A', B', C')$.

Equivalently, there exists a $\lambda \in k$ such that

$$\overrightarrow{AB} = \lambda\overrightarrow{AC},$$

$$\overrightarrow{A'B'} = \lambda\overrightarrow{A'C'}.$$

Proof[4]

As r, s and t are parallel, there exist $\mu, \nu \in k$ such that $\overrightarrow{BB'} = \mu\overrightarrow{CC'}$, and $\overrightarrow{AA'} = \nu\overrightarrow{CC'}$ (see Figure 1.9).

---

[4]   *Proof of Thales' theorem given in the famous song*
             "El teorema de Tales", by **Le Luthiers**:
a paralela a b, b paralela a c, a paralela a b, paralela a c, paralela a d, op es a pq, m es a nt, op es a pq como mn es a mt; a paralela a b, b paralela a c, op es a pq como mn es a nt.
–La bisectriz yo trazaré, y a cuatro planos intersectaré.
–Una igualdad yo encontraré, op más pq es igual a st.
–Usaré la hipotenusa.
–Ay, no te compliques: nadie la usa,
–Trazaré pues un cateto.
–Yo no me meto, yo no me meto.
Triángulo, tetrágono, pentágono, exágono, eptágono, octógono, son todos polígonos. Seno, coseno, tangente y secante, y la cosecante y la cotangente.
Tales... Tales de Mileto, Tales... Tales de Mileto, Tales... Tales de Mileto, Tales... Tales de Mileto. Que es lo que queríamos demostrar, Que es lo que, que es lo que, queríamos demos, demos, demostrar.

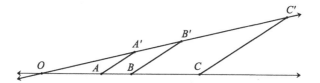

Figure 1.9. Thales' theorem

Let $(A, B, C) = \lambda$ and $(A', B', C') = \sigma$, so that we have $\overrightarrow{AB} = \lambda\overrightarrow{AC}$ and $\overrightarrow{A'B'} = \sigma\overrightarrow{A'C'}$.

We have

$$C' = C + \overrightarrow{CC'} = A + \overrightarrow{AC} + \overrightarrow{CC'}. \tag{1.11}$$

On the other hand

$$
\begin{aligned}
C' &= A + \overrightarrow{AA'} + \overrightarrow{A'C'} \\
&= A + \nu\overrightarrow{CC'} + \sigma^{-1}\overrightarrow{A'B'} \\
&= A + \nu\overrightarrow{CC'} + \sigma^{-1}(\overrightarrow{A'A} + \overrightarrow{AB} + \overrightarrow{BB'}) \\
&= A + \nu\overrightarrow{CC'} + \sigma^{-1}(-\nu\overrightarrow{CC'} + \lambda\overrightarrow{AC} + \mu\overrightarrow{CC'}).
\end{aligned}
\tag{1.12}
$$

Comparing (1.11) and (1.12), and equating the coefficients of $\overrightarrow{AC}$ and $\overrightarrow{CC'}$, one obtains

$$1 = \sigma^{-1}\lambda,$$
$$1 = \nu - \sigma^{-1}\nu + \sigma^{-1}\mu. \tag{1.13}$$

Thus, $(A, B, C) = \lambda = \sigma = (A', B', C')$, completing the proof. □

Corollary 1.34

Let us assume that two parallel straight lines r, s cut two concurrent straight lines respectively in points B, B' (on r) and C, C' (on s). Let A be the point of concurrence. Then the triangles $\triangle ABB'$ and $\triangle ACC'$ are similar, that is, there exists a $\lambda \in k$ such that

$$\overrightarrow{AB} = \lambda\overrightarrow{AC},$$
$$\overrightarrow{AB'} = \lambda\overrightarrow{AC'},$$
$$\overrightarrow{BB'} = \lambda\overrightarrow{CC'}.$$

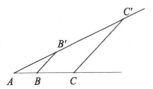

Figure 1.10. Similar triangles

Proof

This is a particular case of Thales' theorem, with $A = A'$ (see Figure 1.10).

It is sufficient to observe that, in this case, formula (1.13) implies $\lambda = \mu$, since $\overrightarrow{AA'} = \vec{0} = \nu\overrightarrow{CC'}$, that is $\nu = 0$. □

Theorem 1.35 (Menelaus' theorem)

Let us assume that a straight line meets the sides of a triangle $\triangle ABC$ in the points P, Q, R, respectively. Then

$$(P, A, B) \cdot (Q, B, C) \cdot (R, C, A) = 1.$$

Proof

Let $(P, A, B) = \lambda$, $(Q, B, C) = \mu$, $(R, C, A) = \nu$. We have $\overrightarrow{PA} = \lambda\overrightarrow{PB}$, $\overrightarrow{QB} = \mu\overrightarrow{QC}$, $\overrightarrow{RC} = \nu\overrightarrow{RA}$ (see Figure 1.11).

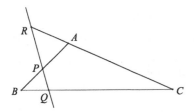

Figure 1.11. Menelaus' theorem

Then

$$\overrightarrow{PA} = \lambda(\overrightarrow{PQ} + \overrightarrow{QB})$$

$$= \lambda\overrightarrow{PQ} + \lambda\mu\overrightarrow{QC}$$

$$= \lambda\overrightarrow{PQ} + \lambda\mu\overrightarrow{QR} + \lambda\mu\nu\overrightarrow{RA}. \tag{1.14}$$

But, since we also have

$$\overrightarrow{PA} = \overrightarrow{PR} + \overrightarrow{RA}, \tag{1.15}$$

and the vectors $\overrightarrow{PR}, \overrightarrow{PQ}$ and $\overrightarrow{QR}$ are proportional, equating (1.14) and (1.15), one obtains

$$\lambda\mu\nu = 1,$$

and this completes the proof.   □

Note that in the statement of Menelaus' theorem it is necessary that $P \neq A, B$, $Q \neq B, C$, $R \neq C, A$, that is, the straight line that cuts the sides of the triangle does not contain any vertex.

The converse of Menelaus' theorem is also true; see Exercise 1.43, page 45.

Observation 1.36

In the axiomatic development of Euclidean Plane Geometry it is not possible to prove directly that if *a straight line intersects one side of a triangle and misses the three vertices, then it must intersect one of the other two sides*.

This must be imposed as an axiom, as was observed by Moritz Pasch and included in D. Hilbert's work of 1899, *Gründlagen der Geometrie*.

From this, and Hilbert's other axioms, it can be proved that if a straight line intersects one side of a triangle and misses the three vertices, then it must intersect one *and only one* of the other two sides.

In our algebraic model, this "only one" part is a direct consequence of Menelaus' theorem and Proposition 1.31, since if a straight line cuts the three sides of a triangle, then the three simple ratios are negative, and so their product cannot be equal to 1.

Theorem 1.37 (Ceva's theorem)

Let $\triangle ABC$ be a triangle. The necessary and sufficient condition for three straight lines, passing respectively through each one of the vertices of the triangle, to be concurrent in a point P, is that the following relationship among simple ratios is satisfied:

$$(P_A, B, C) \cdot (P_B, C, A) \cdot (P_C, A, B) = -1,$$

where P_A, P_B, P_C denote the intersection points of the three given straight lines with the sides BC, AC and AB, respectively.

Proof

First we assume that the three straight lines meet in a point P. Let us consider the straight line s through the point A, parallel to the straight line BC, see Figure 1.12. Let $M = s \cap BP$ and $N = s \cap CP$.

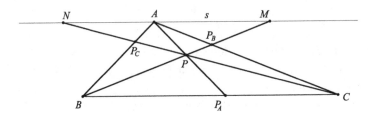

Figure 1.12. Ceva's theorem

Applying Corollary 1.34 of Thales' theorem twice, with vertex P, one obtains

$$(A, M, N) = (P_A, B, C). \tag{1.16}$$

Applying the same corollary, with vertex P_B, one obtains

$$\overrightarrow{CB} = (P_B, C, A)\overrightarrow{AM}.$$

Applying the corollary again, now with vertex P_C, one obtains

$$\overrightarrow{AN} = (P_C, A, B)\overrightarrow{BC}.$$

These two last equalities imply

$$(A, M, N) = -\frac{1}{(P_B, C, A)(P_C, A, B)},$$

which, together with (1.16), gives the result.

To see that the condition is sufficient, let us suppose that there are three points P_A, P_B, P_C satisfying

$$(P_A, B, C) \cdot (P_B, C, A) \cdot (P_C, A, B) = -1.$$

Let $Q = BP_B \cap CP_C$ and denote by P'_A the point $P'_A = AQ \cap BC$.

Then, from the first part of the theorem applied to the three concurrent straight lines in Q, we have

$$(P'_A, B, C) \cdot (P_B, C, A) \cdot (P_C, A, B) = -1.$$

Thus, $(P'_A, B, C) = (P_A, B, C)$, and therefore $P_A = P'_A$. Hence the straight line AP_A goes through Q, and this completes the proof.   □

Because of the influence of this theorem, straight lines through a vertex of a triangle, contained in the plane of the triangle, are called *cevians*, a term apparently introduced by M.A. Poulain in 1888. For instance, the altitudes, the angle bisectors, the perpendicular bisectors and the medians are examples of concurrent cevians; see Exercise 5.11 of Chapter 5, page 171. The intersection of a cevian through a vertex with the opposite side to this vertex is called the *foot* of the cevian. Thus Ceva's theorem says that *three cevians are concurrent if the product of the simple ratios of their feet and corresponding vertices is* -1.

EXERCISES

1.1. Prove that the following actions induce an affine space structure on $\mathbb{A}$:

(a)

$$\mathbb{A} \times E \longrightarrow \mathbb{A}$$
$$P, v \longmapsto P + v$$

where

$$\mathbb{A} = \{P = (x, y, z) \in \mathbb{R}^3 : x + y + z = 1\} \quad \text{and}$$
$$E = \{v = (x, y, z) \in \mathbb{R}^3 : x + y + z = 0\}.$$

(b)

$$\mathbb{A} \times E \longrightarrow \mathbb{A}$$
$$P, (\alpha, \beta) \longrightarrow P + \alpha(1, -1, 0) + \beta(1, 0, -1),$$

where $\mathbb{A} = \{P = (x, y, z) \in \mathbb{R}^3 : x + y + z = 1\}$ and $E = \mathbb{R}^2$.

(c)

$$\mathbb{A} \times E \longrightarrow \mathbb{A}$$
$$(x, y, z), (\alpha, \beta) \longrightarrow (x + \alpha z + \beta y + \alpha \beta, y + \alpha, z + \beta),$$

where $\mathbb{A} = \{(x, y, z) \in \mathbb{R}^3 : x = yz\}$ and $E = \mathbb{R}^2$.

1.2. Let $\mathbb{A}$ be an affine space over a vector space E, defined by the action $\varphi : \mathbb{A} \times E \to \mathbb{A}$, and let $f : E' \to E$ be an isomorphism of vector spaces. Prove that the map

$$\mathbb{A} \times E' \longrightarrow \mathbb{A}$$
$$a, u \longrightarrow \varphi(a, f(u))$$

gives a structure of an affine space over E' on $\mathbb{A}$.

1.3. Let $\mathbb{A}$ be an affine space over a vector space E, defined by the action $\varphi : \mathbb{A} \times E \to \mathbb{A}$. Consider a set $\mathbb{A}'$ and a bijection $g : \mathbb{A} \to \mathbb{A}'$. Prove that the map

$$\mathbb{A}' \times E \longrightarrow \mathbb{A}'$$
$$a', u \longrightarrow g(\varphi(g^{-1}(a'), u))$$

gives a structure of an affine space over E on $\mathbb{A}'$.

1.4. Let $\mathbb{A}$ be an affine space over a vector space E. Find a subset $\mathbb{B}$ of $\mathbb{A}$ and a vector subspace F of E such that conditions $0, 1$ and 2 of Definition 1.5 are fulfilled, but not condition 3.

1.5. Let $\mathbb{A}$ be an affine space of dimension two. Prove that Axioms 1, 2 and 3 of Affine Geometry given in the introduction, page viii, are fulfilled.

1.6. Prove that the set

$$\mathbb{A} = \{(x, y, z) \in \mathbb{R}^3 : x^2 + y^2 - z = 0\}$$

with the action of $\mathbb{R}^2$ given by

$$(x, y, z) + (u, v) = (x + u, y + v, (x + u)^2 + (y + v)^2),$$

is an affine space.

1.7. Let $P_1, \ldots, P_r$ be points of an affine space. Prove that the vectors $\overrightarrow{P_1 P_2}, \ldots, \overrightarrow{P_1 P_r}$ are linearly independent if and only if the vectors $\overrightarrow{P_i P_1}, \ldots, \overrightarrow{P_i P_{i-1}}, \overrightarrow{P_i P_{i+1}}, \ldots, \overrightarrow{P_i P_r}$, $i = 2, \ldots, r - 1$, are linearly independent.

1.8. Find the equation and draw approximately the straight line parallel to $r : (0, 1) + \langle (1, 1) \rangle$, through the point $(0, 2)$, in the affine space of Example 3, page 3.

1.9. Consider the linear varieties of the affine space $\mathbb{R}^4$ given respectively by the following equations:

$$\begin{cases} x + y - z - 2t = 0, \\ 3x - y + z + 4t = 1, \\ 2y - 2z - 5t = -1/2. \end{cases} \qquad \begin{cases} -z + t = 1, \\ 2x + y + z - t = 0, \\ 4x + 2y + 2z + t = 3. \end{cases}$$

$$\begin{cases} 2x - y + t = -1, \\ 2x - y + t = -1, \\ -x + 2y + z - 2t = 2. \end{cases} \qquad \{ 3x + z = 0.$$

$$\begin{cases} -x + 2y + z - 2t = 2, \\ 3x + z = 0. \end{cases} \qquad \begin{cases} 2x - y + t = -1, \\ -x + 2y + z - 2t = 2, \\ 3x + z + t = 4. \end{cases}$$

Write each of them in the form $P + [V]$, where $P \in \mathbb{R}^4$ and V is a vector subspace of the vector space $\mathbb{R}^4$, giving explicitly the point P and a basis of V.

1.10. Find, in an affine space of dimension 3, Cartesian equations for the linear varieties given, in some affine frame, by:

$$\begin{cases} x = 2 - a, \\ y = -1, \\ z = 2 + a, \end{cases} \qquad \begin{cases} x = a - b, \\ y = a - 5, \\ z = 2a + 3b, \end{cases}$$

where $a, b \in k$ are parameters.

1.11. Find, in an affine space of dimension 4, a system of Cartesian and parametric equations for the linear varieties given in some affine frame by:

(a) The straight line through the points $(2,1,0,1)$ and $(1,1,1,2)$.

(b) The plane through the points $(2,1,0,1)$, $(1,1,1,2)$ and $(3,-1,2,3)$.

(c) The linear variety of dimension 3 through the points $(2,1,0,1)$, $(1,1,1,2)$, $(3,-1,2,3)$ and $(0,0,-2,-1)$.

1.12. Find, in an affine space of dimension 4, the dimension and parametric equations of each of the linear varieties given, in some affine frame, by:

$$L: \quad \{ -2x + 3y + 4z + t = 5.$$

$$M: \quad \begin{cases} x - y + 2z - 2t = 7, \\ 3x + z + t = 7, \\ x - y + 5z + 6t = 0, \\ -2x - y + z - 3t = 0. \end{cases}$$

$$N: \quad \begin{cases} -2x + 3y + 4z + t = 5, \\ -x + 4y + z - 5t = 8. \end{cases}$$

Find $L \cap M$, $M \cap N$ and $M + N$.

1.13. Given a linear variety L and a point $P \notin L$, prove that there is a unique linear variety of the same dimension as L, parallel to L and passing through P.

Find, in the affine space $\mathbb{R}^3$, a system of Cartesian and parametric equations of: the plane containing the points $(1,-1,0)$, $(2,1,1)$ and $(2,0,1)$; the plane parallel to the above plane passing through

(0,0,1); a straight line passing through $(1,1,1)$ and parallel to the above two planes.

1.14. Let $L = P + [F]$ be a linear variety of dimension r in an affine space A. Let $Q \notin L$. Prove that the set of points of the straight lines through Q containing some point of L, together with the points of the linear variety $L' = Q + [F]$, is a linear variety of dimension $r + 1$.

1.15. Fix an affine frame in an affine space of dimension 4. Find a system of Cartesian and parametric equations of the following:

(a) The plane containing the point $(1,2,3,4)$ and parallel to the plane:
$$x - y + z + t = 3; \qquad 2x + y - 5t = 10.$$

(b) The plane containing the point $(1,2,3,4)$ and parallel to the plane:
$$x = 2u + v; \qquad y = u + v; \qquad z = u - 2v + 1; \qquad t = v - 2.$$

1.16. Fix an affine frame in an affine space of dimension 4. Find a system of Cartesian equations of the linear varieties given by:
(a) $(x,y,z,t) = (2,3,0,-4) + \lambda(0,2,1,-5)$;
(b) $(x,y,z,t) = (2,3,0,-4) + \lambda(0,2,1,-5) + \mu(3,-1,\frac{1}{2},0)$; and
(c) $(x,y,z,t) = (2,3,0,-4) + \lambda(0,2,1,-5) + \mu(3,-1,\frac{1}{2},0) + \rho(1,1,1,1)$.

1.17. Fix an affine frame in an affine space of dimension 4. Find a system of Cartesian and parametric equations of the plane containing the straight line r and parallel to the straight line s, where:

$$r: \begin{cases} 2x - y + t = -1, \\ -x + 2y + z - 2t = 2, \\ 3x + z + t = 4. \end{cases} \qquad s: \begin{cases} x - y + t = 3, \\ x + 2y + z - 2t = 2, \\ x + z + t = 0. \end{cases}$$

1.18. Fix an affine frame in an affine space of dimension 4. Find a system of Cartesian and parametric equations of a plane Π satisfying:
(a) Π is parallel to the hyperplane $x + y + z + t = 0$.
(b) Π contains the straight line $(1,1,1,1) + \lambda(2,-1,0,-1)$.
(c) Some plane parallel to Π meets in a straight line the plane:
$$\begin{cases} 2x + y - z = 2, \\ 4x + t = 5. \end{cases}$$

1.19. Find, in the affine space $\mathbb{R}^4$, a system of Cartesian and parametric equations of the plane Π generated by the straight line $(1,1,1,1) +$

$\lambda(2,-1,0,-1)$ and the point $(0,0,-2,3)$. Is there a plane Π' passing through the point $(0,1,0,1)$ and such that the intersection $\Pi \cap \Pi'$ is the point $(0,0,-2,3)$?

1.20. Let Π be the plane of the affine space $\mathbb{R}^4$ given by:

$$\begin{cases} 2x + y - z = 2, \\ 4x + t = 5. \end{cases}$$

Determine all the straight lines L passing through the point $(0,1,0,1)$ and such that $\Pi + L = \mathbb{R}^4$.

1.21. Let L_1 and L_2 be the linear varieties of the affine space $\mathbb{R}^4$ given by

$$L_1 = \{(a + 3\lambda + 2\mu, 1 - \lambda - \mu, 4 + \lambda, 6 + 5\lambda + 2\mu) : \lambda, \mu \in \mathbb{R}\},$$

$$L_2 = \{(2 + \alpha + 2\beta, 1, 1 + \alpha + \beta, 3\alpha) : \alpha, \beta \in \mathbb{R}\}.$$

Find $a \in \mathbb{R}$ such that $L_1 \cap L_2 \neq \emptyset$. For this value of a, determine $L_1 \cap L_2$ and $L_1 + L_2$.

1.22. Determine the relative positions of two straight lines in an affine space of dimension n.

1.23. Consider the hyperplane Π of the affine space $\mathbb{R}^4$ given by $x + y + mz = n$ and the straight line r given by

$$x = t, \qquad y = 2 - t, \qquad z = 2t.$$

Study, according to $m \in \mathbb{R}$ and $n \in \mathbb{R}$, the relative position between the plane Π and the straight line r.

1.24. Prove that, in an affine space of dimension 3, the intersection of two different and non-parallel planes is a straight line.

1.25. Let $A_i = (a_i, b_i, c_i)$, $i = 1, 2, 3, 4$, be points in an affine space of dimension 3. Prove that the points A_i are coplanar if and only if

$$\begin{vmatrix} a_1 & b_1 & c_1 & 1 \\ a_2 & b_2 & c_2 & 1 \\ a_3 & b_3 & c_3 & 1 \\ a_4 & b_4 & c_4 & 1 \end{vmatrix} = 0.$$

1.26. Let $A_1, \ldots, A_n$ be points in an affine space. Prove that the straight lines joining each point A_i with the barycenter B_i of the other points (see Figure 1.13) are concurrent (in the barycenter G of the points $A_1, \ldots, A_n$). Find the simple ratio (A_i, B_i, G).

1.27. Let P, Q, R be three points in an affine space of dimension 2. Let M_{PQ} be the midpoint of P and Q, M_{QR} the midpoint of Q and R, and M_{RP} the midpoint of R and P. Prove that the barycenter of P, Q and R coincides with the barycenter of M_{PQ}, M_{QR} and M_{RP}.

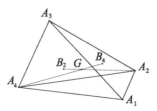

Figure 1.13. Collinear barycenters

1.28. (Barycenter with weights) Given a set of r points P_i, $i = 1, \ldots, r$, in an affine space $\mathbb{A}$, its *barycenter with weights* is the point $\tilde{G}$ given by

$$\tilde{G} = P_1 + \sum_{i=1}^{r} \lambda_i \overrightarrow{P_1 P_i}, \quad \sum_{i=1}^{r} \lambda_i = 1.$$

Prove that for all points $Q \in \mathbb{A}$ we have

$$Q + \sum_{i=1}^{r} \lambda_i \overrightarrow{QP_i} = \tilde{G}, \quad \text{and,} \quad \text{hence,}$$

$$\sum_{i=1}^{r} \lambda_i \overrightarrow{\tilde{G}P_i} = \vec{0}.$$

1.29. Given, in the affine space $\mathbb{R}^4$, the linear varieties

$$L_1 = \{(x, y, z, t) \in \mathbb{R}^4 : x + y = 4, z + t = a\},$$

$$L_2 = \{(3 + \lambda, 2 - 2\lambda, 2\lambda, -1 + \lambda) : \lambda \in \mathbb{R}\},$$

find $a \in \mathbb{R}$ such that the affine space generated by L_1 and L_2 has minimum dimension.

1.30. Let L_1 and L_2 be two straight lines in an affine space $\mathbb{A}$ of dimension 3.
   (a) What are the possible values of $\dim(L_1 + L_2)$?
   (b) Prove that L_1 and L_2 do not meet and are not parallel if and only if $L_1 + L_2 = \mathbb{A}$.

1.31. In the affine plane $\mathbb{A} = \mathbb{Z}/3\mathbb{Z} \times \mathbb{Z}/3\mathbb{Z}$ over the field $\mathbb{Z}/3\mathbb{Z}$:
   (a) How many points are there?
   (b) How many straight lines are there?
   (c) How many points are there on each straight line?
   (d) How many straight lines are there parallel to a given straight line?
   (e) How many different families of parallel straight lines are there?

1.32. Given, in the affine plane $\mathbb{R}^2$, the three straight lines

$$3x + 2y = 1, \qquad y = 5, \qquad 6x + y = -13,$$

find a triangle $\triangle ABC$ such that its medians are on these straight lines, the vertex A is on the first straight line, and the point $(1/3, 0)$ is the midpoint of the side BC.

1.33. Consider, in the affine plane $\mathbb{R}^2$, the points $A = (2, 3)$, $G = (1, -1)$ and the straight lines $r : x - 3y + 1 = 0$, $s : 2x + 5y - 1 = 0$. Determine the unique triangle having A as one of its vertices, G as barycenter, and such that the other two vertices are on the straight lines r and s, respectively.

1.34. Let $\mathcal{C} = (e_1, e_2, e_3)$ be the canonical basis of the vector space $\mathbb{R}^3$. Find the equations of the change of coordinates between the affine frames $\mathcal{R}_1$ and $\mathcal{R}_2$ of the affine space $\mathbb{R}^3$ given by

$$\mathcal{R}_1 = \{(1, 1, 1); (e_3 - e_1, e_1 + e_3, e_1 + e_2)\},$$

$$\mathcal{R}_2 = \{(0, 1, 1); (e_1, 2e_2 - e_3, 3e_1 - e_3)\}.$$

1.35. Consider the affine frame $\mathcal{R}$ of the affine space $\mathbb{R}^3$ given in the canonical affine frame by

$$\mathcal{R} = \{(1, 1, 1); ((1, 1, 1), (0, 1, 0), (2, 1, 0))\}.$$

What are the coordinates of the point $(0, 0, 0) \in \mathbb{R}^3$ in this new affine frame? And those of the point $(1, 1, 1) \in \mathbb{R}^3$? Is there a point with the same coordinates in $\mathcal{R}$ as in the canonical affine frame $\mathcal{C}$? Find the equations of the change of coordinates between $\mathcal{R}$ and $\mathcal{C}$.

1.36. Consider, in the affine space $\mathbb{R}^3$, the affine frames $\mathcal{R}$ and $\mathcal{R}'$ given by

$$\mathcal{R} = \{(0, 0, 0); ((1, 0, 0), (0, 1, 0), (0, 0, 1))\},$$

$$\mathcal{R}' = \{(-1, 0, 0); ((1, 1, 0), (0, -1, 0), (0, 0 - 1))\}.$$

(a) Given the point P with coordinates $(1, 2, -1)$ in $\mathcal{R}$, determine the coordinates of P in $\mathcal{R}'$.

(b) Find, with respect to $\mathcal{R}'$, the equation of the plane Π, given, with respect to $\mathcal{R}$, by the equation $2x - y + z + 2 = 0$.

(c) Find, with respect to $\mathcal{R}'$, the equations of the straight line r given, with respect to $\mathcal{R}$, by the equations

$$\begin{cases} 2x + y = 0, \\ x - 2y + z = 1. \end{cases}$$

1.37. (Gauss' straight line) A straight line cuts the sides AB, BC and AC of the triangle $\triangle ABC$ in the points D, E and F respectively (see Figure 1.14). Prove that the midpoints X, Y, Z of the line segments DC, AE and BF, respectively, are collinear.

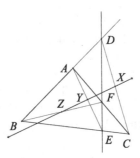

Figure 1.14. Gauss' straight line

1.38. Consider, in a real affine plane, a triangle $\triangle ABC$ and a point P on the straight line AB not belonging to the line segment AB. Let l be a straight line through P that cuts the line segments AC and BC. Let $R = l \cap AC$, $Q = l \cap BC$ and $G = BR \cap AQ$. Prove that the point P', the intersection of the straight lines CG and AB, see Figure 1.15, does not depend on the straight line l.
Hint: Prove that $(P, A, B) = -(P', A, B)$.

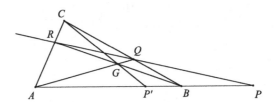

Figure 1.15. P' is called the *harmonic conjugate* of P

1.39. Let $\triangle ABC$ be a triangle in a real affine plane. Let P be the point on the straight line AB with $(B, P, A) = \frac{3}{2}$; Q the point on the straight line BC with $(C, Q, B) = \frac{3}{2}$; and R the point on the straight line AC with $(A, R, C) = -\frac{1}{8}$. Prove that the points P, Q, R are collinear and give, with respect to the affine frame $\mathcal{R} = \{A; (\overrightarrow{AB}, \overrightarrow{AC})\}$, the equation of the straight line that they determine.

1.40. Let us consider a trapezium A, B, C, D, with the side AB parallel to the side DC. Let P be the intersection point of the diagonals AC and BD (see Figure 1.16). Prove that P is the midpoint of the line segment containing P, parallel to the side AB and with endpoints in BC and AD.

1.41. Let A, B, C, D be the vertices of a quadrilateral in an affine plane, that is, four points of the affine plane such that no three of them

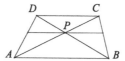

Figure 1.16. Diagonals of a trapezium

are collinear. Prove that the midpoints of the line segments AB, BC, CD, DA are the vertices of a parallelogram (see Figure 1.17). Prove also that a quadrilateral is a parallelogram if and only if the diagonals intersect at their midpoint.

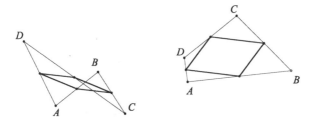

Figure 1.17. The parallelogram associated to a quadrilateral

1.42. Given a triangle $\triangle ABC$ in the real affine plane, consider the points $P_1 = A + \frac{2}{3}\overrightarrow{AB}$, $P_2 = A + \frac{1}{3}\overrightarrow{AB}$, $Q_1 = B + \frac{1}{3}\overrightarrow{BC}$, $Q_2 = B + \frac{2}{3}\overrightarrow{BC}$. Find (A, C, R), where R is the point at which the side AC cuts the straight line P_1Q_2 (see Figure 1.18).

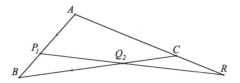

Figure 1.18. Simple ratio

1.43. Prove the converse of Menelaus' theorem. Explicitly, prove that if P, Q, R are points on the sides of a triangle $\triangle ABC$ such that

$$(P, A, B) \cdot (Q, B, C) \cdot (R, C, A) = 1,$$

then P, Q, R are collinear.

1.44. Study the converse of Thales' theorem. Explicitly, given three collinear points A, B, C and three collinear points A', B', C' (on another straight line) such that $(A, B, C) = (A', B', C')$, is it true that the straight lines AA', BB' and CC' are parallel? Prove the converse of Corollary 1.34 of Thales' theorem. That is, prove that if two straight lines r, r' meet two straight lines s, t (concurrent in a point A) in points $B \in r \cap s$, $B' \in r' \cap s$, $C \in r \cap t$, $C' \in r' \cap t$ and the triangles $\triangle ABC$ and $\triangle AB'C'$ are similar, then r and r' are parallel.

1.45. (Desargues' theorem) Let r, r', r'' be straight lines in an affine space, concurrent in a point O; and let $A, B \in r$, $A', B' \in r'$, $A'', B'' \in r''$ be points different to each other and different from O. Prove that if the points $I = AA' \cap BB'$, $J = AA'' \cap BB''$, $K = A'A'' \cap B'B''$ are defined, then they are collinear (see Figure 1.19).

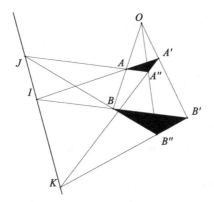

Figure 1.19. Desargues' theorem

1.46. (Pappus' theorem) Let A, B, C be points on a straight line r and let A', B', C' be points on another straight line r', in an affine plane. Prove that if the points $I = AB' \cap BA'$, $J = AC' \cap A'C$, $K = BC' \cap B'C$ are defined, then they are collinear (see Figure 1.20).

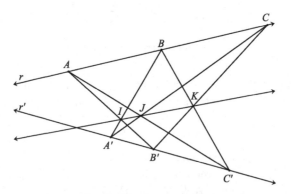

Figure 1.20. Pappus' theorem

2.1 Introduction

From the geometrical point of view, the most natural maps between affine spaces are those taking collinear points to collinear points. We shall see that these maps are also the most natural from the algebraic point of view: they essentially coincide with the affinities.

2.2 Definition of Affinity

Let $\mathbb{A}_1$ and $\mathbb{A}_2$ be affine spaces over the k-vector spaces E_1 and E_2, respectively.

Let us fix a point $P \in \mathbb{A}_1$. Then every map $f : \mathbb{A}_1 \longrightarrow \mathbb{A}_2$ induces a map

$$\tilde{f}_P : E_1 \longrightarrow E_2$$

defined by the formula

$$\tilde{f}_P(v) = \overrightarrow{f(P)f(Q)},$$

where $Q \in \mathbb{A}_1$ is the unique point such that $\overrightarrow{PQ} = v$.

This map $\tilde{f}_P$ between the vector spaces E_1 and E_2 is not in general linear. We shall say that $\tilde{f}_P$ is the map induced by the map f and the point P.

A. Reventós Tarrida, *Affine Maps, Euclidean Motions and Quadrics*,
Springer Undergraduate Mathematics Series,
DOI 10.1007/978-0-85729-710-5_2, © Springer-Verlag London Limited 2011

Definition 2.1

A map $f : \mathbb{A}_1 \longrightarrow \mathbb{A}_2$ between two affine spaces is called an *affinity* if the map $\tilde{f}_P : E_1 \longrightarrow E_2$ induced by f and a point $P \in \mathbb{A}_1$ on the corresponding k-vector spaces is a linear map.

In this case $\tilde{f}_P$ does not depend on the point, that is, $\tilde{f}_P = \tilde{f}_Q$ for all $P, Q \in \mathbb{A}_1$. In fact, if $v = \overrightarrow{QR}$, we have

$$\tilde{f}_Q(v) = \overrightarrow{f(Q)f(R)}$$

$$= \overrightarrow{f(Q)f(P)} + \overrightarrow{f(P)f(R)}$$

$$= -\tilde{f}_P(\overrightarrow{PQ}) + \tilde{f}_P(\overrightarrow{PR})$$

$$= -\tilde{f}_P(\overrightarrow{PQ}) + \tilde{f}_P(\overrightarrow{PQ} + \overrightarrow{QR})$$

$$= \tilde{f}_P(\overrightarrow{QR})$$

$$= \tilde{f}_P(v).$$

Note that we have only used the fact that $\tilde{f}_P$ preserves vector addition. If $\tilde{f}_P$ preserves vector addition but does not preserve scalar multiplication, i.e., $\tilde{f}_P(\lambda v) \neq \lambda \tilde{f}_P(v)$, we still have $\tilde{f}_P = \tilde{f}_Q$ for all $P, Q \in \mathbb{A}_1$, but f is not an affinity.

Since all of the linear maps $\tilde{f}_P$ are equal for an affinity f, i.e. they do not depend on the point P, we shall denote this map simply by $\tilde{f}$.

Thus, if $f : \mathbb{A}_1 \longrightarrow \mathbb{A}_2$ is an affinity, there is a linear map $\tilde{f} : E_1 \longrightarrow E_2$ such that, for every pair of points $P, Q \in \mathbb{A}_1$,

$$\boxed{\tilde{f}(\overrightarrow{PQ}) = \overrightarrow{f(P)f(Q)}}$$

Since

$$f(Q) = f(P) + \overrightarrow{f(P)f(Q)},$$

we have

$$f(P + \overrightarrow{PQ}) = f(Q) = f(P) + \tilde{f}(\overrightarrow{PQ}),$$

and, since $\overrightarrow{PQ}$ is an arbitrary vector, we have, for every point $P \in \mathbb{A}_1$ and for every vector $v \in E_1$,

$$\boxed{f(P + v) = f(P) + \tilde{f}(v)}$$

This equality is equivalent to the commutativity of the following diagram

$$
\begin{array}{ccc}
\mathbb{A}_1 \times E_1 & \longrightarrow & \mathbb{A}_1 \\
{\scriptstyle f \times \tilde{f}}\downarrow & & \downarrow{\scriptstyle f} \\
\mathbb{A}_2 \times E_2 & \longrightarrow & \mathbb{A}_2
\end{array}
$$

We summarize these comments in the following proposition.

Proposition 2.2

A map between two affine spaces $f : \mathbb{A}_1 \longrightarrow \mathbb{A}_2$ is an affinity if and only if there exists a linear map $\tilde{f} : E_1 \longrightarrow E_2$ between the corresponding k-vector spaces such that

$$f(P + v) = f(P) + \tilde{f}(v) \quad \text{for all } P \in \mathbb{A}_1 \text{ and } v \in E_1.$$

Proof

If f is an affinity, we take $\tilde{f} = \tilde{f}_P$, for some P, and we are done.

Conversely, if there exists a linear map $\tilde{f}$ with this property, since $Q = P + \overrightarrow{PQ}$, for all $P, Q \in \mathbb{A}_1$, we have

$$f(Q) = f(P) + \tilde{f}(\overrightarrow{PQ}),$$

and hence, $\tilde{f}(\overrightarrow{PQ}) = \overrightarrow{f(P)f(Q)} = \tilde{f}_P(\overrightarrow{PQ})$, that is, $\tilde{f}_P = \tilde{f}$. Thus, $\tilde{f}_P$ is linear, and f is an affinity. $\qquad\square$

It is clear that if such an $\tilde{f}$ exists, it is unique.

Note that if $f = \mathrm{id}$, that is, $f(P) = P$, for all $P \in \mathbb{A}$, then f is an affinity with $\tilde{f} = \mathrm{id}$, that is, $\tilde{f}(v) = v$, for all $v \in E$. This can only happen when we have $f : \mathbb{A}_1 \longrightarrow \mathbb{A}_1$, that is, when the source and target affine spaces are the same: the same set, the same associated vector space and the same action. For instance, the identity map $\mathrm{id} : \mathbb{R}(x) \longrightarrow \mathbb{R}(x)$ is not an affinity between the affine spaces k_1 and k_2 considered in Observation 1.3, page 5.

2.3 First Properties

In this section $\mathbb{A}_1$ and $\mathbb{A}_2$ are affine spaces over the k-vector spaces E_1 and E_2, respectively.

Proposition 2.3 (Uniqueness)

Let $f, g : \mathbb{A}_1 \longrightarrow \mathbb{A}_2$ be affinities that coincide on some point $P \in \mathbb{A}_1$, that is $f(P) = g(P)$, and which have the same associated linear map, $\tilde{f} = \tilde{g}$. Then $f = g$.

Proof

Let $Q \in \mathbb{A}_1$. Then

$$f(Q) = f(P + \overrightarrow{PQ})$$

$$= f(P) + \tilde{f}(\overrightarrow{PQ})$$

$$= g(P) + \tilde{g}(\overrightarrow{PQ})$$

$$= g(P + \overrightarrow{PQ})$$

$$= g(Q).$$

Hence, $f = g$.   □

Proposition 2.4 (Existence)

Let $\phi : E_1 \longrightarrow E_2$ be a linear map and suppose given two points $P \in \mathbb{A}_1$ and $Q \in \mathbb{A}_2$. Then there exists a unique affinity $f : \mathbb{A}_1 \longrightarrow \mathbb{A}_2$ such that $f(P) = Q$ and $\tilde{f} = \phi$.

Proof

Uniqueness follows from the above proposition.

Let us prove the existence. Define $f : \mathbb{A}_1 \longrightarrow \mathbb{A}_2$ by

$$f(X) = Q + \phi(\overrightarrow{PX}) \quad \text{for all } X \in \mathbb{A}_1.$$

Then, clearly $f(P) = Q$. Moreover, $\tilde{f}_P = \phi$. In fact,

$$\tilde{f}_P(\overrightarrow{PX}) = \overrightarrow{f(P)f(X)} = \overrightarrow{Qf(X)} = \phi(\overrightarrow{PX}).$$

In particular, $\tilde{f}_P$ is linear, and hence f is an affinity with associated linear map $\tilde{f} = \phi$.   □

Theorem 2.5 (Transitivity)

Let $P_1, \ldots, P_r$ be affinely independent points in an affine space $\mathbb{A}_1$. Let $Q_1, \ldots, Q_r$ be points in an affine space $\mathbb{A}_2$. Then there exists an affinity $f : \mathbb{A}_1 \longrightarrow \mathbb{A}_2$ such that $f(P_i) = Q_i$, for $i = 1, \ldots, r$. If $r = \dim \mathbb{A}_1 + 1$, then this affinity is unique.

Proof

Since the points $P_1, \ldots, P_r$ are affinely independent, the vectors $\overrightarrow{P_1 P_2}, \ldots, \overrightarrow{P_1 P_r}$ are linearly independent. We know (see, for instance, [8], page 288) that there exists a linear map $\phi : E_1 \longrightarrow E_2$ such that

$$\phi(\overrightarrow{P_1 P_i}) = \overrightarrow{Q_1 Q_i}, \quad i = 1, \ldots, r.$$

(Notice that the points Q_i are neither necessarily affinely independent nor distinct.)

We take, using Proposition 2.4, the unique affinity $f : \mathbb{A}_1 \longrightarrow \mathbb{A}_2$ such that $f(P_1) = Q_1$ and such that $\tilde{f} = \phi$. Clearly

$$f(P_i) = f(P_1 + \overrightarrow{P_1 P_i}) = f(P_1) + \tilde{f}(\overrightarrow{P_1 P_i}) = Q_1 + \overrightarrow{Q_1 Q_i} = Q_i.$$

If $r = \dim \mathbb{A}_1 + 1$, the vectors $\overrightarrow{P_1 P_2}, \ldots, \overrightarrow{P_1 P_r}$ form a basis of E_1. In this case there exists (see, for instance, [8], page 288) a unique linear map $\phi : E_1 \longrightarrow E_2$ such that

$$\phi(\overrightarrow{P_1 P_i}) = \overrightarrow{Q_1 Q_i}, \quad i = 1, \ldots, r.$$

But any affinity taking the points P_i to the points Q_i has the above linear map ϕ as associated linear map. Hence, by Proposition 2.3, this affinity is unique. $\qquad \square$

Proposition 2.6

Let $f : \mathbb{A}_1 \longrightarrow \mathbb{A}_2$ be an affinity with associated linear map $\tilde{f}$. Then f is injective if and only if $\tilde{f}$ is injective and f is surjective if and only if $\tilde{f}$ is surjective.

Proof

Let us assume that f is injective, and suppose $\tilde{f}(v) = \vec{0}$. Let $v = \overrightarrow{PQ}$. We have

$$0 = \tilde{f}(\overrightarrow{PQ}) = \overrightarrow{f(P)f(Q)},$$

and hence $f(P) = f(Q)$. This implies $P = Q$ and $v = \overrightarrow{PQ} = \vec{0}$, that is, $\tilde{f}$ is injective.

Assume now that $\tilde{f}$ is injective and suppose $f(P) = f(Q)$. Then $\tilde{f}(\overrightarrow{PQ}) = \overrightarrow{f(P)f(Q)} = \vec{0}$, and hence, $\overrightarrow{PQ} = \vec{0}$, that is, $P = Q$, and f is injective.

Assume now that f is surjective and let $v \in E_2$. Let $v = \overrightarrow{P'Q'}$ and choose P, Q such that $f(P) = P'$ and $f(Q) = Q'$. Then $\tilde{f}(\overrightarrow{PQ}) = \overrightarrow{P'Q'} = v$, and hence $\tilde{f}$ is surjective.

Assume now that $\tilde{f}$ is surjective and let $Q \in \mathbb{A}_2$. Take any point $P \in \mathbb{A}_1$ and a vector $v \in E_1$ such that $\tilde{f}(v) = \overrightarrow{f(P)Q}$. Then $f(P + v) = f(P) + \tilde{f}(v) = f(P) + \overrightarrow{f(P)Q} = Q$, and hence f is surjective. $\square$

In particular, f is bijective if and only if $\tilde{f}$ is bijective.

Observation 2.7

Let $P_1, \ldots, P_{n+1} \in \mathbb{A}$ and $Q_1, \ldots, Q_{n+1} \in \mathbb{A}$ be, respectively, affinely independent points. We know, from Theorem 2.5, that there exists a unique affinity f such that $f(P_i) = Q_i$ and a unique affinity g such that $g(Q_i) = P_i$, $i = 1, \ldots, n$.

But, as we saw in the proof of Theorem 2.5, $\tilde{f} = \tilde{g}^{-1}$. Hence, by Proposition 2.6, f is bijective and $f = g^{-1}$.

Observation 2.8

Let $\mathbb{A}$ be an affine space over a k-vector space E. To give an affine frame $\mathcal{R} = \{P; (e_1, \ldots, e_n)\}$ in $\mathbb{A}$ is equivalent to giving a bijective affinity between $\mathbb{A}$ and k^n. For this reason we say that an affine space is "essentially" k^n.

In fact, this bijective affinity is given simply by *taking coordinates*:

$$\mathbb{A} \xrightarrow{f} k^n$$
$$Q \longmapsto (q_1, \ldots, q_n),$$

where $\overrightarrow{PQ} = q_1 e_1 + \cdots + q_n e_n$.

It is clear that f is a bijective map. To see that f is an affinity we compute $\tilde{f}_P$. Let $v \in E$ and let $Q \in \mathbb{A}$ be the unique point such that $v = \overrightarrow{PQ}$. Then

$$\tilde{f}_P(v) = \tilde{f}_P(\overrightarrow{PQ}) = \overrightarrow{f(P)f(Q)} = (q_1, \ldots, q_n),$$

since $f(Q) = (q_1, \ldots, q_n)$ and $f(P) = (0, \ldots, 0)$.

That is, $\tilde{f}_P$ sends the vector of E with components $(q_1, \ldots, q_n)$ in the basis $(e_1, \ldots, e_n)$ to the vector $(q_1, \ldots, q_n) \in k^n$.

Hence,

$$\tilde{f}_P(u+v) = \tilde{f}_P(u) + \tilde{f}_P(v), \quad u,v \in E,$$

$$\tilde{f}_P(\lambda u) = \lambda \tilde{f}_P(u), \quad u \in E, \lambda \in k,$$

since the components of $u+v$ are the components of u plus the components of v, and the components of λu are the components of u multiplied by λ.

Hence, $\tilde{f}_P$ is linear and f is an affinity.

2.4 The Affine Group

We shall see that the set of all bijective affinities from an affine space into itself has the structure of a group. The group operation is, of course, composition of affinities. We begin with some slightly more general results.

Proposition 2.9

Let $f : \mathbb{A}_1 \longrightarrow \mathbb{A}_2$ and $g : \mathbb{A}_2 \longrightarrow \mathbb{A}_3$ be affinities. Then their composition $g \circ f : \mathbb{A}_1 \longrightarrow \mathbb{A}_3$ is an affinity with associated linear map $\tilde{g} \circ \tilde{f}$.

Proof

For each $P \in \mathbb{A}_1$ and each $v \in E_1$ we have

$$(g \circ f)(P+v) = g(f(P) + \tilde{f}(v)) = (g \circ f)(P) + \tilde{g} \circ \tilde{f}(v).$$

By Proposition 2.2, $g \circ f$ is an affinity with

$$\widetilde{g \circ f} = \tilde{g} \circ \tilde{f}.$$

$\square$

Proposition 2.10

Let $f : \mathbb{A}_1 \longrightarrow \mathbb{A}_2$ be a bijective affinity. Then its inverse $f^{-1} : \mathbb{A}_2 \longrightarrow \mathbb{A}_1$ is an affinity with associated linear map $\tilde{f}^{-1}$.

Proof

For each $P \in \mathbb{A}_1$ and each $v \in E_1$ we have

$$f^{-1}(P+v) = f^{-1}(P) + \tilde{f}^{-1}(v),$$

as can be seen directly by applying f to both sides of this equality. By Proposition 2.2, f^{-1} is an affinity with

$$\widetilde{f^{-1}} = \tilde{f}^{-1}.$$

$\square$

Theorem 2.11 (Affine Group)

The set of all bijective affinities from an affine space $\mathbb{A}$ into itself is a group with respect to composition of maps, called the *affine group* or *group of affinities*, and is denoted $\mathbb{G}A$.

Proof

This is an immediate consequence of the above Propositions 2.9 and 2.10. Note that composition of affinities is associative and that the unit element is the identity. $\square$

The natural action of the affine group $\mathbb{G}A$ on the space $\mathbb{A}$ itself is given by

$$\mathbb{G}A \times \mathbb{A} \longrightarrow \mathbb{A}$$
$$f, P \longmapsto f(P).$$

Thus, as noted in Observation 2.7, the group of affinities of the straight line acts simply transitively over ordered pairs of points (given two ordered pairs of points, each pair formed by different points, there exists a unique affinity taking one pair onto the other), the group of affinities of the plane acts simply transitively over ordered triples of points (given two ordered triples of points, each triple formed by different non-collinear points, there exists a unique affinity taking one triple onto the other), etc. Recall the concept of a *simply transitive action* on points (not on pairs, triples, etc.) on page 2.

Now, following F. Klein, we can say that *Affine Geometry is the study of the properties of the figures of $\mathbb{A}$ which are invariant under the action of the affine group $\mathbb{G}A$.*

Observation 2.12

When there is a bijective affinity between two affine spaces we say that these affine spaces are isomorphic. Recall that the concept of affinity is only meaningful when the vector spaces associated to the corresponding affine spaces are modeled on the same field k. Two affine spaces are isomorphic if and only if they have the same dimension. Concretely we have:

Proposition 2.13

Let $\mathbb{A}_1$ and $\mathbb{A}_2$ be affine spaces over the k-vector spaces E_1 and E_2, respectively. Then $\mathbb{A}_1$ and $\mathbb{A}_2$ are isomorphic if and only if they have the same dimension.

Proof

It is clear, by Proposition 2.6, that if the spaces are isomorphic then they have the same dimension.

Conversely, if the spaces have the same dimension, we take an isomorphism $\phi : E_1 \longrightarrow E_2$ between the associated vector spaces, which have, by definition, the same dimension. By Proposition 2.4, there exists an affinity $f : \mathbb{A}_1 \longrightarrow \mathbb{A}_2$ with $\tilde{f} = \phi$, which, by Proposition 2.6, is bijective. $\square$

2.5   Affinities and Linear Varieties

Proposition 2.14

Affinities take linear varieties to linear varieties. Concretely, let $f : \mathbb{A}_1 \longrightarrow \mathbb{A}_2$ be an affinity, $L_1 = P + [F]$ a linear variety of $\mathbb{A}_1$, and $L_2 = Q + [G]$ a linear variety of $\mathbb{A}_2$. Then

$$f(P + [F]) = f(P) + [\tilde{f}(F)],$$
$$f^{-1}(Q + [G]) = P + [\tilde{f}^{-1}(G)], \quad \text{if } P \in f^{-1}(Q + [G]).$$

Proof

Since $f(P + v) = f(P) + \tilde{f}(v)$ for all $P \in \mathbb{A}_1$ and $v \in E_1$, the first equality is evident.

To prove the second equality we observe that P is any point such that $\overrightarrow{Qf(P)} \in G$. Note that in some cases this point P does not exist; in these cases we have

$$f^{-1}(Q + [G]) = \emptyset.$$

Note also that a point $X \in \mathbb{A}_1$ belongs to $f^{-1}(Q + [G])$ if and only if $\overrightarrow{Qf(X)} \in G$.

Finally, a point $X \in \mathbb{A}_1$ belongs to $P + [\tilde{f}^{-1}(G)]$ if and only if $\tilde{f}(\overrightarrow{PX}) \in G$. Since

$$\tilde{f}(PX) = \overrightarrow{f(P)f(X)} = \overrightarrow{f(P)Q} + \overrightarrow{Qf(X)},$$

we have $X \in f^{-1}(Q + [G])$ if and only if $X \in P + \tilde{f}^{-1}(G)$. $\qquad \square$

Corollary 2.15

Injective affinities take linear varieties to linear varieties of the same dimension. In particular, they take straight lines to straight lines.

Proof

Injectivity ensures $\dim \tilde{f}(F) = \dim F$. $\qquad \square$

The Fundamental Theorem of Affine Geometry, which we shall meet later (Theorem 2.46, page 81), deals with the converse of this proposition. The question is: is a map taking straight lines to straight lines necessarily an affinity?

Proposition 2.16

Injective affinities preserve the simple ratio.

Proof

Let $A, B, C \in \mathbb{A}_1$ be three different collinear points such that $\overrightarrow{AB} = \lambda \overrightarrow{AC}$. Let $f : \mathbb{A}_1 \longrightarrow \mathbb{A}_2$ be an injective affinity. By the previous corollary, the three points $f(A), f(B), f(C) \in \mathbb{A}_2$, which are distinct, are collinear. Applying $\tilde{f}$ to the above equality one obtains

$$\tilde{f}(\overrightarrow{AB}) = \overrightarrow{f(A)f(B)} = \tilde{f}(\lambda \overrightarrow{AC}) = \lambda \overrightarrow{f(A)f(C)},$$

and hence

$$(f(A), f(B), f(C)) = \lambda = (A, B, C).$$

$\square$

2.6   Equations of Affinities

Let $f : \mathbb{A}_1 \longrightarrow \mathbb{A}_2$ be an affinity, $\mathcal{R}_1 = \{P_1; (e_1, \dots, e_n)\}$ an affine frame in $\mathbb{A}_1$ and $\mathcal{R}_2 = \{P_2; (v_1, \dots, v_m)\}$ an affine frame in $\mathbb{A}_2$.

The aim of this section is to relate the coordinates $(x_1, \dots, x_n)$ of a point $X \in \mathbb{A}_1$ to the coordinates $(y_1, \dots, y_m)$ of the point $f(X) \in \mathbb{A}_2$.

For this we write

$$\overrightarrow{P_2 f(P_1)} = \sum_{j=1}^{m} a_j v_j,$$

$$\overrightarrow{P_1 X} = \sum_{i=1}^{n} x_i e_i,$$

$$\overrightarrow{P_2 f(X)} = \sum_{j=1}^{m} y_j v_j.$$

That is,

$$f(P_1) = (a_1, \dots, a_m),$$

$$X = (x_1, \dots, x_n),$$

$$f(X) = (y_1, \dots, y_m).$$

We also write

$$\tilde{f}(e_i) = \sum_{j=1}^{m} a_{ji} v_j.$$

That is, $A = (a_{ij})$ is the matrix of $\tilde{f}$ with respect to the bases $\mathcal{B}_1 = (e_1, \dots, e_n)$ and $\mathcal{B}_2 = (v_1, \dots, v_m)$, which is usually denoted by $M(\tilde{f}, \mathcal{B}_1, \mathcal{B}_2)$.

Then

$$\overrightarrow{P_2 f(X)} = \overrightarrow{P_2 f(P_1)} + \overrightarrow{f(P_1) f(X)}$$

$$= \overrightarrow{P_2 f(P_1)} + \tilde{f}(\overrightarrow{P_1 X}),$$

that is,

$$\sum_{j=1}^{m} y_j v_j = \sum_{j=1}^{m} a_j v_j + \sum_{i=1}^{n} x_i \sum_{j=1}^{m} a_{ji} v_j.$$

Equating coefficients we get

$$y_j = a_j + \sum_{i=1}^{n} x_i a_{ji}, \quad j = 1, \dots, m.$$

Matricially

$$
\begin{pmatrix} y_1 \\ \vdots \\ y_m \end{pmatrix} = \begin{pmatrix} a_1 \\ \vdots \\ a_m \end{pmatrix} + \begin{pmatrix} a_{11} & \cdots & a_{1n} \\ \vdots & \vdots & \vdots \\ a_{m1} & \cdots & a_{mn} \end{pmatrix} \begin{pmatrix} x_1 \\ \vdots \\ x_n \end{pmatrix},
$$

or, in a more compact form,

$$
\begin{pmatrix} y_1 \\ \vdots \\ y_m \\ 1 \end{pmatrix} = \left(\begin{array}{ccc|c} a_{11} & \cdots & a_{1n} & a_1 \\ \vdots & \vdots & \vdots & \vdots \\ a_{m1} & \cdots & a_{mn} & a_m \\ \hline 0 & \cdots & 0 & 1 \end{array} \right) \begin{pmatrix} x_1 \\ \vdots \\ x_n \\ 1 \end{pmatrix}.
$$

Or, in an even more abbreviated form,

$$
\boxed{\begin{pmatrix} y \\ 1 \end{pmatrix} = \begin{pmatrix} A & a \\ 0 & 1 \end{pmatrix} \begin{pmatrix} x \\ 1 \end{pmatrix}} \tag{2.1}
$$

or

$$
\boxed{y = Ax + a}
$$

with

$$
y = \begin{pmatrix} y_1 \\ \vdots \\ y_m \end{pmatrix}, \qquad A = (a_{ij}), \qquad x = \begin{pmatrix} x_1 \\ \vdots \\ x_n \end{pmatrix}, \qquad a = \begin{pmatrix} a_1 \\ \vdots \\ a_m \end{pmatrix}.
$$

We shall use the notation

$$
M(f, \mathcal{R}_1, \mathcal{R}_2) = \begin{pmatrix} A & a \\ 0 & 1 \end{pmatrix}
$$

to indicate the matrix of f in the affine frames $\mathcal{R}_1$ and $\mathcal{R}_2$. When $\mathcal{R}_1 = \mathcal{R}_2$ (and thus $\mathbb{A}_1 = \mathbb{A}_2$) we shall simply write $M(f, \mathcal{R}_1)$.

Sometimes, when these affine frames are implicitly given, we shall simply write

$$
M(f) = \begin{pmatrix} A & a \\ 0 & 1 \end{pmatrix}.
$$

Now it is very easy to construct examples of affinities: we simply give the two matrices A and a.

Observation 2.17

Given the affine frames $\mathcal{R}_1 = \{P_1; \mathcal{B}_1\}$ of $\mathbb{A}_1$, $\mathcal{R}_2 = \{P_2; \mathcal{B}_2\}$ of $\mathbb{A}_2$, and a matrix

$$M = \begin{pmatrix} A & a \\ 0 & 1 \end{pmatrix},$$

there exists a unique affinity $f : \mathbb{A}_1 \longrightarrow \mathbb{A}_2$ such that

$$M(f, \mathcal{R}_1, \mathcal{R}_2) = \begin{pmatrix} A & a \\ 0 & 1 \end{pmatrix}.$$

In fact, we define f giving its associated linear map $\tilde{f}$ by the condition

$$M(\tilde{f}, \mathcal{B}_1, \mathcal{B}_2) = A$$

and its value on a point by the condition $f(P_1) = a^{\mathsf{T}}$. That is, $f(P_1)$ is the point with coordinates $(a_1, \ldots, a_n)$ in $\mathcal{R}_2$. From Proposition 2.4, such an affinity exists and is unique.

Observation 2.18

We have seen that, once we fix affine frames, affinities are given by affine equations of the form $y = Ax + a$. If we change these affine frames, the equations will change, but there will still be an affine relationship between the coordinates of a point and those of its transformed image.

For instance, affinities from k^n to k^m are the maps $f : k^n \longrightarrow k^m$, with $f(x_1, \ldots, x_n) = (y_1, \ldots, y_m)$, such that

$$\begin{pmatrix} y_1 \\ \vdots \\ y_m \end{pmatrix} = A \begin{pmatrix} x_1 \\ \vdots \\ x_n \end{pmatrix} + \begin{pmatrix} a_1 \\ \vdots \\ a_m \end{pmatrix},$$

with $A \in \mathcal{M}_{m \times n}(k)$. It suffices to identify the components $(x_1, \ldots, x_n)$ of a point with its coordinates in the canonical affine frame. Here, recall, $\mathcal{M}_{m \times n}(k)$ is the set of matrices with m rows and n columns with elements in the field k.

Observation 2.19

Given two affine frames $\mathcal{R}$ and $\mathcal{R}'$ in an affine space $\mathbb{A}$, we have

$$M(\mathrm{id}, \mathcal{R}', \mathcal{R}) = M(\mathcal{R}', \mathcal{R}),$$

where $M(\mathcal{R}', \mathcal{R})$ is the matrix of the change of coordinates introduced on page 19.

Equation (2.1), page 58, tells us that the relationship between the coordinates x of a point and the coordinates y of the image of this point by an affinity is given by only one matrix, as is the case for linear maps between vector spaces. So we can manipulate affinities "as if" they are linear maps.

In particular, by arguments similar to those given in the composition of linear maps (see [8], page 300), and by the change of basis formula (see [8], page 302), we obtain the next two propositions.

Proposition 2.20

The matrix of the composition of affinities is the product of the matrices of these affinities. Concretely, if $f : \mathbb{A}_1 \longrightarrow \mathbb{A}_2$ and $g : \mathbb{A}_2 \longrightarrow \mathbb{A}_3$ are affinities and we fix affine frames $\mathcal{R}_i$ on $\mathbb{A}_i$, $i = 1, 2, 3$, then

$$M(g \circ f, \mathcal{R}_1, \mathcal{R}_3) = M(g, \mathcal{R}_2, \mathcal{R}_3) \cdot M(f, \mathcal{R}_1, \mathcal{R}_2).$$

Proof

If

$$\begin{pmatrix} y \\ 1 \end{pmatrix} = \begin{pmatrix} A & a \\ 0 & 1 \end{pmatrix} \begin{pmatrix} x \\ 1 \end{pmatrix}$$

are the equations of f, and

$$\begin{pmatrix} z \\ 1 \end{pmatrix} = \begin{pmatrix} B & b \\ 0 & 1 \end{pmatrix} \begin{pmatrix} y \\ 1 \end{pmatrix}$$

are the equations of g, then

$$\begin{pmatrix} z \\ 1 \end{pmatrix} = \begin{pmatrix} B & b \\ 0 & 1 \end{pmatrix} \begin{pmatrix} A & a \\ 0 & 1 \end{pmatrix} \begin{pmatrix} x \\ 1 \end{pmatrix}$$

are the equations of $g \circ f$. Since the product of matrices is associative, we have the result.   □

We summarize this result by simply writing

$$\boxed{M(g \circ f) = M(g) \cdot M(f)}$$

Corollary 2.21

The matrix of the inverse of an affinity is the inverse of the matrix of this affinity. Concretely, if $f : \mathbb{A}_1 \longrightarrow \mathbb{A}_2$ is a bijective affinity and we fix affine frames $\mathcal{R}_i$ on $\mathbb{A}_i$, $i = 1, 2$, then

$$M(f^{-1}, \mathcal{R}_2, \mathcal{R}_1) = M(f, \mathcal{R}_1, \mathcal{R}_2)^{-1}.$$

Proof

We apply the above proposition with $g = f^{-1}$ and $\mathcal{R}_1 = \mathcal{R}_3$. □

We summarize this result by simply writing

$$\boxed{M(f^{-1}) = M(f)^{-1}}$$

Proposition 2.22

Let $f : \mathbb{A}_1 \longrightarrow \mathbb{A}_2$ be an affinity, $\mathcal{R}_1, \mathcal{R}_1'$ affine frames in $\mathbb{A}_1$, and $\mathcal{R}_2, \mathcal{R}_2'$ affine frames in $\mathbb{A}_2$. Then

$$M(f, \mathcal{R}_1, \mathcal{R}_2) = M(\mathrm{id}_2, \mathcal{R}_2, \mathcal{R}_2')^{-1} M(f, \mathcal{R}_1', \mathcal{R}_2') M(\mathrm{id}_1, \mathcal{R}_1, \mathcal{R}_1').$$

Proof

This is a consequence of Proposition 2.20 and the equality

$$f \circ \mathrm{id}_1 = \mathrm{id}_2 \circ f,$$

where id_i denotes the identity map of $\mathbb{A}_i$, $i = 1, 2$. □

In the particular case in which $\mathbb{A}_1 = \mathbb{A}_2$, we can take $\mathcal{R}_1 = \mathcal{R}_2$ and $\mathcal{R}_1' = \mathcal{R}_2'$ and we have

$$\boxed{M(f, \mathcal{R}_1, \mathcal{R}_1) = C^{-1} M(f, \mathcal{R}_1', \mathcal{R}_1') C}$$

where $C = M(\mathrm{id}, \mathcal{R}_1, \mathcal{R}_1') = M(\mathcal{R}_1, \mathcal{R}_1')$ is the matrix of the change of coordinates.

We have already stated that we will write this equation as

$$\boxed{M(f, \mathcal{R}_1) = C^{-1} M(f, \mathcal{R}_1') C} \tag{2.2}$$

Example 2.23

Find, in the affine plane $\mathbb{R}^2$, the equations of an affinity sending the triangle $\triangle ABC$ onto the triangle $\triangle A'B'C'$, with $A = (1,1)$, $B = (3,2)$, $C = (4,4)$, $A' = (-1,0)$, $B' = (-5,2)$, $C' = (7,4)$.

Solution

It follows, from Theorem 2.5, that such an affinity exists and is unique if we assume that the points A, B, C are mapped, respectively, to the points A', B', C'. We can also consider, for example, the case where A, B, C are mapped, respectively, to A', C', B', or to any other permutation of A', B', C'. Here we only study the first case.

Let us consider the affine frames

$$\mathcal{R} = \{A; (\overrightarrow{AB}, \overrightarrow{AC})\} = \{(1,1); ((2,1),(3,3))\},$$

$$\mathcal{R}' = \{A'; (\overrightarrow{A'B'}, \overrightarrow{A'C'})\} = \{(-1,0); ((-4,2),(8,4))\}.$$

By definition of the matrix associated to an affinity, we have

$$M(f, \mathcal{R}, \mathcal{R}') = I_2,$$

where f is the affinity we are looking for.

We shall use the notation I_n for the $n \times n$ identity matrix, i.e. the diagonal matrix with 1s on the diagonal and zeros elsewhere.

By Proposition 2.22 we have

$$M(f, \mathcal{C}, \mathcal{C}) = M(\text{id}, \mathcal{C}, \mathcal{R}')^{-1} M(f, \mathcal{R}, \mathcal{R}') M(\text{id}, \mathcal{C}, \mathcal{R})$$

$$= M(\mathcal{R}', \mathcal{C}) M(\mathcal{R}, \mathcal{C})^{-1},$$

where $\mathcal{C}$ is the canonical affine frame.

These two matrices are easily computable. Concretely we have

$$M(\mathcal{R}', \mathcal{C}) = \begin{pmatrix} -4 & 8 & -1 \\ 2 & 4 & 0 \\ 0 & 0 & 1 \end{pmatrix} \quad \text{and} \quad M(\mathcal{R}, \mathcal{C}) = \begin{pmatrix} 2 & 3 & 1 \\ 1 & 3 & 1 \\ 0 & 0 & 1 \end{pmatrix}$$

and the affinity we are looking for is

$$\begin{cases} x' = -\dfrac{20}{3}x + \dfrac{28}{3}y - \dfrac{11}{3}, \\[2mm] y' = \dfrac{2}{3}x + \dfrac{2}{3}y - \dfrac{4}{3}. \end{cases}$$

$\square$

2.7 Invariant Varieties

In this section we only consider affinities from an affine space $\mathbb{A}$ into itself.

A point $P \in \mathbb{A}$ is a *fixed point* of an affinity $f : \mathbb{A} \longrightarrow \mathbb{A}$ if and only if $f(P) = P$.

A linear variety $L = P + [F] \subset \mathbb{A}$ is *invariant* under an affinity $f : \mathbb{A} \longrightarrow \mathbb{A}$ if and only if $f(L) \subset L$.

In particular, fixed points are invariant linear varieties of dimension zero.

Proposition 2.24

A linear variety $L = P + [F]$ of the affine space $\mathbb{A}$ is invariant under an affinity $f : \mathbb{A} \to \mathbb{A}$ if and only if

(1) $\overrightarrow{Pf(P)} \in F$; and
(2) $\tilde{f}(F) \subset F$.

Proof

By Proposition 2.14 we have

$$f(L) = f(P) + [\tilde{f}(F)] = P + \overrightarrow{Pf(P)} + [\tilde{f}(F)],$$

and hence $f(L) \subset L$ if and only if $\overrightarrow{Pf(P)} \in F$ and $\tilde{f}(F) \subset F$.   □

In the next corollary we use the notion of an eigenvector, which is reviewed on page 333 of the Appendix.

Corollary 2.25

A straight line $L = P + \langle v \rangle$ of an affine space $\mathbb{A}$ is invariant under an affinity $f : \mathbb{A} \to \mathbb{A}$ if and only if there exist $\lambda, \mu \in k$ such that

(1) $\overrightarrow{Pf(P)} = \lambda v$; and
(2) $\tilde{f}(v) = \mu v$ (that is, v *is an eigenvector of* $\tilde{f}$).

Proof

This is the above proposition in dimension 1.   □

Notice that the condition that the direction vector of a straight line L is an eigenvector of $\tilde{f}$ is a necessary but not sufficient condition for L to be invariant under f. For instance, the straight line $y = 1$ of $\mathbb{R}^2$ is not invariant under the affinity

$$\begin{cases} x' = x, \\ y' = y + 1, \end{cases}$$

although its direction vector is an eigenvector of $\tilde{f}$, since $\tilde{f} = \mathrm{id}$. However, the straight line $x = 0$ is invariant, since its direction vector is $v = (0, 1)$ and, taking $P = (0, 0)$, we have $\overrightarrow{Pf(P)} = v$.

Proposition 2.26

If f is a bijective affinity and $L = P + [F]$ is an invariant linear variety, then $f(L) = L$.

Proof

In this case $\tilde{f}$ is also bijective, and hence $\tilde{f}(F) = F$. Thus,

$$f(L) = f(P) + [\tilde{f}(F)] = P + \overrightarrow{Pf(P)} + [F] = P + [F] = L.$$

$\square$

Note that we can have $f(L) = L$, but $f(P) \neq P$ for all $P \in L$. For instance, the translation T_u (see Section 2.8) leaves invariant any straight line with direction vector u, but it does not have any fixed points.

Proposition 2.27

The set $\mathrm{Fix}(f)$ of fixed points of an affinity f is either a linear variety directed by $\ker(\tilde{f} - \mathrm{id})$ or it is the empty set.

Proof

Let us assume that there exists a $P \in \mathbb{A}$ such that $f(P) = P$. Then, for each $u \in E$, $P + u$ is a fixed point of f if and only if

$$P + u = f(P + u) = f(P) + \tilde{f}(u) = P + \tilde{f}(u),$$

that is, if and only if $\tilde{f}(u) = u$, or, equivalently, $u \in \ker(\tilde{f} - \mathrm{id})$.

Hence,

$$\text{Fix}(f) = P + [\ker(\tilde{f} - \text{id})].$$

□

The study of fixed points will be of great interest in what follows, so we begin with the following result.

Proposition 2.28

An affinity $f : \mathbb{A} \longrightarrow \mathbb{A}$ has a unique fixed point if and only if the associated linear map $\tilde{f} : E \longrightarrow E$ does not have eigenvalue 1.

Proof

Let us assume that P is the unique fixed point. Since $\text{Fix}(f) = P + [\ker(\tilde{f} - \text{id})]$, it follows from the above proposition that $\ker(\tilde{f} - \text{id}) = \{\vec{0}\}$, and hence there is no eigenvector with eigenvalue 1.

Conversely, let us assume that $\tilde{f}$ does not have eigenvalue 1.

This means

$$\ker(\tilde{f} - \text{id}) = \{\vec{0}\}$$

and hence, by the above proposition, it only remains to prove that $\text{Fix}(f) \neq \emptyset$.

For this, we look for a point Q and a vector v such that $f(Q + v) = Q + v$. That is, such that $f(Q) + \tilde{f}(v) = Q + v$. Equivalently, we are looking for a point Q and a vector v such that

$$\overrightarrow{Qf(Q)} = -(\tilde{f} - \text{id})(v). \tag{2.3}$$

But, since the kernel of $(\tilde{f} - \text{id})$ is zero, $(\tilde{f} - \text{id})$ is invertible. Hence, given any point Q, there exists a (unique) vector v satisfying (2.3), and so the point $P = Q + v$, with v thus constructed, is fixed. □

Using the equations of an affinity, we can give a very simple proof of Proposition 2.28.

Proposition 2.29 (Equations of fixed points)

Let us assume that the equations of an affinity f in some frame $\mathcal{R} = \{P; \mathcal{B}\}$ of an affine space $\mathbb{A}$ of dimension n is $x' = Ax + a$. Then f has a unique fixed point if and only if $\det(A - I_n) \neq 0$.

Proof

The point with coordinates x is a fixed point if and only if

$$x = Ax + a, \tag{2.4}$$

that is,

$$(A - I_n)x = -a.$$

But this system has a unique solution if and only if $\det(A - I_n) \neq 0$. $\square$

Note that the condition $\det(A - I_n) \neq 0$ is equivalent to the condition that 1 is not an eigenvalue of A. But A is the matrix of $\tilde{f}$ in $\mathcal{B}$, and hence f has a unique fixed point if and only if $\tilde{f}$ does not have eigenvalue 1.

Equation (2.4) is the *equation of fixed points* of f.

2.8 Examples of Affinities

2.8.1 Translations

Definition 2.30

A *translation* is an affinity $f : \mathbb{A} \longrightarrow \mathbb{A}$ such that $\tilde{f} = \mathrm{id}$.

Proposition 2.31

An affinity f is a translation if and only if there is a $u \in E$ such that

$$f(Q) = Q + u \quad \text{for all } Q \in \mathbb{A}.$$

Proof

Since

$$\overrightarrow{Pf(P)} = \overrightarrow{PQ} + \overrightarrow{Qf(Q)} + \overrightarrow{f(Q)f(P)} = \overrightarrow{PQ} + \overrightarrow{Qf(Q)} + \tilde{f}(\overrightarrow{QP}),$$

every affinity satisfies the fundamental relation

$$\boxed{\overrightarrow{Pf(P)} + (\tilde{f} - \mathrm{id})\overrightarrow{PQ} = \overrightarrow{Qf Q}}$$

Hence, if $\tilde{f} = \mathrm{id}$, we have $\overrightarrow{Pf(P)} = \overrightarrow{Qf Q}$ for all $P, Q \in \mathbb{A}$. Let $u = \overrightarrow{Pf(P)}$. Then, for all $Q \in \mathbb{A}$, we have

$$f(Q) = Q + \overrightarrow{Qf(Q)} = Q + \overrightarrow{Pf(P)} = Q + u.$$

Conversely, if there is a $u \in E$ such that $u = \overrightarrow{Pf(P)} = \overrightarrow{Qf(Q)}$, for all $P, Q \in \mathbb{A}$, the *fundamental relation* directly implies $\tilde{f} = \mathrm{id}$. $\qquad\square$

For this reason we shall denote translations by T_u and we will say that T_u is *the translation by vector u*, where u is called the *translation vector*.

Equations of Translations   Let $\mathcal{R} = \{P; \mathcal{B}\}$ be an arbitrary affine frame and assume that the translation vector u has components $u = (u_1, \ldots, u_n)$ in $\mathcal{B}$. In particular, $T_u(P) = P + u = (u_1, \ldots, u_n)$. Then, by Section 2.6, we have

$$M(T_u, \mathcal{R}) = \begin{pmatrix} I_n & u \\ 0 & 1 \end{pmatrix},$$

or, equivalently,

$$\begin{cases} x'_1 = x_1 + u_1, \\ \quad \vdots \\ x'_n = x_n + u_n. \end{cases}$$

If $u \neq 0$, we can complete the translation vector to a basis $\mathcal{B} = (u, e_2, \ldots, e_n)$ of E, and then, in the affine frame $\mathcal{R}' = \{P; \mathcal{B}\}$, with $P \in \mathbb{A}$ arbitrary, we have

$$\begin{cases} x'_1 = x_1 + 1, \\ x'_2 = x_2, \\ \quad \vdots \\ x'_n = x_n. \end{cases}$$

Observation 2.32

The set of all translations of an affine space $\mathbb{A}$ is a group with respect to composition of maps. The identity element is translation by the vector $\vec{0}$. The group properties follow from the equalities $T_u \circ T_v = T_{u+v}$ (the composition of translations is a translation) and $T_u^{-1} = T_{-u}$ (the inverse of a translation is a translation). This group, denoted by $\mathbb{T}$, is a subgroup of the group of affinities $\mathbb{G}A$. Let

$$\Phi : \mathbb{G}A \longrightarrow \text{End E}$$
$$f \longmapsto \tilde{f}$$

be the map sending each affinity f to its associated endomorphism $\tilde{f}$.

By definition of translation we have

$$\mathbb{T} = \ker \Phi.$$

In particular, $\mathbb{T}$ is a normal subgroup of $\mathbb{G}A$, and, from the Isomorphism Theorem (see [8], page 284) and because Φ is surjective, we have

$$\mathbb{G}A/\mathbb{T} \cong \text{End E}.$$

2.8.2 Homotheties

Definition 2.33

A *homothety* is an affinity $f : \mathbb{A} \longrightarrow \mathbb{A}$ such that $\tilde{f} = \lambda \, \text{id}$, $\lambda \neq 0, 1$. λ is called the *similitude ratio* of the homothety.

Proposition 2.34

Homotheties have a unique fixed point.

Proof

Let us fix $P \in \mathbb{A}$ and assume that $Q \in \mathbb{A}$ is a fixed point. Then,

$$P + \overrightarrow{PQ} = Q$$
$$= f(Q)$$
$$= f(P + \overrightarrow{PQ})$$
$$= f(P) + \lambda \overrightarrow{PQ}$$
$$= P + \overrightarrow{Pf(P)} + \lambda \overrightarrow{PQ}.$$

Hence, Q is a fixed point if and only if

$$\overrightarrow{Pf(P)} = (1 - \lambda)\overrightarrow{PQ},$$

that is, if and only if

$$Q = P + \frac{1}{1 - \lambda} \overrightarrow{Pf(P)}. \tag{2.5}$$

Moreover, it is clear that Q is the unique fixed point. To see this let us assume that there are two different fixed points Q, Q'. Then

$$\tilde{f}(\overrightarrow{QQ'}) = \overrightarrow{f(Q)f(Q')} = \overrightarrow{QQ'},$$

implying $\tilde{f} = \mathrm{id}$, a contradiction since $\tilde{f} = \lambda\,\mathrm{id}$ with $\lambda \neq 1$. $\qquad\square$

Note that, in particular, we have proved that for all $P, R \in \mathbb{A}$ we have

$$P + \frac{1}{1 - \lambda} \overrightarrow{Pf(P)} = R + \frac{1}{1 - \lambda} \overrightarrow{Rf(R)}.$$

Since homotheties are determined by the fixed point, called the *center* of the homothety, and by the similitude ratio λ, we shall denote by $h_{P,\lambda}$ the homothety with center P and similitude ratio λ.

Equations of homotheties   Let $\mathcal{R} = \{P; (e_1, \ldots, e_n)\}$ be an affine frame with origin the unique fixed point P of the homothety and with $(e_1, \ldots, e_n)$ an arbitrary basis of E. It follows from Section 2.6 that

$$M(h_{P,\lambda}, \mathcal{R}) = \begin{pmatrix} \lambda I_n & 0 \\ 0 & 1 \end{pmatrix},$$

or, equivalently,

$$\begin{cases} x'_1 = \lambda x_1, \\ \quad\vdots \\ x'_n = \lambda x_n. \end{cases}$$

Observation 2.35

The set of all homotheties of an affine space $\mathbb{A}$ is *not* a group with respect to composition of maps. The identity translation is not a homothety.

Even if we add the identity to the set of all homotheties, we still don't have a group, since *the composition of homotheties with different centers and inverse similitude ratios is a translation.*

Indeed, if we denote by $h_{X,\nu}$ the homothety with center $X \in \mathbb{A}$ and similitude ratio $\nu \in k$, it follows from Proposition 2.9 that

$$h_{P,\lambda} \circ h_{Q,\mu} = \begin{cases} T_{QQ'} & \lambda\mu = 1, \\ h_{R,\lambda\mu} & \lambda\mu \neq 1, \end{cases}$$

where $Q' = h_{P,\lambda}(Q)$ and

$$R = Q + \frac{1}{1 - \lambda\mu}\overrightarrow{Qh_{P,\lambda}(Q)}.$$

We can arrive at the same conclusion using coordinates. Indeed, if we take an affine frame $\mathcal{R}$ with origin at Q, by Proposition 2.20 we have

$$M(h_{P,\lambda} \circ h_{Q,\mu}, \mathcal{R}) = M(h_{P,\lambda}, \mathcal{R}) \circ M(h_{Q,\mu}, \mathcal{R})$$

$$= \begin{pmatrix} \lambda I_n & a \\ 0 & 1 \end{pmatrix}\begin{pmatrix} \mu I_n & 0 \\ 0 & 1 \end{pmatrix}$$

$$= \begin{pmatrix} \lambda\mu I_n & a \\ 0 & 1 \end{pmatrix},$$

which gives the above result, that is, $h_{P,\lambda} \circ h_{Q,\mu}$ is a homothety if $\lambda\mu \neq 1$, or a translation if $\lambda\mu = 1$. Here $n = \dim \mathbb{A}$.

Affinities such that $\tilde{f} = \lambda \operatorname{id}$ (homotheties and translations) are called *dilations*, and they constitute a group (see the definition of dilation given in the introduction, page viii).

2.8.3 Symmetries

Definition 2.36

A *symmetry* is an affinity $f : \mathbb{A} \longrightarrow \mathbb{A}$ such that $f^2 = \operatorname{id}$.

Note first that, for all points $P \in \mathbb{A}$, the point

$$Q = P + \frac{1}{2}\overrightarrow{Pf(P)}$$

is fixed.

Hence, the linear variety $\operatorname{Fix}(f)$ is non-empty. Let us assume $\dim \operatorname{Fix}(f) = r$. Recall $\operatorname{Fix}(f) = Q + [\ker(\tilde{f} - \operatorname{id})]$, where Q is any fixed point.

Since

$$u = \frac{1}{2}(u + \tilde{f}(u)) + \frac{1}{2}(u - \tilde{f}(u)), \quad \text{for all } u \in E,$$

we have

$$E = \ker(\tilde{f} - \mathrm{id}) \oplus \ker(\tilde{f} + \mathrm{id}),$$

since $\frac{1}{2}(u + \tilde{f}(u)) \in \ker(\tilde{f} - \mathrm{id})$, $\frac{1}{2}(u - \tilde{f}(u)) \in \ker(\tilde{f} + \mathrm{id})$ and, obviously, the intersection of these subspaces is the zero vector.

This is the decomposition given by the annihilating polynomial, see [8], page 361. Note that $f^2 = \mathrm{id}$ implies $\tilde{f}^2 = \mathrm{id}$, and hence $(\tilde{f} - \mathrm{id}) \circ (\tilde{f} + \mathrm{id}) = 0$. That is, the polynomial $(x - 1)(x + 1)$ is a multiple of the minimal polynomial.

Every point $P \in \mathbb{A}$ can be written as (see Figure 2.1)

$$P = Q + \overrightarrow{QP} = Q + u + v, \quad u \in \ker(\tilde{f} - \mathrm{id}), \ v \in \ker(\tilde{f} + \mathrm{id}),$$

where Q is the above fixed point.

Thus,

$$f(P) = f(Q + u + v) = Q + u - v = P - 2v.$$

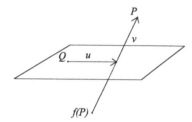

Figure 2.1. Symmetry

Note that if $r = 0$, the linear variety $\mathrm{Fix}(f)$ is a point. In this case f is said to be a *central symmetry*. In particular, $\ker(\tilde{f} - \mathrm{id}) = \{\vec{0}\}$ and $P = Q + v$ with $v \in \ker(\tilde{f} + \mathrm{id})$. The image of a point is given by $f(P) = P - 2v = P - 2\overrightarrow{QP}$.

If $r = 1$, $\mathrm{Fix}(f)$ is a straight line, and we say that f is an *axial symmetry*; if $r = 2$, $\mathrm{Fix}(f)$ is a plane, and we say that f is a *mirror symmetry*.

Equations of symmetries   Let $\mathcal{R} = \{P; (e_1, \ldots, e_n)\}$ be an affine frame with origin a fixed point P of the given symmetry, and let $(e_1, \ldots, e_n)$ be a basis with $e_i \in \ker(\tilde{f} - \mathrm{id})$, $i = 1, \ldots, r$, and $e_i \in \ker(\tilde{f} + \mathrm{id})$, $i = r + 1, \ldots, n$. Then, from Section 2.6, we have

$$M(f, \mathcal{R}) = \left(\begin{array}{ccc|c} 1 & & & 0 \\ & \ddots & & \vdots \\ & & -1 & 0 \\ \hline 0 & \cdots & 0 & 1 \end{array} \right),$$

or, equivalently,

$$\begin{cases} x'_1 = x_1, \\ \quad \vdots \\ x'_r = x_r, \\ x'_{r+1} = -x_{r+1}, \\ \quad \vdots \\ x'_n = -x_n. \end{cases}$$

Observation 2.37

Each decomposition of the vector space E as a direct sum $E = F \oplus G$, together with a point P, gives rise to a symmetry $s_L : \mathbb{A} \longrightarrow \mathbb{A}$ defined by

$$s_L(P + v) = P + v_1 - v_2, \quad \text{for all } v = v_1 + v_2 \in E, \text{ with } v_1 \in F, v_2 \in G,$$

and called the *symmetry with respect to* $L = P + [F]$ *in the direction* G. Note that $s_F^2 = \mathrm{id}$.

If $\dim L = 0$, we have a *central symmetry*; if $\dim L = 1$, we have an *axial symmetry*; if $\dim L = 2$, we have a *mirror symmetry*.

2.8.4 Projections

Definition 2.38

A *projection* is an affinity $f : \mathbb{A} \longrightarrow \mathbb{A}$ such that $f^2 = f$.

In this case it is clear that $\mathrm{Fix}(f) = \Im(f)$. Hence, the linear variety $\mathrm{Fix}(f)$ is non-empty. Let us assume $\dim \mathrm{Fix}(f) = r$ and set $\mathrm{Fix}(f) = Q + [\ker(\tilde{f} - \mathrm{id})]$, where Q is any fixed point.

Since

$$u = \tilde{f}(u) + (u - \tilde{f}(u)), \quad \text{for all } u \in E,$$

we have

$$E = \ker(\tilde{f} - \mathrm{id}) \oplus \ker(\tilde{f}),$$

because $\tilde{f}(u) \in \ker(\tilde{f} - \mathrm{id})$, $(u - \tilde{f}(u)) \in \ker(\tilde{f})$ and, obviously, the intersection of these subspaces is the zero vector.

This is the decomposition induced by the annihilating polynomial of f, $x(x-1)$, see [8], page 361. In fact, $f^2 = f$ implies $\tilde{f}^2 = \tilde{f}$, and hence, $(\tilde{f} - \mathrm{id}) \circ \tilde{f} = 0$. That is, the polynomial $x(x-1)$ is a multiple of the minimal polynomial.

Every point $P \in \mathbb{A}$ can be written in a unique way as (see Figure 2.2)

$$P = Q + \overrightarrow{QP} = Q + u + v, \quad u \in \ker(\tilde{f} - \mathrm{id}), \ v \in \ker(\tilde{f}),$$

where Q is the above fixed point.

Thus

$$f(P) = f(Q + u + v) = Q + u = P - v.$$

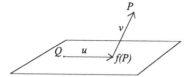

Figure 2.2. Projection

Note that if $r = 0$, the linear variety $\mathrm{Fix}(f)$ reduces to the point P. Then $f(X) = P$ for all $X \in \mathbb{A}$. If $r = \dim E$, then $f = \mathrm{id}$.

Equations of projections   Let $\mathcal{R} = \{P; (e_1, \ldots, e_n)\}$ be an affine frame with origin a fixed point P of the symmetry, and let $(e_1, \ldots, e_n)$ be a basis with $e_i \in \ker(\tilde{f} - \mathrm{id})$, $i = 1, \ldots, r$ and $e_i \in \ker(\tilde{f})$, $i = r + 1, \ldots, n$. Then, by Section 2.6, we have

$$M(f, \mathcal{R}) = \left(\begin{array}{ccc|c} 1 & & & 0 \\ & \ddots & & \vdots \\ & & 0 & 0 \\ \hline 0 & \cdots & 0 & 1 \end{array} \right),$$

or, equivalently,

$$\begin{cases} x_1' = x_1, \\ \quad \vdots \\ x_r' = x_r, \\ x_{r+1}' = 0, \\ \quad \vdots \\ x_n' = 0. \end{cases}$$

Observation 2.39

Each decomposition of the vector space E as a direct sum $E = F \oplus G$, together with a point P, gives rise to a projection $p_L : \mathbb{A} \longrightarrow \mathbb{A}$ defined by

$$p_L(P + v) = P + v_1, \text{ for all } v = v_1 + v_2 \in E, \text{ with } v_1 \in F, v_2 \in G,$$

called the *projection on* $L = P + [F]$ *in the direction* G.

2.9 Characterization of Affinities of the Line

Theorem 2.40

Let $\mathbb{A}_1$, $\mathbb{A}_2$ be two affine spaces over the k-vector spaces E_1, E_2, respectively, of dimension 1, and with $k \neq \mathbb{Z}/2\mathbb{Z}$. Let $f : \mathbb{A}_1 \longrightarrow \mathbb{A}_2$ be a map preserving the simple ratio. Then f is an affinity.

Proof

Fix $P \in \mathbb{A}_1$ and let us study $\tilde{f}_P$. First we want to prove that $\tilde{f}_P(u + v) = \tilde{f}_P(u) + \tilde{f}_P(v)$ for all $u, v \in E_1$. Since this formula is true for $u = \vec{0}$ or $v = \vec{0}$, we can assume given two vectors $u, v \in E_1$ different from zero. We know that there are points (unique) Q, R, S, with $P \neq Q, P \neq R, Q \neq S$ and $R \neq S$, such that

$$u = \overrightarrow{PQ},$$

$$v = \overrightarrow{PR},$$

$$u + v = \overrightarrow{PS}.$$

Note that

$$\overrightarrow{QS} = \overrightarrow{QP} + \overrightarrow{PS} = -u + u + v = \overrightarrow{PR}.$$

Analogously

$$\overrightarrow{RS} = \overrightarrow{RP} + \overrightarrow{PQ} + \overrightarrow{QS} = -v + u + v = \overrightarrow{PQ}.$$

If the three points P, Q, R are distinct (that is, $u \neq v$) we can compute their simple ratio. Put $(P, Q, R) = \lambda$, that is, $\overrightarrow{PQ} = \lambda \overrightarrow{PR}$. Since the simple ratio is preserved, the three points $P' = f(P)$, $Q' = f(Q)$ and $R' = f(R)$ are also distinct and

$$(P', Q', R') = \lambda,$$

that is,

$$\overrightarrow{P'Q'} = \lambda \overrightarrow{P'R'}. \tag{2.6}$$

Note, on the other hand, that if the three points P, Q, S are distinct (that is, $u \neq -v$) we can compute their simple ratio, obtaining

$$(Q, P, S) = -\lambda,$$

since

$$\overrightarrow{QP} = -\lambda \overrightarrow{PR} = -\lambda \overrightarrow{QS}.$$

Since the simple ratio is preserved, the three points P', Q' and $S' = f(S)$ are also distinct and

$$(Q', P', S') = -\lambda,$$

that is,

$$\overrightarrow{Q'P'} = -\lambda \overrightarrow{Q'S'}. \tag{2.7}$$

From (2.6) and (2.7) we directly deduce that $\overrightarrow{P'R'} = \overrightarrow{Q'S'}$.
Thus,

$$\tilde{f}_P(u + v) = \tilde{f}_P(\overrightarrow{PS})$$

$$= \overrightarrow{P'S'}$$

$$= \overrightarrow{P'Q'} + \overrightarrow{Q'S'}$$

$$= \overrightarrow{P'Q'} + \overrightarrow{P'R'}$$

$$= \tilde{f}_P(u) + \tilde{f}_P(v).$$

This proves that $\tilde{f}_P$ preserves vector addition (with the hypothesis that these vectors are neither equal nor opposite). It remains to prove that $\tilde{f}_P$ preserves scalar multiplication. Since the formula that we want to prove, $\tilde{f}_P(\lambda v) = \lambda \tilde{f}_P(v)$, is clearly true for $\lambda = 0$, $\lambda = 1$ or $v = \vec{0}$, we can assume from now on that $\lambda \in k$ and $v \in E_1$, with $\lambda \neq 0$, $\lambda \neq 1$ and $v \neq 0$.

We know that there exists a unique point $Q \in \mathbb{A}_1$ such that $v = \overrightarrow{PQ}$, and a unique point $T \in \mathbb{A}_1$ such that $\lambda v = \overrightarrow{PT}$.

Then it is clear that the points P, Q, T are distinct and that

$$(P, T, Q) = \lambda,$$

hence

$$(P', T', Q') = \lambda,$$

where $P' = f(P)$, $T' = f(T)$, $Q' = f(Q)$.

Thus,

$$\tilde{f}_P(\lambda v) = \tilde{f}_P(\overrightarrow{PT}) = \overrightarrow{P'T'} = \lambda\overrightarrow{P'Q'} = \lambda\tilde{f}_P(v).$$

Finally we remark that, since $\tilde{f}_P$ preserves scalar multiplication, the formula $\tilde{f}_P(u+v) = \tilde{f}_P(u) + \tilde{f}_P(v)$ is also true for $u = \pm v$. This completes the proof. $\square$

For a slightly different proof of this theorem, see Exercise 2.11 of this chapter, page 88.

Note that when $k = \mathbb{Z}/2\mathbb{Z}$ the straight lines have only two points and the hypothesis on the simple ratio doesn't make sense. For this reason, there are maps between affine spaces over $\mathbb{Z}/2\mathbb{Z}$ which are not affinities. For instance, considering the field $k = \mathbb{Z}/2\mathbb{Z}$ as an affine space, the map $f : \mathbb{Z}/2\mathbb{Z} \longrightarrow \mathbb{Z}/2\mathbb{Z}$ given by $f(0) = f(1) = 1$ is not an affinity because $f(1+1) \neq f(1) + f(1)$.

2.10 The Fundamental Theorem of Affine Geometry

First let us recall a definition from linear algebra.

Definition 2.41

Let E_1, E_2 be two k-vector spaces. A map $\tilde{f} : E_1 \longrightarrow E_2$ is called *semi-linear* if there exists an *automorphism* σ of the field k such that

$$\tilde{f}(u+v) = \tilde{f}(u) + \tilde{f}(v), \quad \text{for all } u, v \in E_1,$$

$$\tilde{f}(\lambda u) = \sigma(\lambda)\tilde{f}(u), \quad \text{for all } \lambda \in k, u \in E_1.$$

Recall that an *automorphism* of the field k is a bijective map $\sigma : k \longrightarrow k$ such that $\sigma(a+b) = \sigma(a) + \sigma(b)$, $\sigma(ab) = \sigma(a)\sigma(b)$, for each $a, b \in k$, and $\sigma(1) \neq 0$. This implies $\sigma(0) = 0$ and $\sigma(1) = 1$.

Let us return our attention to affine spaces.

Definition 2.42

A map $f : \mathbb{A}_1 \longrightarrow \mathbb{A}_2$ between two affine spaces is called a *semi-linear affine transformation* if the map $\tilde{f}_P : E_1 \longrightarrow E_2$, induced by f and by a point $P \in \mathbb{A}_1$ on the corresponding k-vector spaces is semi-linear.

In this case we also say that f is a *semi-affinity*.

Recall that, using only the fact that $\tilde{f}_P$ preserves vector addition, one can prove $\tilde{f}_P = \tilde{f}_Q$ for all $Q \in \mathbb{A}$. Hence, it is natural to denote simply by $\tilde{f}$ the semi-linear map associated with the semi-affinity f.

Equivalently, we have the following.

Proposition 2.43

A map $f : \mathbb{A}_1 \longrightarrow \mathbb{A}_2$ is a semi-affinity if and only if there is a semi-linear map $\tilde{f} : E_1 \longrightarrow E_2$ such that

$$f(P + u) = f(P) + \tilde{f}(u), \quad \text{for all } P \in \mathbb{A} \text{ and } u \in E_1.$$

Proof

Compare the proof of Proposition 2.2.   □

If such a map $\tilde{f}$ exists, it is unique.

The Fundamental Theorem of Affine Geometry, which we are going to prove in this section, states that *if a bijective map f takes collinear points to collinear points then it is a semi-affinity.*

Let us first show that the bijective map f maps collinear points to collinear points if and only if it maps straight lines to straight lines.

It could be the case that the image of a straight line is only a proper part of a straight line. For instance, it is easy to construct a map $f : \mathbb{R} \longrightarrow \mathbb{R}$ with $f(\mathbb{R}) \subset (0, 1)$. Such a map sends collinear points to collinear points, but it does not send straight lines to straight lines. However, this f is not bijective. Does there exist a bijective map $f : \mathbb{A} \longrightarrow \mathbb{A}$ sending collinear points to collinear points such that the image $f(L)$ of a straight line L is properly contained in a straight line L'?

Before answering this question, we carefully study the special case of the plane. Readers who wish to proceed directly to the general case may prefer to skip to Proposition 2.45.

Proposition 2.44

Let $\mathbb{A}_1$ and $\mathbb{A}_2$ be two affine planes over the k-vector spaces E_1, E_2, respectively. Assume $k \neq \mathbb{Z}/2\mathbb{Z}$.

Then every bijective map $f : \mathbb{A}_1 \longrightarrow \mathbb{A}_2$ sending collinear points to collinear points bijectively sends straight lines onto straight lines.

Proof

Let L be a straight line in $\mathbb{A}_1$ and P_1, P_2 be distinct points of L. Let L' be the straight line determined by the distinct points $P_1' = f(P_1)$, $P_2' = f(P_2)$.

Since f maps collinear points to collinear points, it is clear that

$$f(L) \subset L'.$$

It remains to prove the opposite inclusion.

Let $Z' \in L'$. There is a $Z \in \mathbb{A}_1$ such that $f(Z) = Z'$. We want to prove that $Z \in L$. Let us assume that $Z \notin L$, see Figure 2.3.

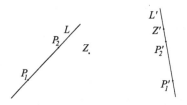

Figure 2.3. Collinear points to collinear points

Take an arbitrary point $X \in \mathbb{A}_1$. If the straight line determined by the points X, Z cuts L in a point Y (see the drawing on the left of Figure 2.4), then $Y' = f(Y) \in L'$; since $Z' \in L'$, and $X' = f(X)$ belongs to the straight line determined by Y' and Z', we must have $X' \in L'$.

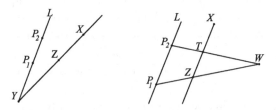

Figure 2.4. Collinear points to collinear points

If the straight line determined by the points X, Z does not cut L (and, therefore, the two lines are parallel, since they lie in a plane, see the drawing on the right of Figure 2.4), we take the point $T = Z + t\overrightarrow{P_1 P_2}$ with $t \neq 0, 1$ (here we use the assumption that $k \neq \mathbb{Z}/2\mathbb{Z}$). Then the straight lines $P_2 T$ and $P_1 Z$ meet in a point W. Since W is collinear with P_1 and Z, we have $W' = f(W) \in L'$.

Since T is collinear with P_2 and W, we have $T' = f(T) \in L'$. Finally, since X is collinear with T and Z, we have $X' \in L'$.

Therefore, in all cases, given $X \in \mathbb{A}_1$ we have $X' \in L'$, that is, we have $f(\mathbb{A}_1) \subset L'$, contradicting the bijectivity of f. Hence, $Z \in L$, and this completes the proof.   □

Let us turn to the general case.

Proposition 2.45

Let $\mathbb{A}_1$ and $\mathbb{A}_2$ be affine spaces of dimension $n \geq 2$ over the k-vector spaces E_1, E_2, respectively. Assume $k \neq \mathbb{Z}/2\mathbb{Z}$.

Then every bijective map $f : \mathbb{A}_1 \longrightarrow \mathbb{A}_2$ sending collinear points to collinear points also bijectively maps linear varieties of dimension r onto linear varieties of dimension r, $r = 1, \dots, n$. In particular, f bijectively maps straight lines onto straight lines.

Proof

The central idea of the proof is to show that f bijectively maps hyperplanes onto hyperplanes. In fact, once we have done this, we can restrict f to a hyperplane. The hypotheses of the theorem will then continue to hold, but in dimension $n - 1$; repeating the argument we eventually prove that f bijectively maps straight lines onto straight lines.

Let L be a hyperplane in $\mathbb{A}_1$. Let $P_1, \dots, P_n$ be points generating this hyperplane, that is, such that

$$L = P_1 + \langle \overrightarrow{P_1 P_2}, \dots, \overrightarrow{P_1 P_n} \rangle, \qquad \dim L = n - 1.$$

Let L' be the linear variety generated by the distinct points $P_1' = f(P_1), \dots,$ $P_n' = f(P_n)$.

Since f takes collinear points to collinear points we have $f(L) \subset L'$. To see this, we define

$$L_i = P_1 + [\langle e_2, \dots, e_i \rangle], \quad e_i = \overrightarrow{P_1 P_i}, \ i = 2, \dots, n,$$

and, since $L = L_n$, we use induction on i. If $i = 2$, since L_1 is the straight line determined by the points P_1 and P_2, it is clear that $f(L_1) \subset L'$.

Let us assume $f(L_k) \subset L'$.

In order to prove that $f(L_{k+1}) \subset L'$ let us take an arbitrary point $X \in L_{k+1}$. Note that X can be written as

$$X = X_0 + xe_{k+1}, \quad X_0 \in L_k.$$

Let $Y = P_1 + \epsilon e_{k+1}$ be a point with $\epsilon \neq 0, x$ (we are assuming that the field has more than two elements). Since Y belongs to the straight line determined by the points P_1 and P_{k+1}, it is clear that $Y' = f(Y) \in L'$.

A short calculation shows that the point

$$T = X + t\overrightarrow{XY}, \quad \text{with } t = \frac{x}{x - \epsilon},$$

belongs to L_k. Hence, $T' = f(T) \in L'$.

Since the points X, Y, T are collinear, the points $X' = f(X), Y', T'$ are also collinear, and hence $X' \in L'$. This proves that $f(L_{k+1}) \subset L'$ and, by induction, that $f(L) \subset L'$.

In order to ensure that *the image of a hyperplane is a hyperplane*, it remains to prove that $f(L) = L'$, and that $\dim L' = n - 1$.

Let us take $Z' \in L'$. We must show that $Z' \in f(L)$. We know that there exists a $Z \in \mathbb{A}_1$ such that $f(Z) = Z'$, but it is not clear a priori that $Z \in L$. Let us assume $Z \notin L$.

Take an arbitrary point $X \in \mathbb{A}_1$. If the straight line determined by the points X, Z cuts L in a point Y, then $Y' = f(Y) \in L'$; since also $Z' \in L'$, and $X' = f(X)$ belongs to the straight line determined by Y' and Z', we must have $X' \in L'$.

If the straight line determined by the points X, Z does not cut L, we have

$$\overrightarrow{ZX} \in \langle \overrightarrow{P_1P_2}, \ldots, \overrightarrow{P_1P_n} \rangle.$$

Let us take a point $T = Z + t\overrightarrow{ZX}$ with $t \neq 0, 1$ (here we use the assumption that $k \neq \mathbb{Z}/2\mathbb{Z}$).

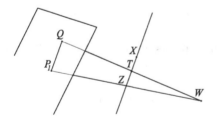

Figure 2.5. The image of a hyperplane is a hyperplane

Let $Q = P_1 + \overrightarrow{ZX} \in L$. Then the straight lines P_1Z and QT meet in a point W, see Figure 2.5. Since W is collinear with P_1 and Z, we have $W' =$

$f(W) \in L'$. Since T is collinear with Q and W, we have $T' = f(T) \in L'$. Finally, since X is collinear with T and Z, we have $X' \in L'$.

Thus, in all cases, given $X \in \mathbb{A}_1$ we have $X' \in L'$, that is, we have $f(\mathbb{A}_1) \subset L'$, contradicting the bijectivity of f. Hence, $Z \in L$, and this proves that $f(L) = L'$.

Finally, note that given a point $P' \notin L'$ it is easy to see, by an argument similar to that used above, that each point of $\mathbb{A}_2$ belongs either to a straight line connecting P' with a point of L' or to the linear variety through P' with the same direction as L. But this implies, see Exercise 1.14 of Chapter 1, page 40, that $\dim L' = n - 1$, and this completes the proof.   □

Theorem 2.46 (Fundamental theorem of affine geometry)

Let $\mathbb{A}_1$, $\mathbb{A}_2$ be two affine spaces over the k-vector spaces E_1, E_2 respectively, of the same dimension n, with $n \geq 2$. Suppose that $k \neq \mathbb{Z}/2\mathbb{Z}$. Let $f : \mathbb{A}_1 \longrightarrow \mathbb{A}_2$ be a bijective map sending collinear points to collinear points.

Then f is a semi-affinity.

Proof

First part: f takes parallel straight lines to parallel straight lines. If $n = 2$, this is evident because f is injective.

Let r_1, r_2 be two distinct parallel straight lines of $\mathbb{A}_1$, with $\dim A_1 = n > 2$. Let us take two distinct points A, B on r_1 and a point C on r_2. Let $D = C + \overrightarrow{AB} \in r_2$.

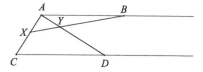

Figure 2.6. Construction of the point X

Take $Y \in AD$ distinct from A and D (here we use $k \neq \mathbb{Z}/2\mathbb{Z}$).

Denote by r_3 the straight line AC, by r_4 the straight line YB and by r_5 the straight line AD.

The straight lines r_3 and r_4 meet in a point X distinct from A, see Figure 2.6. This point X exists because all straight lines of the above diagram are in a plane (concretely in the plane $\Sigma : A + \langle \overrightarrow{AB}, \overrightarrow{AC} \rangle$) and the direction vectors of r_3 and r_4, $\overrightarrow{AC}$ and $\overrightarrow{BY}$, are linearly independent (since $\overrightarrow{AC} = \overrightarrow{BD}$).[1]

---

[1] If the characteristic of the field is different from 2, we can take $Y = A + \frac{1}{2}\overrightarrow{AD}$, and then $X = C$.

Since f is injective and maps straight lines to straight lines, $f(r_1)$ and $f(r_2)$ are non-intersecting straight lines. It still remains to prove that they are in the same plane.

The straight lines $f(r_1)$, $f(r_5)$ meet in $f(A)$ and, therefore, they determine a plane Π. Since $f(Y)$ and $f(B)$ belong to this plane, we have $f(r_4) \subset \Pi$ and in particular $f(X) \in \Pi$. But this implies $f(r_3) \subset \Pi$, and hence $f(C) \in \Pi$. Since also $f(D) \in \Pi$, we have $f(r_2) \subset \Pi$ and $f(r_1)$, $f(r_2)$ are coplanar. Since they do not meet, $f(r_1)$ and $f(r_2)$ are parallel, and this completes the first part of the proof.

Second part: $\tilde{f}_P$ is additive. Let us fix a point $P \in \mathbb{A}_1$. We must show that the map $\tilde{f}_P : E_1 \longrightarrow E_2$ given by

$$\tilde{f}_P(\overrightarrow{PX}) = \overrightarrow{f(P)f(X)}, \quad \text{for all } X \in \mathbb{A}_1,$$

preserves vector addition.

Let $u = \overrightarrow{PQ}$, $v = \overrightarrow{PR}$ be linearly independent vectors and let $S = Q + \overrightarrow{PR}$. We have $u + v = \overrightarrow{PQ} + \overrightarrow{PR} = \overrightarrow{PQ} + \overrightarrow{QS} = \overrightarrow{PS}$.

By the first part of the proof, and by Proposition 1.17, we know that the points $f(P), f(Q), f(R), f(S)$ are the vertices of a parallelogram, that is, $\overrightarrow{f(P)f(R)} = \overrightarrow{f(Q)f(S)}$. Thus,

$$\tilde{f}_P(u) + \tilde{f}_P(v) = \overrightarrow{f(P)f(Q)} + \overrightarrow{f(P)f(R)}$$
$$= \overrightarrow{f(P)f(Q)} + \overrightarrow{f(Q)f(S)}$$
$$= \overrightarrow{f(P)f(S)}$$
$$= \tilde{f}_P(u + v).$$

Hence, $\tilde{f}_P$ preserves the addition of linearly independent vectors.

If $w = \lambda u$, we have

$$\tilde{f}_P(u + w) + \tilde{f}_P(v) = \tilde{f}_P(u + w + v)$$
$$= \tilde{f}_P(u) + \tilde{f}_P(w + v)$$
$$= \tilde{f}_P(u) + \tilde{f}_P(w) + \tilde{f}_P(v).$$

Hence, $\tilde{f}_P$ preserves the addition of vectors, linearly independent or not; that is,

$$\tilde{f}(u + v) = \tilde{f}(u) + \tilde{f}(v), \quad \text{for all } u, v \in E_1.$$

Third part: the behavior of $\tilde{f}_P$ with scalars. Since $\tilde{f}_P$ is additive we have $\tilde{f}_P = \tilde{f}_Q$, for all $P, Q \in \mathbb{A}_1$, and so from now on we shall write $\tilde{f}_P = \tilde{f}$.

Fix $u \in E_1$, $u \neq 0$. Let $\lambda \in k$, $\lambda \neq 0$. There are unique points Q and R such that $u = \overrightarrow{PQ}$ and $\lambda u = \overrightarrow{PR}$.

Since P, Q, R are collinear, the points $f(P), f(Q), f(R)$ are also collinear, and hence there exists a $\mu_u(\lambda) \in k$ such that

$$\tilde{f}(\lambda u) = \overrightarrow{f(P)f(R)} = \mu_u(\lambda)\overrightarrow{f(P)f(Q)} = \mu_u(\lambda)\tilde{f}(u). \qquad (2.8)$$

Moreover, since f is bijective, $\tilde{f}(u) \neq 0$ and, therefore, the scalar $\mu_u(\lambda)$ is unique. Thus we have a map

$$\mu_u : k \longrightarrow k$$
$$\lambda \mapsto \mu_u(\lambda).$$

It is also easy to see that, since f is bijective, μ_u is also bijective.

Note that this map μ_u depends, a priori, on the chosen vector $u \in E_1$. Also note that $\mu_u(0) = 0$ and $\mu_u(1) = 1$.

By definition of semi-affinity, we must prove that $\tilde{f}$ is semi-linear, that is, that the scalars are transformed via an automorphism of the field. Hence, by (2.8), we must verify the following two properties:
(a) $\mu_u = \mu_v$ for all $v \in E_1$; and
(b) μ_u is an automorphism of k.

Proof of (a) *for every* $v \in E_1$ *linearly independent with* u. Let $v \in E_1$ be linearly independent with u. Set $u = \overrightarrow{PQ}$, $v = \overrightarrow{PR}$, $\lambda u = \overrightarrow{PQ'}$ and $\lambda v = \overrightarrow{PR'}$.

From the converse of the corollary of Thales' theorem, see Exercise 1.44 of Chapter 1, page 45, the straight lines RQ and $R'Q'$ are parallel. From the first part of the present proof, page 81, the straight lines $f(R)f(Q)$ and $f(R')f(Q')$ are also parallel, see Figure 2.7.

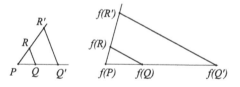

R'
R
P Q Q'

$f(R')$
$f(R)$
$f(P)$ $f(Q)$ $f(Q')$ **Figure 2.7.** Thales' theorem

By Thales' theorem, if the scalar $\tau \in k$ is such that

$$\overrightarrow{f(P)f(Q')} = \tau\overrightarrow{f(P)f(Q)},$$

then

$$\overrightarrow{f(P)f(R')} = \tau\overrightarrow{f(P)f(R)}.$$

Equivalently, if

$$\tilde{f}(\lambda u) = \tau\tilde{f}(u), \qquad (2.9)$$

then

$$\tilde{f}(\lambda v) = \tau \tilde{f}(v). \tag{2.10}$$

From (2.9) we deduce $\tau = \mu_u(\lambda)$ and from (2.10) we deduce $\tau = \mu_v(\lambda)$; hence, $\mu_u = \mu_v$ for all $v \in E_1$ linearly independent with u.

Proof of (b). The equality $\mu_u = \mu_v$ for all $v \in E_1$ linearly independent with u enables us to prove that μ_u is an automorphism of k.

Indeed, since f is bijective, there exists a $v \in E_1$ linearly independent with u such that $\tilde{f}(u) \neq 0$. If $\lambda' \in k$, $\lambda' \neq 0$, then setting $\mu = \mu_u$, we have

$$\mu(\lambda\lambda')\tilde{f}(v) = \tilde{f}(\lambda\lambda' v) = \mu(\lambda)\tilde{f}(\lambda' v) = \mu(\lambda)\mu(\lambda')\tilde{f}(v),$$

since, as $\lambda' v$ is linearly independent with u, $\mu_u = \mu_{\lambda' v}$.

Thus,

$$\mu(\lambda\lambda') = \mu(\lambda)\mu(\lambda'), \quad \lambda, \lambda' \in k \backslash \{0\}.$$

Since $\mu(0) = 0$, this equality holds for all $\lambda, \lambda' \in k$.

Since $\tilde{f}((\lambda + \lambda')u) = \tilde{f}(\lambda u) + \tilde{f}(\lambda' u)$, we have $\mu(\lambda + \lambda') = \mu(\lambda) + \mu(\lambda')$ and clearly $\mu(1) = 1$. Hence $\mu : k \longrightarrow k$ is an automorphism of k.

End of the proof of (a). It remains only to prove that

$$\mu_u = \mu_v \quad \text{when } v = \nu u, \ \nu \in k.$$

But we have

$$\tilde{f}(\lambda v) = \tilde{f}(\lambda \nu u) = \mu_u(\lambda \nu)\tilde{f}(u) = \mu_u(\lambda)\mu_u(\nu)\tilde{f}(u),$$

and also

$$\tilde{f}(\lambda v) = \mu_v(\lambda)\tilde{f}(\nu u) = \mu_v(\lambda)\mu_u(\nu)\tilde{f}(u).$$

Comparing these two equalities we obtain the result. □

Corollary 2.47

Let $\mathbb{A}_1, \mathbb{A}_2$ be two affine spaces over the k-vector spaces E_1, E_2, respectively, of the same dimension n, with $n \geq 2$. Let us assume that $k \neq \mathbb{Z}/2\mathbb{Z}$. Let $f : \mathbb{A}_1 \longrightarrow \mathbb{A}_2$ be a bijective map that takes straight lines to straight lines and preserves the simple ratio.

Then f is an affinity.

Proof

Every element λ of the field can be written as the simple ratio $\lambda = (A, B, C)$ of three points.

Since f is a semi-affinity we have

$$\tilde{f}(\overrightarrow{AB}) = \tilde{f}(\lambda \overrightarrow{AC}) = \sigma(\lambda)\tilde{f}(\overrightarrow{AC}).$$

Hence,

$$\lambda = (A, B, C) = (f(A), f(B), f(C)) = \sigma(\lambda),$$

where σ is the automorphism of the field associated to f. Thus, $\sigma(\lambda) = \lambda$ for all λ, that is, σ is the identity, and hence f is an affinity. $\square$

Example 2.48

Find a bijective map of an affine space $\mathbb{A}$ into itself taking straight lines onto straight lines and such that it is not a semi-affinity.

Solution

By the Fundamental Theorem, we must look for an affine space over a $\mathbb{Z}/2\mathbb{Z}$-vector space. We take $\mathbb{A} = E = (\mathbb{Z}/2\mathbb{Z})^3$ and

$$f : \mathbb{A} \longrightarrow \mathbb{A}$$
$$(0,0,0) \mapsto (0,0,0)$$
$$(1,0,0) \mapsto (1,0,1)$$
$$(1,1,0) \mapsto (1,1,1)$$
$$(0,1,0) \mapsto (0,1,1)$$
$$(0,0,1) \mapsto (0,\overset{.}{0},1)$$
$$(1,0,1) \mapsto (1,0,0)$$
$$(1,1,1) \mapsto (1,1,0)$$
$$(0,1,1) \mapsto (0,1,0)$$

f is bijective and takes straight lines onto straight lines, but it is not a semi-affinity. Indeed, take $P = (0,0,0)$, $Q = (1,0,0)$, $R = (0,1,0)$ and denote $\tilde{f}_P$ by $\tilde{f}$. Notice that $\overrightarrow{PQ} = (1,0,0)$ and $\overrightarrow{PR} = (0,1,0)$. We have

$$\tilde{f}(\overrightarrow{PQ} + \overrightarrow{PR}) = \tilde{f}(1,1,0) = (1,1,1).$$

But,

$$\tilde{f}(\overrightarrow{PQ}) + \tilde{f}(\overrightarrow{PR}) = \tilde{f}(1,0,0) + \tilde{f}(0,1,0)$$

$$= (1,0,1) + (0,1,1)$$

$$= (1,1,0) \neq (1,1,1).$$

$\square$

2.10.1 A Note on Automorphisms of Fields

It is useful to know that the fields $\mathbb{Q}$, $\mathbb{R}$ and $\mathbb{Z}/p\mathbb{Z}$ have a unique automorphism: the identity. Hence, when we work over one of these fields (with $p \neq 2$), the Fundamental Theorem of Affine Geometry states that *every bijective map taking straight lines to straight lines is an affinity.*

The only automorphism of $\mathbb{Z}/p\mathbb{Z}$ is the identity. Let σ be an automorphism of $\mathbb{Z}/p\mathbb{Z}$. Since $\sigma(1) = 1$, the other elements are also fixed, because $\mathbb{Z}/p\mathbb{Z}$ is additively generated by 1. That is, $\sigma(m) = m\sigma(1) = m$, and hence $\sigma = \text{id}$.

The only automorphism of $\mathbb{Q}$ is the identity. Let σ be an automorphism of $\mathbb{Q}$. Since $\sigma(1) = 1$, all integers are fixed, because $\mathbb{Z}$ is additively generated by 1. Moreover, $\sigma(n \cdot \frac{1}{n}) = \sigma(n) \cdot \sigma(\frac{1}{n}) = 1$; hence, $\sigma(\frac{1}{n}) = \frac{1}{\sigma(n)} = \frac{1}{n}$.

Thus,

$$\sigma\left(\frac{m}{n}\right) = m \cdot \sigma\left(\frac{1}{n}\right) = \frac{m}{n},$$

and so $\sigma = \text{id}$.

The only automorphism of $\mathbb{R}$ is the identity. Let σ be an automorphism of $\mathbb{R}$. We know, by the above argument, that σ is the identity on $\mathbb{Q}$.

Let $a \in \mathbb{R}$. Let us assume $a > 0$. Then $a = b^2$, with $b \in \mathbb{R}$. Thus, $\sigma(a) = \sigma(b)^2$, and hence $\sigma(a) > 0$. Since σ is bijective, we have $a > 0$ if and only if $\sigma(a) > 0$.

Let us assume that there is a real number c such that $\sigma(c) - c > 0$ (a very similar argument can be applied if $\sigma(c) - c < 0$). Choose $r \in \mathbb{Q}$ such that $\sigma(c) < r < c$. Then

$$\sigma(c - r) = \sigma(c) - \sigma(r) = \sigma(c) - r < 0,$$

a contradiction, since $c - r > 0$. Hence, $\sigma(c) = c$ for all $c \in \mathbb{R}$, and $\sigma = \text{id}$.

See, for instance, [30], page 49.

The field $\mathbb{C}$, however, has infinitely many automorphisms[2] and the fields of p^r elements have exactly r automorphisms; see for instance [20], page 184.

---

[2] It seems impossible (or very difficult) to explicitly describe these automorphisms (except, of course, for the identity and conjugation). Nevertheless, given any permutation P of a family of complex numbers S, algebraically independent over the

EXERCISES

2.1. Verify if the following maps are affinities or not:

$$\begin{array}{ccc} \mathbb{C} \longrightarrow \mathbb{C} & \mathbb{R}^2 \longrightarrow \mathbb{R}^2 & \mathbb{R}^3 \longrightarrow \mathbb{R}^2 \\ z \mapsto \overline{z} & (x,y) \mapsto (x,-y) & (x,y,z) \mapsto (1+x, \sqrt{\pi}) \end{array}$$

In the first case consider $\mathbb{C}$ as an affine space over $\mathbb{C}$ and also over $\mathbb{R}$.

2.2. Consider the affinity $f : \mathbb{R}^3 \longrightarrow \mathbb{R}^2$ given in the respective canonical affine frames by

$$\begin{pmatrix} x' \\ y' \end{pmatrix} = \begin{pmatrix} 2 & 1 & 0 \\ -1 & 1 & 3 \end{pmatrix} \begin{pmatrix} x \\ y \\ z \end{pmatrix} + \begin{pmatrix} 2 \\ 1 \end{pmatrix}.$$

Find the equations of f in the affine frames $\mathcal{R}_1$ and $\mathcal{R}_2$ given by

$$\mathcal{R}_1 = \{(1,0,0); ((2,-1,0), (0,2,-1), (-1,0,2))\},$$

$$\mathcal{R}_2 = \{(2,1); ((1,1), (1,-1))\}.$$

2.3. Find, in the affine plane $\mathbb{R}^2$, the equations of the projection on the straight line $r: x + y = 1$ in the direction of the straight line $s: x - y = 2$.

2.4. Find, in the affine space $\mathbb{R}^3$, the equations of the symmetry with respect to $L = (1,1,0) + [F]$ in the direction G in the following cases:
   1. $F = \langle (1,2,3), (0,0,1) \rangle$, $G = \langle (3,1,1) \rangle$.
   2. $F = \langle (3,1,1) \rangle$, $G = \langle (1,2,3), (0,0,1) \rangle$.
   3. $F = \{\vec{0}\}$.
   4. $G = \{\vec{0}\}$.

2.5. Find, in the affine space $\mathbb{R}^3$, the equations of the projection on $L = (1,1,0) + [F]$ in the direction G in the same four cases considered in the previous exercise.

2.6. Consider, in an affine space of dimension 3, the linear varieties given in some affine frame $\mathcal{R}$ by

$$L = \{(x,y,x) : x + y + z = 3, x - 4y + 2z = 1\},$$

$$M = \{(x,y,x) : 2x - 3y - z = 1\}.$$

Find the equations of the symmetry with respect to L in the direction of M and the equations of the symmetry with respect to M in the direction of L.

---

field of algebraic numbers $\bar{Q}$, and any automorphism σ of $\bar{Q}$, there exists an automorphism of $\mathbb{C}$ that restricted to S acts as P and restricted to $\bar{Q}$ acts as σ. See, for instance, [20], page 270, Exercise 1.

2.7. Let $f : \mathbb{R}^3 \longrightarrow \mathbb{R}^3$ be an affinity of the affine space $\mathbb{R}^3$ such that the points of the plane $\varPi : x + z = 1$ are fixed and such that $f(0,0,0) = (0,-3,0)$.
   (a) Find the matrix of f in the canonical affine frame $\mathcal{C}$ of the affine space $\mathbb{R}^3$.
   (b) Prove that on the planes parallel to $\varPi$, f acts as a translation, and find the translation vector.

2.8. Let $\mathcal{R} = \{P; (e_1, e_2)\}$ be an affine frame of a given affine plane. Find the equations of the axial symmetries, and their compositions, with respect to the straight lines given in $\mathcal{R}$ by the equations $2x + 3y = 2$ and $x + 3y = 2$, in the directions $e_1 + e_2$ and e_1 respectively.

2.9. Prove that dilations of an affine plane satisfy the Axioms 4 and 5 of Affine Geometry given in the introduction, page viii.

2.10. Prove that the image under an affinity of the barycenter of r points is equal to the barycenter of the images of these points. The same is true for the barycenter with weights. Is a map preserving the barycenter with weights necessarily an affinity?

2.11. Prove Theorem 2.40 in the following three steps:
   (1) Reduce the problem to the case $A_1 = A_2 = k$.
   (2) Further reduce the problem to the case $f(0) = 0$ (composing with a translation).
   (3) Observe that $f(b) = a \cdot b$ where $f(1) = a$.
   Thus, f is multiplication by a, which is linear. This exercise was suggested by F. Cedó.

2.12. Let f be an affinity of an affine plane $\mathbb{A}$, given in the affine frame $\mathcal{R} = \{P; (e_1, e_2)\}$ by the equations $f(x,y) = (x - y + 1, y + 2)$. Find the equations of f in the affine frame $\mathcal{R}' = \{(3,2); (e_1 - e_2, 2e_1 + e_2)\}$.

2.13. Consider the affinity of the affine space $\mathbb{R}^3$ given by

$$T(x,y,z) = (1 + x + 2y + z, 2 + y - z, -1 + x + 9z).$$

Is it bijective? Find the associated endomorphism. Find the image of the straight line $x = a$, $y = 2 - a$, $z = -1$ and the image of the plane $2x + y - z = 1$. Find the preimage of these linear varieties. Find the fixed points. Find also the equations of T in the affine frame $\{(1,0,4); ((1,1,0), (2,0,1), (8,0,7))\}$.

2.14. Find the affinity f of the affine space $\mathbb{R}^3$ leaving the plane $\varPi : x + y = 1$ invariant, acting on this plane as a translation by the vector $v = (0,0,1)$, and such that $f(1,1,0) = (0,0,0)$.

2.15. Let f be the affinity of an affine space of dimension 3 given in an affine frame $\mathcal{R}$ by

$$\begin{cases} x' = 1 - \dfrac{1}{3}x + \dfrac{2}{3}y + \dfrac{2}{3}z, \\[2mm] y' = \dfrac{2}{3}x - \dfrac{1}{3}y + \dfrac{2}{3}z, \\[2mm] z' = -1 + \dfrac{2}{3}x + \dfrac{2}{3}y - \dfrac{1}{3}z. \end{cases}$$

Find, with respect to $\mathcal{R}$:
(a) The fixed points and the invariant straight lines.
(b) The image of the straight line

$$\frac{x}{1} = \frac{y-2}{2} = \frac{z+1}{3}.$$

(c) The preimage of the plane $x' - y' - z' = 2$.

2.16. (a) Construct all the affinities of the affine plane $\mathbb{R}^2$ which leave invariant the figure formed by a point P and a straight line r with $P \notin r$.

(b) Give, in the affine plane $\mathbb{R}^2$, the expression of all affinities f such that $f(P) = P$ and $f(r) = r$ where $P = (1,0)$ and r: $y - x = 0$.

2.17. Prove that, in an affine plane, given two intersecting straight lines r, s and a point P not belonging to them, and given another analogous configuration, that is, two intersecting straight lines r', s' and a point P' not belonging to them, there exists a unique affinity f such that $f(r) = r'$, $f(s) = s'$ and $f(P) = P'$. Find this affinity, in the affine plane $\mathbb{R}^2$, where

$$r: x - y = 2, \qquad s: x - 2y = -1, \qquad P = (0,0);$$
$$r': x = 4, \qquad s': x - y = 1, \qquad P' = (1,2).$$

2.18. (a) Find, with respect to the canonical affine frame of the affine plane $\mathbb{R}^2$, the equations of a homothety of center $(2,3)$ and similitude ratio -1.

(b) Is it true that the image of a straight line that does not contain the center of a homothety is another straight line parallel to it?

(c) Find a homothety which, when composed with the homothety of part a), yields a translation, and give the translation vector.

2.19. Let $\mathbb{A}$ be an affine plane. Find the affinities of $\mathbb{A}$ fixing one vertex of a triangle and permuting the other two. Find the affinities of $\mathbb{A}$ leaving invariant a given straight line.

2.20. Let $\mathbb{A}$ be an affine space of dimension 3. Find the affinities of $\mathbb{A}$ with one straight line of fixed points r and leaving invariant another straight line parallel to r.

2.21. Consider the map $f : \mathbb{R}^4 \to \mathbb{R}^4$ given by

$$f(x,y,z,t) = (1 + x + y, -1 + y + z, 2 + z + t, 1 + \alpha x + t), \quad \alpha \in \mathbb{R}.$$

Find the value of α for which f is not bijective. Let $\Pi : ax + by + cz + dt = 1$ be a hyperplane in the affine space $\mathbb{R}^4$. Find, for the previous value of α, the condition that must be fulfilled by a, b, c, d in order that $f(\Pi)$ be a linear variety of dimension smaller than 3.

2.22. Study the affinities of the affine space $\mathbb{R}^2$ preserving the hyperbola $xy = 1$.

2.23. Let f be an affine map. Prove:

(a) If f^2 has a fixed point, then f also has a fixed point.

(b) If there exists an $n \in \mathbb{N}$ such that f^n has a fixed point, then f also has a fixed point.

2.24. Prove that given two triangles T_1 and T_2 of an affine plane $\mathbb{A}$, there exists a bijective affinity $f : \mathbb{A} \to \mathbb{A}$ such that $f(T_1) = T_2$. How many maps with this property are there? Is the previous statement true if we replace triangles by quadrilaterals, parallelograms or triples of points?

2.25. Let $\triangle ABC$ and $\triangle A'B'C'$ be triangles whose sides are respectively parallel. Prove that there exists a translation or a homothety mapping one of them onto the other.

2.26. (a) Find, with respect to the canonical affine frame of the affine space $\mathbb{R}^2$, the equations of an affinity mapping the points $A = (0,0)$, $B = (1,0)$, $C = (0,1)$, respectively onto the points B, C, A. What is the image under this affinity of the barycenter of the triangle $\triangle ABC$?

(b) Find, with respect to the canonical affine frame of the affine space $\mathbb{R}^2$, the equations of an affinity mapping the points $A = (a_1, a_2)$, $B = (b_1, b_2)$, $C = (c_1, c_2)$, respectively onto the points B, C, A. What is the image under this affinity of the barycenter of the triangle $\triangle ABC$?

Classification of Affinities

3.1 Introduction

In this chapter we only consider affinities from an affine space $\mathbb{A}$ into itself. The aim is to see how many affinities there are. We give a complete answer to this question, modulo an equivalence relation, in dimensions 1 and 2. We also give a geometric interpretation of the affinities of the real affine plane. Arbitrary dimensions are considered in the next chapter.

3.2 Similar Endomorphisms

First let us recall the following definitions from linear algebra. Let E be a k-vector space.

Definition 3.1

Two endomorphisms $\tilde{f}$ and $\tilde{g}$ of E are *similar* if and only if there exists an invertible endomorphism $\tilde{h}$ of E such that $\tilde{f} = \tilde{h}^{-1} \circ \tilde{g} \circ \tilde{h}$.

In this case we say that $\tilde{h}$ conjugates $\tilde{f}$ and $\tilde{g}$. Analogously we have the following.

A. Reventós Tarrida, *Affine Maps, Euclidean Motions and Quadrics*,
Springer Undergraduate Mathematics Series,
DOI 10.1007/978-0-85729-710-5_3, © Springer-Verlag London Limited 2011

Definition 3.2

Two matrices $A, B \in \mathcal{M}_{n \times n}(k)$ are *similar* if and only if there exists an invertible matrix $C \in \mathcal{M}_{n \times n}(k)$ such that $A = C^{-1}BC$.

In this case we say that C conjugates A and B.

The relationship between these two definitions is given in the next proposition.

Proposition 3.3

Let $\tilde{f}$ and $\tilde{g}$ be endomorphisms of E. Then:
  (i) The matrices $M(\tilde{f}, \mathcal{B}_1)$ and $M(\tilde{f}, \mathcal{B}_2)$ are similar, where $\mathcal{B}_1$ and $\mathcal{B}_2$ are bases of E.
  (ii) $\tilde{f}$ and $\tilde{g}$ are similar if and only if $M(\tilde{f}, \mathcal{B})$ and $M(\tilde{g}, \mathcal{B})$ are similar, where $\mathcal{B}$ is a basis of E.
  (iii) $\tilde{f}$ and $\tilde{g}$ are similar if and only if there are bases $\mathcal{B}_1$ and $\mathcal{B}_2$ of E such that $M(\tilde{f}, \mathcal{B}_1) = M(\tilde{g}, \mathcal{B}_2)$.

Proof

See [8], page 321.   □

From this, we easily deduce that similar endomorphisms have the same characteristic polynomial and, in particular, the same trace and the same determinant, see [8], page 332.

3.3 Similar Affinities

For affinities we have an analogous definition.

Definition 3.4

Two affinities $f, g : \mathbb{A} \longrightarrow \mathbb{A}$ are *similar* if and only if there exists a bijective affine map $h : \mathbb{A} \longrightarrow \mathbb{A}$ such that $f = h^{-1} \circ g \circ h$.

In this case we say that h conjugates f and g.

The relationship between similar endomorphisms and similar affinities is given in the following proposition.

Proposition 3.5

If two affinities are similar, then their associated linear maps are also similar.

Proof

Since the linear map associated with a composition of affinities is equal to the composition of the corresponding linear maps, the equality

$$f = h^{-1} \circ g \circ h$$

implies

$$\tilde{f} = \tilde{h}^{-1} \circ \tilde{g} \circ \tilde{h}.$$

Hence, if f is similar to g, then $\tilde{f}$ is similar to $\tilde{g}$. $\qquad\square$

The converse is not true. For instance, the identity and a translation T_u, with $u \neq 0$, have the same associated linear map, but they are not similar.

Proposition 3.6

Let T_u and T_v be two translations, different from the identity, of an affine space $\mathbb{A}$. Then T_u and T_v are similar.

Proof

Let φ be any bijective linear map such that $\varphi(u) = v$. We know, from Proposition 2.4, that there is an affinity h such that $\tilde{h} = \varphi$ (in fact, there are many such maps).

Then h conjugates T_u and T_v. In fact,

$$(h \circ T_u)(P) = h(P + u) = h(P) + \varphi(u) = h(P) + v = (T_v \circ h)(P).$$

$\qquad\square$

3.4 Computations in Coordinates

Proposition 3.3, which refers to endomorphisms, can be directly generalized to affinities.

For convenience, we define *affinely similar* matrices.

Definition 3.7

We say that the matrices $\left(\begin{smallmatrix} A & a \\ 0 & 1 \end{smallmatrix}\right)$ and $\left(\begin{smallmatrix} B & b \\ 0 & 1 \end{smallmatrix}\right)$ are *affinely similar* if there is an invertible matrix $\left(\begin{smallmatrix} C & c \\ 0 & 1 \end{smallmatrix}\right)$ such that

$$\begin{pmatrix} A & a \\ 0 & 1 \end{pmatrix} = \begin{pmatrix} C & c \\ 0 & 1 \end{pmatrix}^{-1} \begin{pmatrix} B & b \\ 0 & 1 \end{pmatrix} \begin{pmatrix} C & c \\ 0 & 1 \end{pmatrix},$$

with $A, B, C \in M_{n \times n}(k); a, b, c \in M_{n \times 1}(k)$.

Then we have the following.

Proposition 3.8

Let f and g be affinities of an affine space $\mathbb{A}$. Then:
  (i) The matrices $M(f, \mathcal{R}_1)$ and $M(f, \mathcal{R}_2)$ are affinely similar, where $\mathcal{R}_1$ and $\mathcal{R}_2$ are affine frames of $\mathbb{A}$.
  (ii) f and g are similar if and only if $M(f, \mathcal{R})$ and $M(g, \mathcal{R})$ are affinely similar, where $\mathcal{R}$ is an affine frame of $\mathbb{A}$.
  (iii) f and g are similar if and only if there exist affine frames $\mathcal{R}_1$ and $\mathcal{R}_2$ of $\mathbb{A}$ such that $M(f, \mathcal{R}_1) = M(g, \mathcal{R}_2)$.

Proof

The proof is an adaptation of that of Proposition 3.3. For convenience, we reproduce it here.
  (i) This is a consequence of the change of basis formula, (2.2) on page 61.
  (ii) ($\Rightarrow$) Let us assume that f and g are similar. Then there is an invertible affinity h of $\mathbb{A}$ such that $f = h^{-1} \circ g \circ h$. Hence, by Proposition 2.20,

$$M(f, \mathcal{R}) = M(h^{-1}, \mathcal{R}) M(g, \mathcal{R}) M(h, \mathcal{R}).$$

Since $M(h^{-1}, \mathcal{R}) = M(h, \mathcal{R})^{-1}$, the matrices $M(f, \mathcal{R})$ and $M(g, \mathcal{R})$ are similar, and the matrix conjugating them is of the form $\left(\begin{smallmatrix} C & c \\ 0 & 1 \end{smallmatrix}\right)$, since it corresponds to the matrix of an affinity.
  ($\Leftarrow$) Let us assume that the matrices $M(f, \mathcal{R})$ and $M(g, \mathcal{R})$ are conjugated by a matrix

$$M = \begin{pmatrix} C & c \\ 0 & 1 \end{pmatrix}.$$

Let h be the affinity of $\mathbb{A}$ such that $M(h, \mathcal{R}) = M$. Since M is invertible, h is bijective and $M(h^{-1}, \mathcal{R}) = M^{-1}$. Now we have

$$M(f, \mathcal{R}) = M(h^{-1}, \mathcal{R})M(g, \mathcal{R})M(h, \mathcal{R})$$
$$= M(h^{-1} \circ g \circ h, \mathcal{R}).$$

Hence, $f = h^{-1} \circ g \circ h$, and this proves that f and g are similar.

(iii) ($\Leftarrow$) Let us assume that there are affine frames $\mathcal{R}_1$ and $\mathcal{R}_2$ of $\mathbb{A}$ such that $M(f, \mathcal{R}_1) = M(g, \mathcal{R}_2)$. From (i), $M(g, \mathcal{R}_2)$ is similar to $M(g, \mathcal{R}_1)$, and from (ii), f and g are similar.

($\Rightarrow$) Let us assume that f and g are similar. Let h be an invertible affinity of $\mathbb{A}$ such that $f = h^{-1} \circ g \circ h$.

Then

$$M(f, \mathcal{R}_1) = M(h, \mathcal{R}_1)^{-1}M(g, \mathcal{R}_1)M(h, \mathcal{R}_1).$$

Since h is bijective, if we put $\mathcal{R}_1 = \{P; (e_1, \ldots, e_n)\}$, then

$$\mathcal{R}_2 = \{h(P); \tilde{h}(e_1), \ldots, \tilde{h}(e_n)\}$$

is an affine frame of $\mathbb{A}$ and

$$M(h, \mathcal{R}_1) = M(\mathcal{R}_2, \mathcal{R}_1).$$

Hence,

$$M(f, \mathcal{R}_1) = M(\mathcal{R}_2, \mathcal{R}_1)^{-1}M(g, \mathcal{R}_1)M(\mathcal{R}_2, \mathcal{R}_1) = M(g, \mathcal{R}_2).$$

$\square$

Note that Proposition 3.6 can now easily be proved:

Corollary 3.9

Let T_u and T_v be two translations, different from the identity, of an affine space $\mathbb{A}$. Then T_u and T_v are similar.

Proof

We have seen, on page 67, that the equations of a translation T_u, with $u \neq 0$, are

$$\begin{cases} x_1' = x_1 + 1, \\ x_2' = x_2, \\ \quad \vdots \\ x_n' = x_n. \end{cases}$$

Now the result is a consequence of part (iii) of Proposition 3.8.   □

Analogously, we have the following.

Corollary 3.10

The homotheties $h_{P,\lambda}$ and $h_{Q,\mu}$ are similar if and only if $\lambda = \mu$.

Proof

If $\lambda = \mu$, the matrix of the first homothety with respect to an affine frame
with origin the fixed point P is equal to the matrix of the second homothety
with respect to an affine frame with origin the fixed point Q. From part (iii) of
Proposition 3.8, these homotheties are similar.

   If the homotheties are similar, it is clear by Proposition 3.5 that $\lambda = \mu$.   □

Corollary 3.11

Two projections f and g of an affine space $\mathbb{A}$ are similar if and only if
$\dim \operatorname{Fix}(f) = \dim \operatorname{Fix}(g)$.

Proof

See the equations of projections on page 73.   □

Corollary 3.12

Two symmetries f and g of an affine space $\mathbb{A}$ are similar if and only if
$\dim \operatorname{Fix}(f) = \dim \operatorname{Fix}(g)$.

Proof

See the equations of symmetries on page 72.   □

Corollary 3.13

Let $\mathcal{R} = \{P; \mathcal{B}\}$ be an affine frame of an affine space $\mathbb{A}$ of dimension n. Let f, g
be affinities with

$$M(f,\mathcal{R}) = \begin{pmatrix} A & a \\ 0 & 1 \end{pmatrix}, \qquad M(g,\mathcal{R}) = \begin{pmatrix} B & b \\ 0 & 1 \end{pmatrix},$$

where $A, B \in \mathcal{M}_{n \times n}(k)$ and $a, b \in \mathcal{M}_{n \times 1}(k)$.

Then, f is similar to g if and only if there exist an invertible matrix $C \in \mathcal{M}_{n \times n}(k)$ and a column matrix $c \in \mathcal{M}_{n \times 1}(k)$ such that

$$\begin{cases} A = C^{-1}BC, \\ (B - I_n)c = Ca - b. \end{cases}$$

Proof

From part (ii) of Proposition 3.8, f is similar to g if and only if there exists a matrix

$$\begin{pmatrix} C & c \\ 0 & 1 \end{pmatrix},$$

with C invertible, such that

$$\begin{pmatrix} A & a \\ 0 & 1 \end{pmatrix} = \begin{pmatrix} C & c \\ 0 & 1 \end{pmatrix}^{-1} \begin{pmatrix} B & b \\ 0 & 1 \end{pmatrix} \begin{pmatrix} C & c \\ 0 & 1 \end{pmatrix}. \tag{3.1}$$

Equivalently,

$$\begin{pmatrix} C & c \\ 0 & 1 \end{pmatrix} \begin{pmatrix} A & a \\ 0 & 1 \end{pmatrix} = \begin{pmatrix} B & b \\ 0 & 1 \end{pmatrix} \begin{pmatrix} C & c \\ 0 & 1 \end{pmatrix},$$

and hence

$$CA = BC, \qquad Ca + c = Bc + b,$$

which can be written as

$$\begin{cases} A = C^{-1}BC, \\ (B - I_n)c = Ca - b. \end{cases}$$

$\square$

Part (i) of Proposition 3.8 allows us to define the determinant and trace of an affinity as the determinant and trace of the matrix of the affinity with respect to any affine frame.[1]

---

[1] Indeed, we can define not only the determinant and trace, but any function of the elementary symmetric polynomials.

Since

$$M(f, \mathcal{R}) = \begin{pmatrix} A & a \\ 0 & 1 \end{pmatrix},$$

we have $\det \tilde{f} = \det f$ and trace $\tilde{f} = $ trace $f - 1$.

Part (ii) of Proposition 3.8 states that the determinant and the trace of an affinity are invariant within each equivalence class.

3.5 Invariance Level

Definition 3.14 (Invariance level)

An affinity f has *invariance level* $\rho(f) = r$, for $r = 0, \dots, \dim \mathbb{A}$, if and only if f has invariant linear varieties of dimension r and all linear varieties of smaller dimension (if $r > 0$) are not invariant.

Thus, for instance, if f has a fixed point, the invariance level is $\rho(f) = 0$, since there are invariant varieties of dimension zero (and it does not make sense to talk about varieties of smaller dimension).

If f is a translation (different from the identity) it has invariance level $\rho(f) = 1$, since translations have invariant straight lines, but they do not have fixed points.

Equivalently, $\rho(f)$ is the *minimum of the dimensions of the invariant varieties*:

$$\boxed{\rho(f) = \min_{L \subset \mathbb{A}} \{\dim L : f(L) \subset L\}}$$

We illustrate this definition in dimensions 1 and 2.

3.5.1 Invariance Level in Dimension 1

Let f be an affinity of an affine space of dimension 1. Then:

(0) f has invariance level $\rho(f) = 0$ if and only if it has a fixed point.

(1) f has invariance level $\rho(f) = 1$ if and only if it does not have a fixed point.

Hence, in order to find $\rho(f)$ it is only necessary to determine the fixed points of f.

3.5.2 Invariance Level in Dimension 2

Let f be an affinity of an affine space of dimension 2. Then:

(0) f has invariance level $\rho(f) = 0$ if and only if it has a fixed point.

(1) f has invariance level $\rho(f) = 1$ if and only if it does not have a fixed point, but it has an invariant straight line.

(2) f has invariance level $\rho(f) = 2$ if and only if it neither has fixed points nor invariant straight lines.

Hence, in order to find $\rho(f)$ it is only necessary to determine the fixed points and invariant straight lines of f.

In the next proposition we prove that similar affinities have the same invariance level.

Proposition 3.15

Let f, g be similar affinities. Then $\rho(f) = \rho(g)$.

Proof

We know that there is an invertible affinity h such that

$$f = h^{-1} \circ g \circ h.$$

Let L be a linear variety of the affine space $\mathbb{A}$ invariant under f, that is, such that $f(L) \subset L$. Then the linear variety $h(L)$, which has the same dimension as L because $\tilde{h}$ is an isomorphism, is invariant under g, since

$$g(h(L)) = h(f(L)) \subset h(L).$$

Hence, if f has invariant varieties of a certain dimension, then g also has invariant varieties of the same dimension, and conversely. Hence, $\rho(f) = \rho(g)$, and this completes the proof. $\square$

3.6 Classification of Affinities of the Line

In this section $\mathbb{A}$ denotes an affine space over a k-vector space E of dimension 1.

We first observe that Proposition 2.29, which refers to the existence of a unique fixed point, can be stated as follows:

Proposition 3.16

Let f be an affinity of an affine space of dimension 1. Let us assume that, in some affine frame, f has equation $x' = Ax + a$. Then, f has a unique fixed point if and only if $A \neq 1$.

Proof

In this case $A \in \mathcal{M}_{1 \times 1}(k)$, that is, $A \in k$, and hence the condition $\det(A - 1) \neq 0$ of Proposition 2.29 reduces to $A - 1 \neq 0$. $\square$

Note that this unique fixed point has coordinate $x = \frac{a}{1-A} \in k$.

Theorem 3.17 (Classification)

Let f be an affinity of an affine space of dimension 1. Then there exists an affine frame $\mathcal{R} = \{P; e\}$ such that the equation of f in this affine frame is one (and only one) of the following:

$$x' = Ax, \quad A \neq 1,$$

$$x' = x + 1,$$

$$x' = x.$$

Proof

If f has two fixed points, it is the identity. In any affine frame the equation is

$$x' = x.$$

In this case $\rho(f) = 0$.

If f has a unique fixed point P, we take the affine frame $\mathcal{R} = \{P; e\}$, where $e \in E$ is any non-zero vector. Since $\dim E = 1$, we have $\tilde{f}(e) = Ae$ for some scalar $A \in k$. By Proposition 3.16, $A \neq 1$. Thus,

$$M(f, \mathcal{R}) = \begin{pmatrix} A & 0 \\ 0 & 1 \end{pmatrix},$$

and hence the equation is

$$x' = Ax, \quad A \neq 1.$$

In this case, $\rho(f) = 0$.

Finally, if f does not have fixed points, we take any point P and define $e = \overrightarrow{Pf(P)}$. Note that $e \neq 0$. Again by Corollary 2.28, we must have $\tilde{f}(e) = e$. In the affine frame $\mathcal{R} = \{P; e\}$ we have

$$M(f, \mathcal{R}) = \begin{pmatrix} 1 & 1 \\ 0 & 1 \end{pmatrix},$$

and hence the equation is

$$x' = x + 1.$$

In this case, $\rho(f) = 1$.

The uniqueness of such expression follows from the invariance of the determinant: if $\det f = A \neq 1$, we are in the first case; if $\det f = 1$, we are either in the second case if $\rho(f) = 1$, or in the third case if $\rho(f) = 0$. □

Summarizing, we have Table 3.1.

Linear	Affine	Name and invariance level	Equation
$1 \neq A \in k$	$\begin{pmatrix} A & 0 \\ 0 & 1 \end{pmatrix}$	Homothety, $\rho = 0$	$x' = Ax$
1	$\begin{pmatrix} 1 & 1 \\ 0 & 1 \end{pmatrix}$	Translation, $\rho = 1$	$x' = x + 1$
1	$\begin{pmatrix} 1 & 0 \\ 0 & 1 \end{pmatrix}$	Identity, $\rho = 0$	$x' = x$

Table 3.1.   Affinities of the line

Theorem 3.18 (Characterization)

Let f and g be affinities of an affine space of dimension 1. Then f is similar to g if and only if $\tilde{f}$ is similar to $\tilde{g}$ and $\rho(f) = \rho(g)$.

Proof

We have seen in Propositions 3.5 and 3.15 that if f is similar to g then $\tilde{f}$ is similar to $\tilde{g}$ and $\rho(f) = \rho(g)$.

Conversely, let us assume given two affinities f and g such that $\tilde{f}$ is similar to $\tilde{g}$, and $\rho(f) = \rho(g)$. The condition "$\tilde{f}$ is similar to $\tilde{g}$" (which in dimension 1 is equivalent to $\tilde{f} = \tilde{g}$) implies $\det(f) = \det(g) = A$.

Looking at the above table we see that if $A \neq 1$, both f and g are homotheties of similitude ratio A, and hence, by Corollary 3.10, f is similar to g.

If $A = 1$, the equality $\rho(f) = \rho(g)$ implies either that f and g are both translations, and so by Proposition 3.6, f is similar to g, or that f and g are equal to the identity.   □

3.7 Classification of Affinities of the Real Plane

In this section $\mathbb{A}$ denotes an affine space over an $\mathbb{R}$-vector space E of dimension 2. First let us recall a result from linear algebra.

3.7.1 Classification of Endomorphisms in Real Dimension Two

Let E be an $\mathbb{R}$-vector space of dimension two. The study of an endomorphism of E requires the study of its characteristic polynomial, but the knowledge of this polynomial for each endomorphism is not enough to determine if two endomorphisms of E are similar or not.

Since we are in dimension two, the characteristic polynomial is of degree two, with real coefficients, and therefore the roots of this polynomial are complex, real simple or real multiple.

It is known, see for instance [8], that *in dimension two*, two endomorphisms $\tilde{f}, \tilde{g}$ are similar if and only if their characteristic polynomials have the same roots and, moreover, $\dim(\ker(\tilde{f} - a\,\mathrm{id})) = \dim(\ker(\tilde{g} - a\,\mathrm{id}))$, where $a \in \mathbb{R}$ is a common real multiple root (if there is such a root).

Concretely, if the characteristic polynomial of $\tilde{f}$, $p_c(\tilde{f}) = x^2 + bx + c$, has complex roots, that is, $b^2 < 4c$, then there exists a basis $\mathcal{B}$ of E such that

$$M(\tilde{f}, \mathcal{B}) = \begin{pmatrix} 0 & -c \\ 1 & -b \end{pmatrix}.$$

If $p_c(\tilde{f}) = (x - a)(x - b)$, with a, b different real numbers, then there exists a basis $\mathcal{B}$ of E such that

$$M(\tilde{f}, \mathcal{B}) = \begin{pmatrix} a & 0 \\ 0 & b \end{pmatrix}.$$

If $p_c(\tilde{f}) = (x - a)^2$ with $a \in \mathbb{R}$, and $\dim(\ker(\tilde{f} - a\,\mathrm{id})) = 1$, then there exists a basis $\mathcal{B}$ of E such that

$$M(\tilde{f}, \mathcal{B}) = \begin{pmatrix} a & 0 \\ 1 & a \end{pmatrix}.$$

If $p_c(\tilde{f}) = (x - a)^2$ with $a \in \mathbb{R}$, and $\dim(\ker(\tilde{f} - a\,\mathrm{id})) = 2$, then there exists a basis $\mathcal{B}$ of E such that

$$M(\tilde{f},\mathcal{B}) = \begin{pmatrix} a & 0 \\ 0 & a \end{pmatrix}.$$

3.7.2 List of Canonical Expressions of Endomorphisms

We have obtained four canonical expressions for the matrix of $\tilde{f}$ as a function of the roots of its characteristic polynomial: complex, simple real or double real (two cases). However, because of Proposition 3.16, we must still separate the cases in which one of the real roots is 1. This increases the four cases to eight.

We organize the above classification in the following way: given an endomorphism $\tilde{f}$ of a real vector space E of dimension 2, there exists a basis $\mathcal{B}$ of E such that $M(\tilde{f},\mathcal{B})$ is equal to one and only one of the matrices of the following list:

$$\tilde{\mathcal{E}}(b,c) = \begin{pmatrix} 0 & -c \\ 1 & -b \end{pmatrix}, \quad b^2 < 4c$$

$$\tilde{\mathcal{H}}(a,b) = \begin{pmatrix} a & 0 \\ 0 & b \end{pmatrix}, \quad a,b \neq 1;\ a < b$$

$$\tilde{h}(a) = \begin{pmatrix} a & 0 \\ 0 & a \end{pmatrix}, \quad a \neq 1$$

$$\tilde{\mathcal{P}}(a) = \begin{pmatrix} a & 0 \\ 1 & a \end{pmatrix}, \quad a \neq 1$$

$$\tilde{h}_g(a) = \begin{pmatrix} 1 & 0 \\ 0 & a \end{pmatrix}, \quad a \neq 1$$

$$\tilde{h}_e = \begin{pmatrix} 1 & 0 \\ 1 & 1 \end{pmatrix},$$

$$\tilde{\mathrm{id}} = \begin{pmatrix} 1 & 0 \\ 0 & 1 \end{pmatrix} = I_2,$$

No two different matrices in this list are conjugate.

Inspired by this, we next give a list of matrices that will allow us to give a more concise statement of the Classification Theorem.

3.7.3 Canonical Expressions of Affinities

$$\mathcal{E}(b,c) = \left(\begin{array}{cc|c} 0 & -c & 0 \\ 1 & -b & 0 \\ \hline 0 & 0 & 1 \end{array}\right) = \left(\begin{array}{c|c} \tilde{\mathcal{E}}(b,c) & 0 \\ & 0 \\ \hline 0 \quad 0 & 1 \end{array}\right), \quad b^2 < 4c,$$

$$\mathcal{H}(a,b) = \left(\begin{array}{cc|c} a & 0 & 0 \\ 0 & b & 0 \\ \hline 0 & 0 & 1 \end{array}\right) = \left(\begin{array}{c|c} \tilde{\mathcal{H}}(a,b) & 0 \\ & 0 \\ \hline 0 \quad 0 & 1 \end{array}\right), \quad a,b \neq 1; \; a < b,$$

$$h(a) = \left(\begin{array}{cc|c} a & 0 & 0 \\ 0 & a & 0 \\ \hline 0 & 0 & 1 \end{array}\right) = \left(\begin{array}{c|c} \tilde{h}(a) & 0 \\ & 0 \\ \hline 0 \quad 0 & 1 \end{array}\right), \quad a \neq 1,$$

$$\mathcal{P}(a) = \left(\begin{array}{cc|c} a & 0 & 0 \\ 1 & a & 0 \\ \hline 0 & 0 & 1 \end{array}\right) = \left(\begin{array}{c|c} \tilde{\mathcal{P}}(a) & 0 \\ & 0 \\ \hline 0 \quad 0 & 1 \end{array}\right), \quad a \neq 1,$$

$$Th_g(a) = \left(\begin{array}{cc|c} 1 & 0 & 1 \\ 0 & a & 0 \\ \hline 0 & 0 & 1 \end{array}\right) = \left(\begin{array}{c|c} \tilde{h}_g(a) & 0 \\ & 1 \\ \hline 0 \quad 0 & 1 \end{array}\right), \quad a \neq 1,$$

$$h_g(a) = \left(\begin{array}{cc|c} 1 & 0 & 0 \\ 0 & a & 0 \\ \hline 0 & 0 & 1 \end{array}\right) = \left(\begin{array}{c|c} \tilde{h}_g(a) & 0 \\ & 0 \\ \hline 0 \quad 0 & 1 \end{array}\right), \quad a \neq 1,$$

$$Th_e = \left(\begin{array}{cc|c} 1 & 0 & 1 \\ 1 & 1 & 0 \\ \hline 0 & 0 & 1 \end{array}\right) = \left(\begin{array}{c|c} \tilde{h}_e & 1 \\ & 0 \\ \hline 0 \quad 0 & 1 \end{array}\right),$$

$$h_e = \left(\begin{array}{cc|c} 1 & 0 & 0 \\ 1 & 1 & 0 \\ \hline 0 & 0 & 1 \end{array}\right) = \left(\begin{array}{c|c} \tilde{h}_e & 0 \\ & 0 \\ \hline 0 \quad 0 & 1 \end{array}\right),$$

$$T = \left(\begin{array}{cc|c} 1 & 0 & 0 \\ 0 & 1 & 1 \\ \hline 0 & 0 & 1 \end{array}\right) = \left(\begin{array}{c|c} I_2 & 0 \\ & 1 \\ \hline 0 \quad 0 & 1 \end{array}\right),$$

$$id = \left(\begin{array}{cc|c} 1 & 0 & 0 \\ 0 & 1 & 0 \\ \hline 0 & 0 & 1 \end{array}\right) = \left(\begin{array}{c|c} I_2 & 0 \\ & 0 \\ \hline 0 \quad 0 & 1 \end{array}\right) = I_3.$$

The matrices of this list are called *canonical matrices*.
The notation comes from the following:

$$\mathcal{E}(b,c) = \text{Elliptic.}$$

$$\mathcal{H}(a,b) = \text{Hyperbolic.}$$

$$\mathcal{P}(a) = \text{Parabolic.}$$

$$h(a) = \text{Homothety.}$$

$$h_g(a) = \text{General homology.}$$

$$h_e = \text{Special homology.}$$

$$Th_g(a) = \text{General homology followed by translation.}$$

$$Th_e = \text{Special homology followed by translation.}$$

$$T = \text{Translation.}$$

$$\text{id} = \text{Identity.}$$

See Section 3.9, page 116, for a geometrical interpretation of these affinities.

Theorem 3.19 (Classification)

Let f be an affinity of a real affine plane. Then there exists an affine frame
$\mathcal{R} = \{P; \mathcal{B}\}$ such that the matrix $M(f, \mathcal{R})$ is a canonical matrix.

Proof

We know that there is a basis $\mathcal{B} = (e_1, e_2)$ of the associated vector space E such
that $M(\vec{f}, \mathcal{B})$ is equal to one (and only one) of the matrices

$$\tilde{\mathcal{E}}(b,c), \tilde{\mathcal{H}}(a,b), \tilde{h}(a), \tilde{\mathcal{P}}(a), \tilde{h}_g(a), \tilde{h}_e, \tilde{\text{id}}.$$

First case: $M(\vec{f}, \mathcal{B}) = \tilde{\mathcal{E}}(b,c), \tilde{\mathcal{H}}(a,b), \tilde{h}(a), \tilde{\mathcal{P}}(a)$. In this case, by Proposition 3.16, there exists a unique fixed point P.

Let us consider the affine frame $\mathcal{R} = \{P; \mathcal{B}\}$ with origin this fixed point.
Then, since

$$M(f, \mathcal{R}) = \left(\begin{array}{c|c} M(\vec{f}, \mathcal{B}) & \begin{array}{c} 0 \\ 0 \end{array} \\ \hline \begin{array}{cc} 0 & 0 \end{array} & 1 \end{array} \right),$$

the matrix $M(f, \mathcal{R})$ is equal to one of the matrices $\mathcal{E}(b,c), \mathcal{H}(a,b), h(a)$ or $\mathcal{P}(a)$.

Second case: $M(\tilde{f}, \mathcal{B}) = \tilde{h}_g(a)$. In this case, we take an affine frame with origin any point Q, $\mathcal{S} = \{Q; \mathcal{B}\}$. Then

$$M(f, \mathcal{S}) = \begin{pmatrix} 1 & 0 & \alpha \\ 0 & a & \beta \\ 0 & 0 & 1 \end{pmatrix}.$$

Let us see if there are any fixed points. The equations of fixed points are

$$x + \alpha = x,$$

$$ay + \beta = y,$$

so we have two cases:

$\alpha = 0$. In this case, there are fixed points (any point with $y = \frac{\beta}{1-a}$). Let P be one of these fixed points, and let $\mathcal{R} = \{P; \mathcal{B}\}$. Then

$$M(f, \mathcal{R}) = \begin{pmatrix} 1 & 0 & 0 \\ 0 & a & 0 \\ 0 & 0 & 1 \end{pmatrix} = h_g(a).$$

$\alpha \neq 0$. In this case, there are no fixed points. We look for invariant straight lines. Since the eigenvectors are the two vectors e_1 and e_2 of the basis $\mathcal{B}$, the candidates to be invariant straight lines must have equations of the form $x = \text{constant}$ or $y = \text{constant}$. Since $y' = ay + \beta$, it is clear that the straight line

$$y = \frac{\beta}{1-a}$$

is invariant. In fact, it is the unique invariant straight line, since $x' = x + \alpha$, and hence, the straight line $x = c$, where c is a constant, is mapped to the straight line $x = c + \alpha$, with $\alpha \neq 0$.

Since the invariant straight line has an equation of the form $y = \text{constant}$, it has direction vector e_1. Let P be a point of this straight line, and let $\mathcal{R} = \{P; \lambda e_1, e_2\}$ with $\overrightarrow{Pf(P)} = \lambda e_1$. Then

$$M(f, \mathcal{R}) = \begin{pmatrix} 1 & 0 & 1 \\ 0 & a & 0 \\ 0 & 0 & 1 \end{pmatrix} = Th_g(a).$$

Third case: $M(\tilde{f}, \mathcal{B}) = \tilde{h}_e$. In this case, we take an affine frame with origin any point Q, $\mathcal{S} = \{Q; \mathcal{B}\}$. Then

$$M(f, \mathcal{S}) = \begin{pmatrix} 1 & 0 & \alpha \\ 1 & 1 & \beta \\ 0 & 0 & 1 \end{pmatrix}.$$

Let us see if there are any fixed points. The equations of fixed points are

$$x + \alpha = x,$$

$$x + y + \beta = y,$$

so we have two cases:

$\alpha = 0$. In this case, there are fixed points (any point with $x = -\beta$). Let P be one of these fixed points, and let $\mathcal{R} = \{P; \mathcal{B}\}$. Then

$$M(f, \mathcal{R}) = \begin{pmatrix} 1 & 0 & 0 \\ 1 & 1 & 0 \\ 0 & 0 & 1 \end{pmatrix} = h_e.$$

$\alpha \neq 0$. In this case, there are no fixed points. Let us see if there are invariant straight lines. Since the unique eigenvector is e_2, the second vector of the basis $\mathcal{B}$, the candidates to be invariant straight lines must have equations of the form $x = \text{constant}$. But, since $x' = x + \alpha$, there are no invariant straight lines.

Therefore, there are no special points which can be considered as the origin. Nevertheless, we can slightly improve the initial expression of f in the following way: Let

$$\mathcal{R} = \{Q; (\alpha e_1 + \beta e_2, \alpha e_2)\},$$

where, as before, $\overrightarrow{Qf(Q)} = \alpha e_1 + \beta e_2$. Then $\tilde{f}(\alpha e_1 + \beta e_2) = \alpha e_1 + \beta e_2 + \alpha e_2$, and $\tilde{f}(\alpha e_2) = \alpha e_2$, and hence

$$M(f, \mathcal{R}) = \begin{pmatrix} 1 & 0 & 1 \\ 1 & 1 & 0 \\ 0 & 0 & 1 \end{pmatrix} = Th_e.$$

Fourth case: $M(\tilde{f}, \mathcal{B}) = \tilde{\text{id}} = I_2$. In this case f is, by definition, a translation T_v.

If $f = T_v$, with $v \neq 0$, we complete v to a basis $\mathcal{B}' = (u, v)$. Let P be any point and $\mathcal{R} = \{P; \mathcal{B}'\}$. Then

$$M(f, \mathcal{R}) = \begin{pmatrix} 1 & 0 & 0 \\ 0 & 1 & 1 \\ 0 & 0 & 1 \end{pmatrix} = T.$$

If $f = T_v$ with $v = \vec{0}$, then $f = \text{id}$, and in any affine frame $\mathcal{R}$ we have

$$M(f, \mathcal{R}) = \begin{pmatrix} 1 & 0 & 0 \\ 0 & 1 & 0 \\ 0 & 0 & 1 \end{pmatrix} = I_3.$$

□

When $M(f,\mathcal{R})$ is a canonical matrix we also say that $M(f,\mathcal{R})$ is the *canonical expression* of f.

It remains to be seen if there can exist affine frames $\mathcal{R}$ and $\mathcal{S}$ such that $M(f,\mathcal{R})$ and $M(f,\mathcal{S})$ are two different canonical matrices. For instance, $h_g(a)$ and $Th_g(a)$ have the same characteristic polynomial, and in particular the same trace and the same determinant, and we could have, in principle, $M(f,\mathcal{R}) = h_g(a)$ and $M(f,\mathcal{S}) = Th_g(a)$ for the same affinity f. Likewise for h_e and Th_e, and for T and id.

We shall answer this question in Proposition 3.23.

Example 3.20

Let f be the affinity of the affine plane $\mathbb{R}^2$ given in the canonical affine frame $\mathcal{C}$ by

$$M(f,\mathcal{C}) = \begin{pmatrix} 2 & 2 & 3 \\ 0 & 1 & 4 \\ 0 & 0 & 1 \end{pmatrix}.$$

Find an affine frame $\mathcal{R}$ in $\mathbb{R}^2$ such that $M(f,\mathcal{R})$ is a canonical matrix. Find the fixed points and the invariant straight lines.

Solution

We first study the fixed points.

Fixed points. The equations of the fixed points are

$$2x + 2y + 3 = x,$$

$$y + 4 = y.$$

Hence, there are no fixed points.

Eigenvectors. The matrix

$$A = \begin{pmatrix} 2 & 2 \\ 0 & 1 \end{pmatrix}$$

has $u = (1,0)$ as eigenvector with eigenvalue $\lambda = 2$ and $v = (-2,1)$ as eigenvector with eigenvalue $\mu = 1$.

Invariant straight lines of direction u. These straight lines have equation $y = $ constant and, hence, they are not invariant $(y' = y + 4)$.

Invariant straight lines of direction v. These straight lines have equation $x + 2y + c = 0$, $c \in \mathbb{R}$, and they are invariant if for each of their points (x,y)

there exists a $\lambda \in \mathbb{R}$ such that

$$\begin{pmatrix} 2-1 & 2 & 3 \\ 0 & 1-1 & 4 \\ 0 & 0 & 1-1 \end{pmatrix} \begin{pmatrix} x \\ y \\ 1 \end{pmatrix} = \lambda \begin{pmatrix} -2 \\ 1 \\ 0 \end{pmatrix}. \tag{3.2}$$

Equivalently,

$$\begin{pmatrix} 1 & 2 & 3 \\ 0 & 0 & 4 \\ 0 & 0 & 0 \end{pmatrix} \begin{pmatrix} -c-2y \\ y \\ 1 \end{pmatrix} = \lambda \begin{pmatrix} -2 \\ 1 \\ 0 \end{pmatrix}.$$

From this we deduce that $\lambda = 4$ and $c = 11$. Hence, the invariant straight line is $x + 2y + 11 = 0$.

Invariant straight lines (second method). The straight line $ax + by + c = 0$ is invariant if

$$\begin{pmatrix} 2-\lambda & 0 & 0 \\ 2 & 1-\lambda & 0 \\ 3 & 4 & 1-\lambda \end{pmatrix} \begin{pmatrix} a \\ b \\ c \end{pmatrix} = \begin{pmatrix} 0 \\ 0 \\ 0 \end{pmatrix},$$

where λ is an eigenvalue associated to one of the eigenvectors (not necessarily to the direction vector of the straight line!). See point (8) on page 111.

For $\lambda = 1$ there is no solution. For $\lambda = 2$ the solution is $b = 2a$, $c = 11a$, and we get the same straight line $x + 2y + 11 = 0$ obtained above.

In particular, we have obtained $\rho(f) = 1$.

Hence, looking at the table on page 104, we see that the given affinity is similar to an affinity of the type $Th_g(a)$, namely

$$Th_g(2) = \begin{pmatrix} 1 & 0 & 1 \\ 0 & 2 & 0 \\ 0 & 0 & 1 \end{pmatrix}.$$

To find the affine frame in which the matrix of f has this expression, we take a point P of the invariant straight line, that is, a point of the form $P = (-11 - 2y, y)$, next we compute $f(P)$ and we put

$$\overrightarrow{Pf(P)} = \lambda v.$$

Since this equation coincides with (3.2) we have $\lambda = 4$. Hence, as we have seen in the proof of Theorem 3.19, the affine frame $\mathcal{R}$ (one for each value of y) such that $M(f, \mathcal{R})$ is the canonical matrix $Th_g(2)$ is

$$\mathcal{R} = \{(-11 - 2y); 4v, u\}.$$

The matrix of the change of basis is

$$C = \begin{pmatrix} -8 & 1 & -11-2y \\ 4 & 0 & y \\ 0 & 0 & 1 \end{pmatrix}.$$

Note that this matrix C satisfies $C^{-1}M(f,\mathcal{C})C = Th_g(2)$.   □

Observation 3.21

We collect here some observations that may be useful in trying to classify affinities of the plane.

1. *General homology.* Let f be an affinity such that

$$M(f,\mathcal{S}) = \begin{pmatrix} 1 & 0 & \alpha \\ 0 & a & \beta \\ 0 & 0 & 1 \end{pmatrix}.$$

If $\alpha = 0$, there exists an affine frame $\mathcal{R}$ such that $M(f,\mathcal{R}) = h_g(a)$. If $\alpha \neq 0$, there exists an affine frame $\mathcal{R}$ such that $M(f,\mathcal{R}) = Th_g(a)$.

2. *Special homology.* Let f be an affinity such that

$$M(f,\mathcal{S}) = \begin{pmatrix} 1 & 0 & \alpha \\ 1 & 1 & \beta \\ 0 & 0 & 1 \end{pmatrix}.$$

If $\alpha = 0$, there exists an affine frame $\mathcal{R}$ such that $M(f,\mathcal{R}) = h_e$. If $\alpha \neq 0$, there exists an affine frame $\mathcal{R}$ such that $M(f,\mathcal{R}) = Th_e$.

3. *Translation.* Let f be an affinity such that

$$M(f,\mathcal{S}) = \begin{pmatrix} 1 & 0 & \alpha \\ 0 & 1 & \beta \\ 0 & 0 & 1 \end{pmatrix}.$$

If $\alpha = \beta = 0$, $M(f,\mathcal{S}) = I_3$. If $\alpha \neq 0$ or $\beta \neq 0$, there exists an affine frame $\mathcal{R}$ such that $M(f,\mathcal{R}) = T$.

4. *Origin.* Note that the origin of any affine frame $\mathcal{R}$, such that $M(f,\mathcal{R})$ is a canonical matrix, is the unique fixed point in cases $\mathcal{E}(b,c)$, $\mathcal{H}(a,b)$, $h(a)$, $\mathcal{P}(a)$; an arbitrary fixed point in cases $h_g(a)$, h_e, id; a point on the invariant straight line in case $Th_g(a)$; and an arbitrary point in cases Th_e and T.

5. *Fixed points. First method.* A point P with coordinates $P = (x,y)$ in an affine frame $\mathcal{R}$ is a *fixed point* of the affinity f if and only if

$$\boxed{(x,y,1) \in \ker(M(f,\mathcal{R}) - I_3)}$$

6. *Fixed points. Second method.* To find the fixed points of an affinity f, we first find its canonical expression, next we find the fixed points in the affine frame in which the canonical expression is written, and finally we obtain the results in the given affine frame.

   Concretely: let us assume given $M(f, \mathcal{R}')$, and let $\mathcal{R}$ be an affine frame such that $M(f, \mathcal{R})$ is a canonical matrix. Let $C = M(\mathcal{R}, \mathcal{R}')$ be the matrix of the change of affine frame.

   Then, if a fixed point has coordinates (x, y) in $\mathcal{R}$, it has coordinates (x', y') in $\mathcal{R}'$, given by (1.3) on page 19, that is,

$$\begin{pmatrix} x' \\ y' \\ 1 \end{pmatrix} = C \begin{pmatrix} x \\ y \\ 1 \end{pmatrix}.$$

7. *Invariant straight lines. First method.* Recall that the straight line $P + \langle v \rangle$ is invariant under an affinity f if and only if v is an eigenvector of $\tilde{f}$ and $\overrightarrow{Pf(P)} \in \langle v \rangle$.

   Let us fix an affine frame $\mathcal{R}$ and set $P = (p_1, p_2)$, $v = (v_1, v_2)$. Then the condition $\overrightarrow{Pf(P)} \in \langle v \rangle$ is equivalent to

$$(M - I_3)p = \mu v,$$

   for some scalar μ, where

$$M = M(f, \mathcal{R}), \qquad p = \begin{pmatrix} p_1 \\ p_2 \\ 1 \end{pmatrix}, \qquad v = \begin{pmatrix} v_1 \\ v_2 \\ 0 \end{pmatrix}.$$

8. *Invariant straight lines. Second method.* A straight line r, given in an affine frame $\mathcal{R}$ by the equation $\alpha x + \beta y + \gamma = 0$ (α or β different from zero), is invariant under the affinity f if and only if

$$\boxed{(\alpha, \beta, \gamma) \in \ker(M^{\mathsf{T}} - \mu \cdot I_3)}$$

   where μ is an eigenvalue of $M = M(f, \mathcal{R})$.

   This follows from the fact that the equation of the straight line can be written matricially as

$$(\alpha, \beta, \gamma) \begin{pmatrix} x \\ y \\ 1 \end{pmatrix} = 0.$$

The condition $(\alpha, \beta, \gamma) \in \ker(M^{\mathsf{T}} - \mu \cdot I_3)$ can be written as

$$M^{\mathsf{T}} \begin{pmatrix} \alpha \\ \beta \\ \gamma \end{pmatrix} = \mu \begin{pmatrix} \alpha \\ \beta \\ \gamma \end{pmatrix},$$

or, transposing,

$$\mu(\alpha, \beta, \gamma) = (\alpha, \beta, \gamma)M,$$

and hence

$$(\alpha, \beta, \gamma) \begin{pmatrix} x' \\ y' \\ 1 \end{pmatrix} = (\alpha, \beta, \gamma)M \begin{pmatrix} x \\ y \\ 1 \end{pmatrix} = \mu(\alpha, \beta, \gamma) \begin{pmatrix} x \\ y \\ 1 \end{pmatrix}.$$

Thus, the point $P = (x, y)$ belongs to the straight line r if and only if the point $f(P) = (x', y')$ belongs to the same straight line r. Hence, this straight line is invariant.

Conversely, if the straight line is invariant, the previous equality is satisfied, and hence, by the same argument as that made on page 25, $(\alpha, \beta, \gamma)M = \mu(\alpha, \beta, \gamma)$, or equivalently $(\alpha, \beta, \gamma) \in \ker(M^{\mathsf{T}} - \mu \cdot I_3)$.

Recall that the eigenvalues of M and M^{T} coincide, but the corresponding eigenvectors do not.

9. *Invariant straight lines. Third method.* To compute the invariant straight lines of an affinity f, we first find its canonical expression, next we find the invariant straight lines in the affine frame in which the canonical expression is written, and finally we obtain the results in the given affine frame.

Concretely: let us assume given $M(f, \mathcal{R}')$, and let $\mathcal{R}$ be the affine frame such that $M(f, \mathcal{R})$ is a canonical matrix. Let $C = M(\mathcal{R}, \mathcal{R}')$ be the matrix of change of affine frame.

Then, if an invariant straight line has equation $\alpha x + \beta y + \gamma = 0$ in $\mathcal{R}$, it has equation $\alpha' x + \beta' y + \gamma' = 0$ in $\mathcal{R}'$, with

$$\begin{pmatrix} \alpha' \\ \beta' \\ \gamma' \end{pmatrix} = (C^{\mathsf{T}})^{-1} \begin{pmatrix} \alpha \\ \beta \\ \gamma \end{pmatrix}.$$

3.8 Invariance Level in the Real Plane

Proposition 3.22

Let f be an affinity of the real affine plane and let $\mathcal{R}$ be an affine frame such that $M(f, \mathcal{R})$ is a canonical matrix. Then

$$\rho(f) = 0, \quad \text{for } M(f, \mathcal{R}) = \mathcal{E}(b, c), \mathcal{H}(a, b), h(a), \mathcal{P}(a), h_g(a), h_e, \text{id}.$$

$$\rho(f) = 1, \quad \text{for } M(f, \mathcal{R}) = Th_g(a), T.$$

$$\rho(f) = 2, \quad \text{for } M(f, \mathcal{R}) = Th_e.$$

Proof

To find the fixed points and the invariant straight lines of f we can fix any affine frame and use coordinates. In particular, we can choose the affine frame $\mathcal{R}$ given in the statement. The proof is, therefore, a simple calculation in each of the ten cases. □

Proposition 3.23

Let f be an affinity of the real affine plane and let us assume that there are affine frames $\mathcal{R} = \{P; \mathcal{B}\}$ and $\mathcal{R}' = \{P'; \mathcal{B}'\}$ such that $M(f, \mathcal{R})$ and $M(f, \mathcal{R}')$ are canonical matrices. Then $M(f, \mathcal{R}) = M(f, \mathcal{R}')$.

Proof

The matrices of a given affinity f in different affine frames are conjugated by an invertible matrix of the form

$$\begin{pmatrix} C & c \\ 0 & 1 \end{pmatrix}.$$

This conjugation between $M(f, \mathcal{R})$ and $M(f, \mathcal{R}')$ induces a conjugation, performed by the matrix C, between the matrices $M(\tilde{f}, \mathcal{B})$ and $M(\tilde{f}, \mathcal{B}')$.

Hence, if $M(f, \mathcal{R})$ and $M(f, \mathcal{R}')$ are canonical matrices, then $M(\tilde{f}, \mathcal{B})$ and $M(\tilde{f}, \mathcal{B}')$ must, respectively, be conjugate to and equal to matrices in the seven families

$$\tilde{\mathcal{E}}(b, c), \tilde{\mathcal{H}}(a, b), \tilde{h}(a), \tilde{\mathcal{P}}(a), \tilde{h}_g(a), \tilde{h}_e, \tilde{\text{id}},$$

but, by the classification of endomorphisms in dimension two given in Section 3.7.1, this is impossible, unless $M(\tilde{f}, \mathcal{B}) = M(\tilde{f}, \mathcal{B}')$.

If $M(\tilde{f},\mathcal{B}) = M(\tilde{f},\mathcal{B}') = \tilde{\mathcal{E}}(b,c), \tilde{\mathcal{H}}(a,b), \tilde{h}(a), \tilde{\mathcal{P}}(a)$, it can be seen directly, just by looking at the list of canonical matrices, that we must have

$$M(f,\mathcal{R}) = M(f,\mathcal{R}') = \mathcal{E}(b,c), \mathcal{H}(a,b), h(a), \mathcal{P}(a),$$

respectively.

If $M(\tilde{f},\mathcal{B}) = M(\tilde{f},\mathcal{B}') = \tilde{h}_g(a)$, it can be seen directly, just by looking at the list of canonical matrices, that we could have, in principle, $M(f,\mathcal{R}) = h_g(a)$ and $M(f,\mathcal{R}') = Th_g(a)$. But the invariance level of f is different for the matrices $h_g(a)$ and $Th_g(a)$, so this cannot happen, and hence either

$$M(f,\mathcal{R}) = M(f,\mathcal{R}') = h_g(a)$$

or

$$M(f,\mathcal{R}) = M(f,\mathcal{R}') = Th_g(a).$$

An analogous argument holds for $M(\tilde{f},\mathcal{B}) = M(\tilde{f},\mathcal{B}') = \tilde{h}_e, \text{id}$. This completes the proof. $\qquad\square$

Theorem 3.24 (Characterization)

Let f and g be affinities of the real affine plane. Then f is similar to g if and only if $\tilde{f}$ is similar to $\tilde{g}$ and $\rho(f) = \rho(g)$.

Proof

It has been proved in Propositions 3.5 and 3.15, respectively, that if f is similar to g then $\tilde{f}$ similar to $\tilde{g}$ and $\rho(f) = \rho(g)$.

Conversely, let us assume that we have two affinities of the plane, f and g, such that $\tilde{f}$ is similar to $\tilde{g}$ and $\rho(f) = \rho(g)$.

Case $\rho(f) = \rho(g) = 0$. Let P be a fixed point of f, $f(P) = P$; Q be a fixed point of g, $g(Q) = Q$; and set

$$\tilde{f} = \tilde{h}^{-1} \circ \tilde{g} \circ \tilde{h}.$$

We then define h as the unique affinity such that $\tilde{h}$ is its associated linear map and such that $h(P) = Q$.

Note that

$$(h \circ f)(P) = h(P) = Q = g(Q) = (g \circ h)(P).$$

Hence,

$$h \circ f = g \circ h,$$

since they have the same associated linear map $\tilde{h} \circ \tilde{f} = \tilde{g} \circ \tilde{h}$, and they coincide at the point P.

Case $\rho(f) = \rho(g) = 1$. In this case there is an affine frame $\mathcal{R} = \{P; \mathcal{B}\}$ such that either $M(f, \mathcal{R}) = Th_g(a)$ or $M(f, \mathcal{R}) = T$; and there is an affine frame $\mathcal{R}' = \{P'; \mathcal{B}'\}$ such that either $M(g, \mathcal{R}') = Th_g(a')$ or $M(g, \mathcal{R}') = T$. In particular, we must have either

$$M(\tilde{f}, \mathcal{B}) = \tilde{h}_g(a) \quad \text{or} \quad M(\tilde{f}, \mathcal{B}) = \tilde{\mathrm{id}},$$

and either

$$M(\tilde{g}, \mathcal{B}') = \tilde{h}_g(a') \quad \text{or} \quad M(\tilde{g}, \mathcal{B}') = \tilde{\mathrm{id}}.$$

But, since $\tilde{h}_g(a)$ and $\tilde{\mathrm{id}}$ are not conjugate, and $\tilde{h}_g(a)$ is conjugate to $\tilde{h}_g(a')$ if and only if $a = a'$, we must have either

$$M(f, \mathcal{R}) = M(g, \mathcal{S}) = Th_g(a) \quad \text{or}$$

$$M(f, \mathcal{R}) = M(g, \mathcal{S}) = T.$$

Hence, f is similar to g.

Case $\rho(f) = \rho(g) = 2$. In this case, there are affine frames $\mathcal{R}$, $\mathcal{S}$ such that

$$M(f, \mathcal{R}) = Th_e,$$

$$M(g, \mathcal{S}) = Th_e.$$

Hence, f is similar to g.   □

Thus, every affinity of the real affine plane is similar to one and only one of the affinities of Table 3.3.

3.8.1 Summary of the Classification of Affinities of the Real Plane

Let us assume an affinity f is given in some affine frame $\mathcal{R}$ by the matrix

$$M(f, \mathcal{R}) = \begin{pmatrix} P & p \\ 0 & 1 \end{pmatrix}.$$

In order to determine which kind of affinity f is (without searching for an affine frame in which f has canonical expression) we will use the following procedure:
(1) First we find the characteristic polynomial of P and compute its roots.
(2) If it has complex roots, we are in case $\tilde{\mathcal{E}}(b, c)$. There is an affine frame with respect to which the matrix of f is equal to $\mathcal{E}(b, c)$.

(3) If it has simple real roots, we are either in case $\tilde{\mathcal{H}}(a,b)$ or in case $\tilde{h}_g(a)$. In the first case there is an affine frame with respect to which the matrix of f is equal to $\mathcal{H}(a,b)$. In the second case we still need to compute ρ.

(4) If it has a multiple real root a, we compute $\dim \ker(P - aI_2)$ in order to determine if P is diagonalizable or not. This tells us in which of the cases $\tilde{h}(a)$, $\tilde{\mathcal{P}}(a)$, $\tilde{h}_e$ or $\tilde{\mathrm{id}}$ we are in. In the first two cases there is an affine frame in which the matrix of f is equal, respectively, to $h(a)$ or $\mathcal{P}(a)$. In the two last cases we must compute ρ.

(5) In order to compute ρ, we compute the eigenvectors of the eigenvalue 1 of $M(f,\mathcal{R})$ (point 5 in Observation 3.21, page 110). If one of them has its third component different from zero, then $\rho = 0$. Otherwise, $\rho = 1$ in cases $\tilde{h}_g(a)$ and $\tilde{\mathrm{id}}$, and $\rho = 2$ in case $\tilde{h}_e$.

3.9 Geometrical Interpretation

3.9.1 Homologies

Affinities with canonical expression given by $h_g(a)$ are called *general homologies* and those with canonical expression given by h_e are called *special homologies*.

More geometrically, an affinity f of the real affine plane with a straight line of fixed points is called a *homology*. This straight line e of fixed points is called the *homology axis* of f. Outside of this axis there are no fixed points.

Then, either for all $P \notin e$ the straight line $Pf(P)$ is parallel to the homology axis or there is a point $P \notin e$ such that the straight line $Pf(P)$ meets e in a point O.

In the first case, all straight lines parallel to the homology axis are invariant (verify this), and we can find an affine frame $\mathcal{R}$ in which $M(f,\mathcal{R}) = h_e$, that is, f is a *special homology*.

In the second case, every straight line parallel to the straight line $Pf(P)$ is invariant (verify this), and we can find an affine frame $\mathcal{R}$ in which $M(f,\mathcal{R}) = h_g(a)$, that is, f is a *general homology*.

The knowledge of the axis and the image $f(P)$ of a given point P allows us to find graphically the image $f(Q)$ of any other point Q.

For *special homologies* we need only cut the straight line $Of(P)$ with the straight line parallel to the axis through Q. The point O is the intersection of the straight line PQ with the homology axis, see Figure 3.1.

For *general homologies* we need only cut the straight line $Of(P)$ with the straight line parallel to the straight line $Pf(P)$ through Q. The point O is the intersection of the straight line PQ with the homology axis, see Figure 3.2.

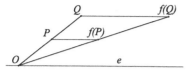

Figure 3.1. Special homology

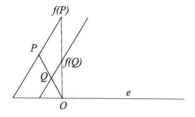

Figure 3.2. General homology

Note that symmetries and projections, studied on pages 70 and 72, are particular cases of general homologies. Indeed, with respect to a suitable affine frame, general homologies have equations

$$x' = ax, \quad a \neq 1,$$
$$y' = y.$$

If $a = 0$, the homology is a *projection* over the straight line $x = 0$, in the direction of the straight line $y = 0$.

If $a = -1$, the homology is a *symmetry* with respect to the straight line $x = 0$, in the direction of the straight line $y = 0$.

An affinity with canonical expression $Th_g(a)$ is a general homology followed by a translation T_u such that the *translation vector u is linearly independent with the direction vector of the invariant straight line*. Recall that the direction vector of the invariant straight line of a general homology is linearly independent with the direction vector of the axis.

An affinity with canonical expression Th_e is a special homology followed by a translation T_u such that the *translation vector u is linearly independent with the direction vector of the invariant straight line*. Recall that the direction vector of the invariant straight line of a special homology is linearly dependent with the direction vector of the axis.

3.9.2 Elliptic Affinities

Affinities with canonical expression given by $\mathcal{E}(b, c)$ are called *elliptic affinities*. They have a unique fixed point and no invariant straight lines.

Note that rotations are elliptic affinities. Concretely, if

$$M(f,\mathcal{R}) = \begin{pmatrix} \cos\alpha & -\sin\alpha & 0 \\ \sin\alpha & \cos\alpha & 0 \\ 0 & 0 & 1 \end{pmatrix},$$

then there exists an affine frame $\mathcal{R}'$ such that $M(f,\mathcal{R}') = \mathcal{E}(-2\cos\alpha,1)$.

On the other hand, since

$$\begin{pmatrix} 0 & -c & 0 \\ 1 & -b & 0 \\ 0 & 0 & 1 \end{pmatrix} = \begin{pmatrix} 0 & \lambda & 0 \\ \frac{1}{\lambda} & 0 & 0 \\ 0 & 0 & 1 \end{pmatrix} \begin{pmatrix} 1 & 2\lambda\alpha & 0 \\ 0 & -1 & 0 \\ 0 & 0 & 1 \end{pmatrix} \begin{pmatrix} \lambda & 0 & 0 \\ 0 & \lambda & 0 \\ 0 & 0 & 1 \end{pmatrix},$$

with $\lambda^2 = c$ and $2\alpha\lambda = -b$, we see that *every elliptic affinity can be considered as the composition of a homothety (if $\lambda \neq 1$) with two general homologies.*

Since

$$\begin{pmatrix} \lambda & 0 & 0 \\ 0 & \lambda & 0 \\ 0 & 0 & 1 \end{pmatrix} = \begin{pmatrix} \lambda & 0 & 0 \\ 0 & 1 & 0 \\ 0 & 0 & 1 \end{pmatrix} \begin{pmatrix} 1 & 0 & 0 \\ 0 & \lambda & 0 \\ 0 & 0 & 1 \end{pmatrix},$$

every homothety is the composition of two general homologies, and hence *every elliptic affinity is a composition of general homologies*; two if $\lambda = 1$, and four if $\lambda \neq 1$.

Notice also that

$$\begin{pmatrix} 0 & \lambda & 0 \\ \frac{1}{\lambda} & 0 & 0 \\ 0 & 0 & 1 \end{pmatrix}^2 = \begin{pmatrix} 1 & 2\lambda\alpha & 0 \\ 0 & -1 & 0 \\ 0 & 0 & 1 \end{pmatrix}^2 = \begin{pmatrix} 1 & 0 & 0 \\ 0 & 1 & 0 \\ 0 & 0 & 1 \end{pmatrix},$$

and hence these two general homologies are symmetries, and we can conclude that *every elliptic affinity is the composition of a homothety (if $\lambda \neq 1$) followed by two symmetries with concurrent axes (concurrent in the center of the homothety).*

The equation of the axis of the first symmetry, with respect to the affine frame we are using, is $x = \lambda y$ (and the direction of symmetry is that of the straight line $x + y\lambda = 0$); and the axis of the second symmetry has equation $y = 0$ (and the direction of symmetry is that of the straight line $\alpha\lambda y + x = 0$).

3.9.3 Parabolic Affinities

Affinities with canonical expression given by $\mathcal{P}(a)$ are called *parabolic affinities*. They have a unique fixed point and a unique invariant straight line, called the *axis* of the affinity.

A characterization of these affinities is given in Exercise 3.1, page 124.

To construct graphically the image of a point under a parabolic affinity f we need to know the fixed point O, the axis e, a pair of corresponding points P and $f(P)$ on the axis, and a pair of corresponding straight lines r and $f(r)$, through O.

Construction of $f(Q)$ when $Q \in r$. We shall follow the following steps (see Figure 3.3):

(1) Let Q_1 be the intersection of the parallel to the axis through Q with r'.
(2) Then Q' is the intersection of the parallel to PQ_1 through P'.

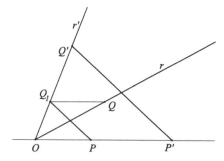

Figure 3.3. Parabolic affinity. First case

Construction of $f(R)$ when $R \notin r$. We shall follow the following steps (see Figure 3.4):

(1) Consider the point $M = QR \cap e$.
(2) Next consider the point R_1 defined as the intersection of the straight line MQ_1 with the parallel to the axis e through R.
(3) The point $R' = f(R)$ we are looking for is the intersection of the straight line OR_1 with the parallel to PR_1 through P'.

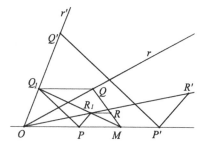

Figure 3.4. Parabolic affinity. Second case

For a justification of the validity of these two constructions, see Exercise 3.2, page 124.

3.9.4 Hyperbolic Affinities

Affinities with canonical expression given by $\mathcal{H}(a,b)$ are called *hyperbolic affinities*. They have a unique fixed point and two invariant straight lines, called *axes* of the affinity.

 If we know the axes e_1, e_2 and a pair of corresponding points, that is, a point P and its image $P' = f(P)$, for some point P that does not belong to the axes, then we can construct $Q' = f(Q)$ graphically for any other point Q.

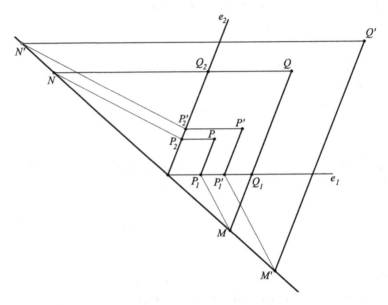

Figure 3.5. Hyperbolic affinity

 We shall follow the following steps (see Figure 3.5):
(1) Construct $P_1, P_2, P_1', P_2', Q_1, Q_2$ as the intersection with the axes of the parallel to them through P, P' and Q, respectively.
(2) Draw an auxiliary straight line r through the intersection point of the two axes, and construct the points $M = r \cap QQ_1$ and $N = r \cap QQ_2$.
(3) Draw the parallel to MP_1 through P_1' and find the point M', the intersection of this parallel with r.
(4) Draw the parallel to NP_2 through P_2' and find the point N', the intersection of this parallel with r.
(5) Obtain the point $Q' = f(Q)$ we are looking for as the intersection of the parallel e_1 through N' with the parallel to e_2 through M'.
For a justification of the validity of this construction, see Exercise 3.3, page 124.

Observation 3.25

An affinity f of the plane is:

(1) Elliptic if and only if the characteristic polynomial $p_{\tilde{f}}(x)$ of the associated linear map has complex roots. Equivalently, if and only if $(\text{trace}(\tilde{f}))^2 < 4 \det \tilde{f}$.

(2) Parabolic or a homothety if and only if $p_{\tilde{f}}(x)$ has a double real root different from 1.

(3) Hyperbolic if and only if $p_{\tilde{f}}(x)$ has two different real roots different from 1.

(4) A general homology or a general homology followed by a translation if and only if $p_{\tilde{f}}(x)$ has two different real roots and one of them is equal to 1.

(5) A special homology, a special homology followed by a translation, a translation or the identity if and only if $p_{\tilde{f}}(x)$ has 1 as a double real root.

3.10   Decomposition of Affinities in the Real Plane

Since

$$
\begin{pmatrix} a & b & h \\ c & d & k \\ 0 & 0 & 1 \end{pmatrix} = \begin{pmatrix} a & -c & h \\ c & a & k \\ 0 & 0 & 1 \end{pmatrix} \begin{pmatrix} 1 & 0 & 0 \\ 0 & \lambda & 0 \\ 0 & 0 & 1 \end{pmatrix} \begin{pmatrix} 1 & \mu & 0 \\ 0 & 1 & 0 \\ 0 & 0 & 1 \end{pmatrix},
$$

with

$$
\lambda = \frac{ad - bc}{a^2 + c^2}, \qquad \mu = \frac{ab + cd}{a^2 + c^2},
$$

we see that *every bijective affinity of the plane is equal to the composition of a special homology (or the identity if $\mu = 0$), followed by a general homology (or the identity if $\lambda = 1$), followed by an elliptic affinity (or a homothety if $c = 0$ and $a \neq 1$, or the identity or a translation if $c = 0$ and $a = 1$).*

Note that the general and special homologies above have the same axis. Note also that any matrix of the form

$$
\begin{pmatrix} a & -c \\ c & a \end{pmatrix}, \quad c \neq 0,
$$

satisfies the ellipticity condition, since $(2a)^2 < 4(a^2 + c^2)$.

Finally we remark that this decomposition of the affinity is not unique.

Table 3.2 summarizes the sixteen different ways in which a bijective affinity can be decomposed according to the above matricial product.

$\mathcal{E} \circ h_e$	$h \circ h_e$	h_e	$T \circ h_e$
$\mathcal{E} \circ h_g \circ h_e$	$h \circ h_g \circ h_e$	$h_g \circ h_e$	$T \circ h_g \circ h_e$
$\mathcal{E} \circ h_g$	$h \circ h_g$	h_g	$T \circ h_g$
$\mathcal{E}$	h	T	id

Table 3.2. Decomposition of affinities

Here $T =$ translation, $h =$ homothety, $h_g =$ general homology, $h_e =$ special homology, $\mathcal{E} =$ elliptic.

Note that

$$\begin{pmatrix} 1 & 0 & 0 \\ 1 & 1 & 0 \\ 0 & 0 & 1 \end{pmatrix} = \begin{pmatrix} \alpha & \beta & 0 \\ -\beta & \alpha & 0 \\ 0 & 0 & 1 \end{pmatrix} \begin{pmatrix} a & 0 & 0 \\ 0 & 1 & 0 \\ 0 & 0 & 1 \end{pmatrix} \begin{pmatrix} \alpha' & \beta' & 0 \\ -\beta' & \alpha' & 0 \\ 0 & 0 & 1 \end{pmatrix},$$

with

$$a = \frac{3+\sqrt{5}}{2}, \qquad \alpha = \frac{5-\sqrt{5}}{20}, \qquad \beta = -\frac{\sqrt{5}}{10}, \qquad \alpha' = 2, \qquad \beta' = \sqrt{5}-1.$$

Thus, every special homology is the composition of an elliptic affinity with a general homology and with another elliptic affinity. We shall write

$$\boxed{h_e = \mathcal{E} \circ h_g \circ \mathcal{E}} \tag{3.3}$$

Summing up, we have the following.

Theorem 3.26

Every affinity is a composition of general homologies.

Proof

This follows from Table 3.2 on page 122, together with formula (3.3) and the fact that homotheties and elliptic affinities are a composition of general homologies, see page 118.

Translations and the identity are a composition of general homologies (with parallel axes), because

$$\begin{pmatrix} 1 & 0 & 0 \\ 0 & 1 & 0 \\ 0 & 0 & 1 \end{pmatrix} = \begin{pmatrix} a & 0 & 0 \\ 0 & 1 & 0 \\ 0 & 0 & 1 \end{pmatrix} \begin{pmatrix} a^{-1} & 0 & 0 \\ 0 & 1 & 0 \\ 0 & 0 & 1 \end{pmatrix}, \quad a \neq 0, 1,$$

$$\begin{pmatrix} 1 & 0 & 0 \\ 0 & 1 & 1 \\ 0 & 0 & 1 \end{pmatrix} = \begin{pmatrix} 1 & 0 & 0 \\ 0 & -1 & 1 \\ 0 & 0 & 1 \end{pmatrix} \begin{pmatrix} 1 & 0 & 0 \\ 0 & -1 & 0 \\ 0 & 0 & 1 \end{pmatrix}.$$

Affinity	Name and properties	Equation
$\mathcal{E}(b,c)$	*Elliptic* One fixed point. No invariant straight lines. $\rho = 0$.	$x' = -cy, \quad b^2 < 4c,$ $y' = x - by.$
$\mathcal{H}(a,b)$	*Hyperbolic* One fixed point. Two invariant straight lines. $\rho = 0$.	$x' = ax, \quad a,b \neq 1,$ $y' = by, \quad a \neq b.$
$h(a)$	*Homothety* Infinite invariant straight lines. One fixed point. $\rho = 0$.	$x' = ax, \quad a \neq 1,$ $y' = ay.$
$\mathcal{P}(a)$	*Parabolic* One fixed point. One invariant straight line. $\rho = 0$.	$x' = ax, \quad a \neq 1,$ $y' = x + ay.$
$h_g(a)$	*General homology* Invariant lines not $\parallel$ to the fixed one. Line of fixed points. $\rho = 0$.	$x' = ax, \quad a \neq 1,$ $y' = y.$
$Th_g(a)$	*Gen. hom. followed by trans.* One invariant straight line. No fixed points. $\rho = 1$.	$x' = ax, \quad a \neq 1,$ $y' = y + 1.$
h_e	*Special homology* Invariant lines $\parallel$ to the fixed one. Line of fixed points. $\rho = 0$.	$x' = x,$ $y' = x + y.$
Th_e	*Sp. hom. followed by trans.* No invariant straight lines. No fixed points. $\rho = 2$.	$x' = x + 1,$ $y' = x + y.$
T	*Translation* $\rho = 1$.	$x' = x,$ $y' = y + 1.$
id	*Identity* $\rho = 0$.	$x' = x,$ $y' = y.$

Table 3.3. Affinities of the plane

Hence, every element in Table 3.2 on page 122 can be reduced to a composition of general homologies.   □

EXERCISES

3.1. Prove that the composition of a special homology of axis e with a homothety of center $O \in e$ is a parabolic affinity.

3.2. Prove that a parabolic affinity $\mathcal{P}(a)$ maps every straight line r parallel to the axis into another straight line r' parallel to the axis, and such that the simple ratio $(e, r, r') = a$ (with the natural definition of the simple ratio of three parallel straight lines).

Justify, following the above comments, the two constructions of the image of a point under a parabolic affinity given on page 119.

3.3. Justify the construction of the image of a point under a hyperbolic affinity given on page 120.

3.4. Give examples of two elliptic affinities whose composition is respectively: elliptic, parabolic, hyperbolic, a homothety, a special homology, a special homology followed by a translation, a general homology, a general homology followed by a translation, a translation or the identity.

3.5. Repeat the previous exercise, but this time consider the composition of two hyperbolic affinities or two parabolic affinities.

3.6. Study the composition of two affinities of different type in the real affine plane.

3.7. Let h be a homology and let u be an arbitrary vector. Prove that there exists a vector v such that

$$T_v \circ h = h \circ T_u.$$

Study the relationship between the following two sets of affinities of the plane:

$$TH = \{T_u \circ h : u \in E, h \text{ homology}\},$$

$$HT = \{h \circ T_u : u \in E, h \text{ homology}\}.$$

3.8. Classify the affinity f of the affine plane $\mathbb{R}^2$ given, in the canonical affine frame, by

$$\begin{cases} x' = x - \frac{1}{8}y - \frac{1}{8}, \\ y' = 2x - \frac{1}{8}. \end{cases}$$

Find the fixed points, the invariant straight lines, and an affine frame $\mathcal{R}$ such that $M(f,\mathcal{R})$ is the canonical expression of f.

3.9. Classify, in each case, the affinity f of a real affine plane given, in some affine frame, by

(a) $\begin{cases} x' = 4y + 1, \\ y' = -x + 4y + 1. \end{cases}$

(b) $\begin{cases} x' = 2x + 9y + 12, \\ y' = x + 4y + 4. \end{cases}$

(c) $\begin{cases} x' = -\frac{1}{2}x + \frac{3}{2}y + \frac{11}{2}, \\ y' = \frac{3}{2}x - \frac{1}{2}y - \frac{7}{2}. \end{cases}$

(d) $\begin{cases} x' = 2x - y - 2, \\ y' = 3y + 2. \end{cases}$

(e) $\begin{cases} x' = \frac{3}{2}x - \frac{1}{2}y - 1, \\ y' = \frac{1}{2}x + \frac{1}{2}y + 1. \end{cases}$

(f) $\begin{cases} x' = x - 3y + 3, \\ y' = x - 2y + 2. \end{cases}$

Find, in each case, the fixed points, the invariant straight lines, and an affine frame $\mathcal{R}$ such that $M(f,\mathcal{R})$ is the canonical expression of f.

3.10. Classify the affinities f_a of the affine plane $\mathbb{R}^2$ given, in the canonical affine frame, by

$$\begin{cases} x' = ax + y + a, \\ y' = x + ay + a. \end{cases}$$

Find, in each case, the fixed points, the invariant straight lines, and an affine frame $\mathcal{R}_a$ such that $M(f_a, \mathcal{R}_a)$ is the canonical expression of f_a.

Find the locus of the images of a point under all these affinities. That is, describe the set $\{f_a(P) : a \in \mathbb{R}\}$ for each point $P \in \mathbb{R}^2$.

3.11. Classify the affinities f_a of the affine plane $\mathbb{R}^2$ given, in the canonical affine frame, by

$$\begin{cases} x' = (1+a)x - ay + 1, \\ y' = a^2x + (1 + 2a - 4a^2 + a^3)y, \end{cases}$$

and such that they do not have fixed points. Find the invariant straight lines, and an affine frame $\mathcal{R}_a$ such that $M(f_a, \mathcal{R}_a)$ is the canonical expression of f_a.

Find the locus of the images of a point under all these affinities. That is, describe the set $\{f_a(P) : a \in \mathbb{R}\}$ for each given point $P \in \mathbb{R}^2$.

3.12. Classify the affinity f of a real affine plane given, in some affine frame, by

$$\begin{pmatrix} x' \\ y' \end{pmatrix} = A \begin{pmatrix} x \\ y \end{pmatrix} + a,$$

in the following cases:

(a) $A = \begin{pmatrix} -17 & 10 \\ 30 & 18 \end{pmatrix}$, $a = \begin{pmatrix} -2 \\ 4 \end{pmatrix}$,

(b) $A = \begin{pmatrix} -4 & 1 \\ -9 & -2 \end{pmatrix}$, $a = \begin{pmatrix} -2 \\ 2 \end{pmatrix}$,

(c) $A = \begin{pmatrix} 11 & -25 \\ 4 & -9 \end{pmatrix}$, $a = \begin{pmatrix} -5 \\ -2 \end{pmatrix}$,

(d) $A = \begin{pmatrix} 11 & -25 \\ 4 & -9 \end{pmatrix}$, $a = \begin{pmatrix} -39 \\ -13 \end{pmatrix}$,

(e) $A = \begin{pmatrix} -7 & -4 \\ 12 & 7 \end{pmatrix}$, $a = \begin{pmatrix} 2 \\ -3 \end{pmatrix}$,

(f) $A = \begin{pmatrix} 4 & 6 \\ -2 & -3 \end{pmatrix}$, $a = \begin{pmatrix} -9 \\ 6 \end{pmatrix}$.

Find, in each case, the fixed points, the invariant straight lines, and an affine frame $\mathcal{R}$ such that $M(f,\mathcal{R})$ is the canonical expression of f.

3.13. Classify the affinity f of a real affine plane given, in some affine frame, by

$$\begin{cases} x' = 3x + 4y + 3, \\ y' = -x - y - 1. \end{cases}$$

Find the fixed points, the invariant straight lines, and an affine frame $\mathcal{R}$ such that $M(f,\mathcal{R})$ is the canonical expression of f.

3.14. Classify the affinity f of a real affine plane given, in some affine frame, by

$$\begin{cases} x' = 5x - 2y + 3, \\ y' = \frac{9}{2}x - y + 4. \end{cases}$$

Find the fixed points, the invariant straight lines, and an affine frame $\mathcal{R}$ such that $M(f,\mathcal{R})$ is the canonical expression of f.

3.15. Consider the two affinities of a real affine plane given, in some affine frame, by

$$f(x,y) = (-2x - y - 1, 9x + 4y + 4),$$

$$g(x,y) = (3x + 4y + 3, -x - y - 1).$$

Are they similar?

3.16. Classify the affinity f of a real affine plane given by

$$M(f,\mathcal{R}) = \begin{pmatrix} 1 & 0 & 0 \\ 0 & -1 & 1 \\ 0 & 0 & 1 \end{pmatrix}.$$

Find the fixed points, the invariant straight lines, and an affine frame $\mathcal{R}'$ such that $M(f,\mathcal{R}')$ is the canonical expression of f.

3.17. Classify the affinity f_a of the affine plane $\mathbb{R}^2$ given by

$$M(f_a,\mathcal{C}) = \begin{pmatrix} 1 & 0 & 0 \\ 5 & a & 0 \\ 0 & 0 & 1 \end{pmatrix},$$

where $\mathcal{C}$ is the canonical affine frame. Find, as a function of the values of the parameter $a \in \mathbb{R}$, the fixed points, the invariant straight lines, and an affine frame $\mathcal{R}_a$ such that $M(f_a,\mathcal{R}_a)$ is the canonical expression of f_a.

3.18. (a) Find, in the canonical affine frame of the affine plane $\mathbb{R}^2$, the equations of a homothety with center $(1,1)$ and similitude ratio 3.

(b) Can two homotheties with different centers be similar?

(c) Can two homotheties with different ratios be similar?

3.19. Let $\mathcal{R} = \{P;(e_1,e_2)\}$ be an affine frame in a real affine plane $\mathbb{A}$. Study the affinity $f = S \circ T_u$ of $\mathbb{A}$ given by the composition of the translation T_u by the vector $u = (3,4)$ with the symmetry S with respect to the straight line $L_1 = P + \langle e_1 \rangle$ in the direction $\langle e_2 \rangle$. Find the fixed points, the invariant straight lines, and an affine frame $\mathcal{R}'$ such that $M(f,\mathcal{R}')$ is the canonical expression of f.

3.20. Let $\mathcal{R} = \{P;(e_1,e_2)\}$ be an affine frame in a real affine plane $\mathbb{A}$. Study the affinity $f = T_u \circ S$ of $\mathbb{A}$ given by the composition of the symmetry S with respect to the straight line $L_1 = P + \langle e_1 \rangle$ in the direction $\langle e_2 \rangle$ and the translation T_u by the vector $u = (2,5)$. Find the fixed points, the invariant straight lines, and an affine frame $\mathcal{R}'$ such that $M(f,\mathcal{R}')$ is the canonical expression of f.

3.21. Study the affinities of a complex affine plane.

Classification of Affinities in Arbitrary Dimension

4.1 Introduction

In this chapter we study affinities in arbitrary dimension. We shall see, using the classification of endomorphisms, that we can give a very general classification of affinities. For this, we shall try to generalize the results obtained in dimension two to arbitrary dimension.

4.2 Jordan Matrices

First let us recall the following results on the classification of endomorphisms. Let $\tilde{f}$ be an endomorphism of a k-vector space E of dimension n and assume that the characteristic polynomial $p_c(\tilde{f})$ factorizes as

$$p_c(\tilde{f}) = (x-1)^s q(x),$$

with $\gcd((x-1)^s, q(x)) = 1$.

Then

$$E = \ker(\tilde{f} - \mathrm{id})^s \oplus \mathrm{Im}(\tilde{f} - \mathrm{id})^s, \tag{4.1}$$

and this decomposition is invariant under $\tilde{f}$, see, for instance [8], page 359. We shall call this the *canonical decomposition* of E induced by $\tilde{f}$.

A. Reventós Tarrida, *Affine Maps, Euclidean Motions and Quadrics*,
Springer Undergraduate Mathematics Series,
DOI 10.1007/978-0-85729-710-5_4, © Springer-Verlag London Limited 2011

From now on we shall use the notation

$$E_1 = \ker(\tilde{f} - \mathrm{id})^s,$$

$$E_2 = \mathrm{Im}(\tilde{f} - \mathrm{id})^s.$$

Thus, equality (4.1) is now written as

$$E = E_1 \oplus E_2.$$

We also know that there is a basis $\mathcal{B}_1$ of E_1, called the *Jordan basis*, such that

$$M(\tilde{f}_{|E_1}, \mathcal{B}_1) = \begin{pmatrix} 1 & 0 & \cdots & & \cdots & 0 \\ \varepsilon_1 & 1 & \ddots & & & \vdots \\ 0 & \ddots & \ddots & & \ddots & \vdots \\ \vdots & \ddots & & \ddots & 1 & 0 \\ 0 & \cdots & & 0 & \varepsilon_{n-1} & 1 \end{pmatrix}, \tag{4.2}$$

with $\varepsilon_i \in \{0, 1\}$, for all $i = 1, \ldots, n-1$. See, for instance [8], page 388.

The matrices

$$(1), \quad \begin{pmatrix} 1 & 0 \\ 1 & 1 \end{pmatrix}, \quad \begin{pmatrix} 1 & 0 & 0 \\ 1 & 1 & 0 \\ 0 & 1 & 1 \end{pmatrix}, \quad \cdots$$

are called *Jordan boxes* of order 1, 2, 3, etc. Thus, the matrix (4.2) is composed of Jordan boxes, and these boxes are uniquely determined, so that the expression (4.2) is unique up to the order of these boxes. (One normally puts the larger Jordan boxes at the top of the matrix, however in the classification theorem for affinities that we are about to prove it is not possible to maintain this convention; we will be forced to permute the boxes.)

Recall that the basis $\mathcal{B}_1$ is formed by certain special vectors e_i, called generators, and their images under $(\tilde{f} - \mathrm{id})$, $(\tilde{f} - \mathrm{id})^2$, $(\tilde{f} - \mathrm{id})^3$, etc. These generators will play a fundamental role in the classification theorem for affinities.

The basis $\mathcal{B}_1$ has, therefore, the following form (ordered by rows):

$$e_1, (\tilde{f} - \mathrm{id})e_1, \ldots, (\tilde{f} - \mathrm{id})^{s_1-1}e_1,$$

$$e_2, (\tilde{f} - \mathrm{id})e_2, \ldots, (\tilde{f} - \mathrm{id})^{s_2-1}e_2,$$

$$\vdots$$

$$e_r, (\tilde{f} - \mathrm{id})e_r, \ldots, (\tilde{f} - \mathrm{id})^{s_r-1}e_r,$$

where $s_1, \ldots, s_r$ are the orders of the Jordan boxes of (4.2). We shall write

$$\boxed{e_{ij} = (\tilde{f} - \mathrm{id})^{j-1} e_i, \quad j = 2, \ldots, s_i}$$

Thus we have

$$\mathcal{B}_1 = (e_1, e_{12}, \ldots, e_{1s_1}, e_2, e_{22}, \ldots, e_{2s_2}, \ldots \ldots, e_r, e_{r2}, \ldots, e_{rs_r}). \tag{4.3}$$

Note that the vector e_i is not the i-th vector of the basis, but the generator of the i-th Jordan box.

Denoting

$$C_j = \langle e_j, e_{j2}, \ldots, e_{js_j} \rangle, \quad j = 1, \ldots, r,$$

we have the following decomposition of E_1 as a direct sum of $\tilde{f}$-invariant subspaces

$$E_1 = C_1 \oplus \cdots \oplus C_t \oplus \cdots \oplus C_r. \tag{4.4}$$

Unlike the decomposition (4.1), this one is not canonical.

In summary, the most important properties of this basis are the following:

$$
\begin{aligned}
s_t &= \text{order of the } t\text{-th Jordan box,} \\
e_t &= \text{generator of the } t\text{-th Jordan box,} \\
e_{tk} &= \text{element of the basis of the } t\text{-th box,} \\
(\tilde{f} - \mathrm{id})(e_t) &= e_{t2}, \quad \text{if } s_t > 1, \\
(\tilde{f} - \mathrm{id})(e_{tk}) &= e_{t(k+1)}, \quad 2 \le k < s_t, \\
(\tilde{f} - \mathrm{id})(e_{ts_t}) &= 0, \\
(\tilde{f} - \mathrm{id})^{s_t}(e_t) &= 0.
\end{aligned}
\tag{4.5}
$$

The sixth equality states that the last vector of the basis of each Jordan box is an eigenvector.

For instance, if

$$
M(\tilde{f}_{|E_1}, \mathcal{B}_1) = \left(\begin{array}{ccc|ccc}
1 & & & & & \\
1 & 1 & & & & \\
\hline
 & & & 1 & & \\
 & & & 1 & 1 & \\
 & & & & 1 & 1
\end{array} \right),
$$

we have a basis $\mathcal{B}_1$ of the form

$$e_1$$
$$e_{12} = (\tilde{f} - \text{id})e_1,$$
$$e_2$$
$$e_{22} = (\tilde{f} - \text{id})e_2,$$
$$e_{23} = (\tilde{f} - \text{id})^2 e_2.$$

Here $s_1 = 2$, $s_2 = 3$, and the generators are e_1 and e_2. Note finally that $(\tilde{f} - \text{id})^2 e_1 = 0$ and that $(\tilde{f} - \text{id})^3 e_2 = 0$. This is the reason why the first box has order two and the second one has order three.

Observation 4.1

We have decomposed $E = E_1 \oplus E_2$ and we have found a Jordan basis $\mathcal{B}_1$ of E_1. If the characteristic polynomial of $\tilde{f}_{|E_2}$ factorizes in linear factors, we can also find a Jordan basis of E_2. Otherwise, we can find a basis of E_2 in which the matrix of $\tilde{f}$ has rational form; see [8], page 428.

Observation 4.2

Note that conjugate Jordan matrices are equal, up to the order of the boxes. This is a consequence of the uniqueness, up to the order, of the Jordan matrix associated to an endomorphism; see [8], page 373. For instance, we have $J' = C^{-1}JC$, with

$$J = \begin{pmatrix} 1 & 0 & 0 \\ 1 & 1 & 0 \\ 0 & 0 & 1 \end{pmatrix}, \qquad J' = \begin{pmatrix} 1 & 0 & 0 \\ 0 & 1 & 0 \\ 0 & 1 & 1 \end{pmatrix}, \qquad C = \begin{pmatrix} 0 & 1 & 0 \\ 0 & 0 & 1 \\ 1 & 0 & 0 \end{pmatrix}.$$

We remark, once again, that the Jordan basis is not unique. For instance,

$$\begin{pmatrix} a & 0 & 0 \\ b & a & 0 \\ c & b & a \end{pmatrix} \begin{pmatrix} 1 & 0 & 0 \\ 1 & 1 & 0 \\ 0 & 1 & 1 \end{pmatrix} = \begin{pmatrix} 1 & 0 & 0 \\ 1 & 1 & 0 \\ 0 & 1 & 1 \end{pmatrix} \begin{pmatrix} a & 0 & 0 \\ b & a & 0 \\ c & b & a \end{pmatrix},$$

for all a, b, c. This equality can be written as $J = C^{-1}JC$, where J is the Jordan matrix. If $a \neq 0$, C is the matrix of a change of basis leaving J invariant.

4.3 Similar Endomorphisms and Canonical Decomposition

Definition 4.3

Let $\tilde{f}$ be an endomorphism of a k-vector space E. Let $E = E_1 \oplus E_2$ be the canonical decomposition of E induced by $\tilde{f}$. A basis $\mathcal{B} = (e_1, \ldots, e_k, e_{k+1}, \ldots, e_n)$ of E is *adapted* to $\tilde{f}$ if $\mathcal{B}_1 = (e_1, \ldots, e_k)$ is a basis of E_1 and $\mathcal{B}_2 = (e_{k+1}, \ldots, e_n)$ is a basis of E_2.

In this case we shall write $\mathcal{B} = (\mathcal{B}_1, \mathcal{B}_2)$.

Proposition 4.4

Let $\tilde{f}$ and $\tilde{g}$ be similar endomorphisms of E, and let $\mathcal{B} = (\mathcal{B}_1, \mathcal{B}_2)$ and $\mathcal{B}' = (\mathcal{B}'_1, \mathcal{B}'_2)$ be bases adapted respectively to $\tilde{f}$ and $\tilde{g}$. Then there exists an invertible matrix

$$C = \begin{pmatrix} P_1 & P_2 \\ 0 & P_4 \end{pmatrix},$$

where P_1 is a $p \times p$ matrix (p=number of elements of $\mathcal{B}_1$=number of elements of $\mathcal{B}'_1$), and P_4 is a $(n-p) \times (n-p)$ matrix, $n = \dim E$, such that

$$M(\tilde{f}, \mathcal{B}) = C^{-1} M(\tilde{g}, \mathcal{B}') C.$$

Proof

Let us assume that $\tilde{f}$ is similar to $\tilde{g}$, and let

$$E = E_1 \oplus E_2$$

$$E = E'_1 \oplus E'_2$$

be the two canonical decompositions induced respectively by $\tilde{f}$ and $\tilde{g}$, given in (4.1).

Let $\tilde{h}$ be an isomorphism conjugating $\tilde{f}$ and $\tilde{g}$, that is, such that $\tilde{f} = \tilde{h}^{-1} \circ \tilde{g} \circ \tilde{h}$.

Note that

$$(\tilde{f} - \mathrm{id})^s = \tilde{h}^{-1}(\tilde{g} - \mathrm{id})^s \tilde{h}. \tag{4.6}$$

Hence,

$$\tilde{h}(E_1) = E'_1.$$

This implies that the matrix $C = M(\tilde{h}, \mathcal{B}, \mathcal{B}')$ is of the required form, and by the formula for the change of basis, the proof is complete. $\square$

The next result is included only to emphasize that, in the context of Jordan bases, not only is there a matrix like the above C conjugating $M(\tilde{f}, \mathcal{B})$ and $M(\tilde{g}, \mathcal{B}')$, but any matrix conjugating them must be of the same type as C. Concretely, we have the following.

Definition 4.5

Let $\tilde{f}$ be an endomorphism of a k-vector space E. Let $E = E_1 \oplus E_2$ be the canonical decomposition of E induced by $\tilde{f}$. A basis $\mathcal{B} = (e_1, \dots, e_k, e_{k+1}, \dots, e_n)$ of E is a *Jordan basis adapted to* $\tilde{f}$ if $\mathcal{B}_1 = (e_1, \dots, e_k)$ is a Jordan basis of E_1 and $\mathcal{B}_2 = (e_{k+1}, \dots, e_n)$ is a basis of E_2.

Note that if $\mathcal{B}$ is a Jordan basis adapted to $\tilde{f}$, then

$$M(\tilde{f}, \mathcal{B}) = \begin{pmatrix} J & 0 \\ 0 & K \end{pmatrix},$$

where J is a Jordan matrix.

Proposition 4.6

Let us assume

$$\begin{pmatrix} P_1 & P_2 \\ P_3 & P_4 \end{pmatrix} \begin{pmatrix} J & 0 \\ 0 & K \end{pmatrix} = \begin{pmatrix} J' & 0 \\ 0 & K' \end{pmatrix} \begin{pmatrix} P_1 & P_2 \\ P_3 & P_4 \end{pmatrix},$$

where J, J' are $p \times p$ Jordan matrices with 1s on the diagonal, K, K' are $q \times q$ matrices such that K' does not admit the eigenvalue 1, P_1 is a $p \times p$ matrix, P_4 is a $q \times q$ matrix, P_3 is a $q \times p$ matrix, and P_2 is a $p \times q$ matrix. Then $P_3 = 0$.

Proof

Using the block product of matrices, see [8], page 182, one obtains

$$P_3 J = K' P_3. \tag{4.7}$$

We write P_3 in columns as $P_3 = (C_1, \dots, C_p)$. The last column of $P_3 J$ (the p-th column) is equal to P_3 multiplied by the last column of J, but we know

that

$$\text{last column of } J = \begin{pmatrix} 0 \\ \vdots \\ 0 \\ 1 \end{pmatrix},$$

and, hence, the last column of $P_3 J$ is equal to the last column, C_p, of P_3. But, at the same time, it must be equal to the last column of $K'P_3$, which is equal to K' multiplied by C_p. Hence, since K' does not have eigenvalue 1, $C_p = 0$.

Equating now the penultimate columns of (4.7) we get either

$$C_{p-1} + C_p = K'C_{p-1}$$

if the penultimate column of J has two 1s, or

$$C_{p-1} = K'C_{p-1}$$

if the penultimate column of J has one 1. But, since $C_p = 0$, in both cases we have the same result: $C_{p-1} = K'C_{p-1}$. Since K' does not have the eigenvalue 1, $C_{p-1} = 0$. We repeat this procedure until we reach the first column of P_3. Therefore all the columns of P_3 are zero. That is, $P_3 = 0$, and this completes the proof. □

4.4 Clarifying Examples

Before initiating the study of the general classification, we provide examples of the many different situations that can occur. With these examples in mind, the proof of the general theorem is much easier to follow.

Example 4.7

Let f be an affinity of an affine space of dimension 4. Let us assume that in some affine frame $\mathcal{R} = \{P; (e_1, e_2, e_3, e_4)\}$ we have

$$M(f, \mathcal{R}) = \left(\begin{array}{cccc|c} 1 & 0 & 0 & 0 & 1 \\ 1 & 1 & 0 & 0 & 0 \\ 0 & 0 & 1 & 0 & 1 \\ 0 & 0 & 1 & 1 & 0 \\ \hline 0 & 0 & 0 & 0 & 1 \end{array} \right).$$

Equivalently,

$$x_1' = x_1 + 1,$$

$$x_2' = x_1 + x_2,$$

$$x_3' = x_3 + 1,$$

$$x_4' = x_3 + x_4.$$

If we make the change of basis

$$e_1' = e_1 + e_3 \quad (\text{the vector } \overrightarrow{Pf(P)}),$$

$$e_2' = (\tilde{f} - \mathrm{id})(e_1') = e_2 + e_4,$$

$$e_3' = e_3,$$

$$e_4' = e_4,$$

we have

$$M(f, \mathcal{R}') = \left(\begin{array}{cc|cc|c} 1 & 0 & 0 & 0 & 1 \\ 1 & 1 & 0 & 0 & 0 \\ \hline 0 & 0 & 1 & 0 & 0 \\ 0 & 0 & 1 & 1 & 0 \\ \hline 0 & 0 & 0 & 0 & 1 \end{array}\right).$$

We remark that, since e_2 and e_4 are eigenvectors, $(\tilde{f} - \mathrm{id})^2(e_1') = 0$.

Thus, we have changed the basis in such a way that the Jordan matrix has not changed (two 2×2 boxes), and the vector $\overrightarrow{Pf(P)}$ has components $(1, 0, 0, 0)$ in this new basis. The origin of the two affine frames, $\mathcal{R}$ and $\mathcal{R}'$, is the same point P.

We recall that the Jordan matrix is unique up to the order of the boxes, but the Jordan basis is not unique, as we have just seen.

Example 4.8

Let f be an affinity of an affine space of dimension 5. Let us assume that in a given affine frame $\mathcal{R} = \{P; (e_1, e_2, e_3, e_4, e_5)\}$ we have

$$M(f, \mathcal{R}) = \left(\begin{array}{ccc|cc|c} 1 & 0 & 0 & 0 & 0 & 1 \\ 1 & 1 & 0 & 0 & 0 & 0 \\ 0 & 1 & 1 & 0 & 0 & 0 \\ \hline 0 & 0 & 0 & 1 & 0 & 1 \\ 0 & 0 & 0 & 1 & 1 & 0 \\ \hline 0 & 0 & 0 & 0 & 0 & 1 \end{array}\right).$$

Equivalently,

$$x_1' = x_1 + 1,$$

$$x_2' = x_1 + x_2,$$

$$x_3' = x_2 + x_3,$$

$$x_4' = x_4 + 1,$$

$$x_5' = x_4 + x_5.$$

If we make the change of basis

$$e_1' = e_1 + e_4 \quad \text{(the vector } \overrightarrow{Pf(P)}),$$

$$e_2' = (\tilde{f} - \mathrm{id})(e_1') = e_2 + e_5,$$

$$e_3' = (\tilde{f} - \mathrm{id})(e_2') = e_3,$$

$$e_4' = e_4,$$

$$e_5' = e_5,$$

we have

$$M(f, \mathcal{R}') = \left(\begin{array}{ccc|cc|c} 1 & 0 & 0 & 0 & 0 & 1 \\ 1 & 1 & 0 & 0 & 0 & 0 \\ 0 & 1 & 1 & 0 & 0 & 0 \\ \hline 0 & 0 & 0 & 1 & 0 & 0 \\ 0 & 0 & 0 & 1 & 1 & 0 \\ \hline 0 & 0 & 0 & 0 & 0 & 1 \end{array} \right).$$

We remark that, since e_3 is an eigenvector, $(\tilde{f} - \mathrm{id})^3(e_1') = 0$.

Thus, we have changed the basis in such a way that the Jordan matrix has not changed (one 3×3 box and one 2×2 box), and the vector $\overrightarrow{Pf(P)}$ has components $(1, 0, 0, 0, 0)$ in this new basis. The origin of the two affine frames, $\mathcal{R}$ and $\mathcal{R}'$, is the same point P.

Example 4.9

Let f be an affinity of an affine space of dimension 5. Let us assume that in a given affine frame $\mathcal{R} = \{P; (e_1, e_2, e_3, e_4, e_5)\}$ we have

$$M(f,\mathcal{R}) = \left(\begin{array}{cc|ccc|c} 1 & 0 & 0 & 0 & 0 & 1 \\ 1 & 1 & 0 & 0 & 0 & 0 \\ \hline 0 & 0 & 1 & 0 & 0 & 1 \\ 0 & 0 & 1 & 1 & 0 & 0 \\ 0 & 0 & 0 & 1 & 1 & 0 \\ \hline 0 & 0 & 0 & 0 & 0 & 1 \end{array}\right).$$

Equivalently,

$$x_1' = x_1 + 1,$$

$$x_2' = x_1 + x_2,$$

$$x_3' = x_3 + 1,$$

$$x_4' = x_3 + x_4,$$

$$x_5' = x_4 + x_5.$$

If we make the change of basis

$$e_1' = e_1 + e_3 \quad \text{(the vector } \overrightarrow{Pf(P)}),$$

$$e_2' = (\tilde{f} - \text{id})(e_1') = e_2 + e_4,$$

$$e_3' = (\tilde{f} - \text{id})(e_2') = e_5,$$

$$e_4' = e_1,$$

$$e_5' = e_2,$$

we have

$$M(f,\mathcal{R}') = \left(\begin{array}{ccc|cc|c} 1 & 0 & 0 & 0 & 0 & 1 \\ 1 & 1 & 0 & 0 & 0 & 0 \\ 0 & 1 & 1 & 0 & 0 & 0 \\ \hline 0 & 0 & 0 & 1 & 0 & 0 \\ 0 & 0 & 0 & 1 & 1 & 0 \\ \hline 0 & 0 & 0 & 0 & 0 & 1 \end{array}\right).$$

We remark that, since e_5 is an eigenvector, $(\tilde{f} - \text{id})^3(e_1') = \vec{0}$.

Thus, we have changed the basis in such a way that the Jordan matrix has not changed (one 3×3 box and one 2×2 box), and the vector $\overrightarrow{Pf(P)}$ has components $(1,0,0,0,0)$ in this new basis. The origin of the two affine frames, $\mathcal{R}$ and $\mathcal{R}'$, is the same point P.

But the order of the Jordan boxes has changed!

Example 4.10

Let f be an affinity of an affine space of dimension 5. Let us assume that in a given affine frame $\mathcal{R} = \{P; (e_1, e_2, e_3, e_4, e_5)\}$ we have

$$
M(f, \mathcal{R}) = \left(\begin{array}{ccc|ccc}
1 & 0 & 0 & 0 & 0 & 0 \\
1 & 1 & 0 & 0 & 0 & a \\
0 & 1 & 1 & 0 & 0 & b \\
\hline
0 & 0 & 0 & 1 & 0 & 1 \\
0 & 0 & 0 & 1 & 1 & c \\
0 & 0 & 0 & 0 & 0 & 1
\end{array}\right).
$$

Equivalently,

$$x_1' = x_1,$$

$$x_2' = x_1 + x_2 + a,$$

$$x_3' = x_2 + x_3 + b,$$

$$x_4' = x_4 + 1,$$

$$x_5' = x_4 + x_5 + c.$$

Let us take as a new origin of the affine frame any point Q with coordinates in $\mathcal{R}$ given by

$$Q = (-a, -b, q_3, -c, q_5).$$

Then

$$\overrightarrow{Qf(Q)} = (q_1' - q_1, q_2' - q_2, q_3' - q_3, q_4' - q_4, q_5' - q_5)$$

$$= (0, 0, 0, 1, 0),$$

and hence

$$
M(f, \mathcal{R}') = \left(\begin{array}{ccc|ccc}
1 & 0 & 0 & 0 & 0 & 0 \\
1 & 1 & 0 & 0 & 0 & 0 \\
0 & 1 & 1 & 0 & 0 & 0 \\
\hline
0 & 0 & 0 & 1 & 0 & 1 \\
0 & 0 & 0 & 1 & 1 & 0 \\
0 & 0 & 0 & 0 & 0 & 1
\end{array}\right),
$$

where $\mathcal{R}' = \{Q; (e_1, e_2, e_3, e_4, e_5)\}$.

We could stop here, but we prefer to permute the boxes $(e'_1 = e_4, e'_2 = e_5,$ $e'_3 = e_1, e'_4 = e_2, e'_5 = e_3)$ in order to obtain

$$M(f, \mathcal{R}'') = \left(\begin{array}{cc|ccc|c} 1 & 0 & 0 & 0 & 0 & 1 \\ 1 & 1 & 0 & 0 & 0 & 0 \\ \hline 0 & 0 & 1 & 0 & 0 & 0 \\ 0 & 0 & 1 & 1 & 0 & 0 \\ 0 & 0 & 0 & 1 & 1 & 0 \\ \hline 0 & 0 & 0 & 0 & 0 & 1 \end{array} \right),$$

where $\mathcal{R}'' = \{Q; (e'_1, e'_2, e'_3, e'_4, e'_5)\}$. Thus, we have changed the initial basis in such a way that the Jordan matrix has not changed (one 3×3 box and one 2×2 box), and the vector $\overrightarrow{Qf(Q)}$ has components $(1, 0, 0, 0, 0)$ in this new basis. The origin has changed and *the order of the Jordan boxes has also changed!*

Example 4.11

Let f be an affinity of an affine space of dimension 4. Let us assume that in a given affine frame $\mathcal{R} = \{P; (e_1, e_2, e_3, e_4)\}$ we have

$$M(f, \mathcal{R}) = \left(\begin{array}{cc|cc|c} 1 & 0 & 0 & 0 & 0 \\ 1 & 1 & 0 & 0 & 1 \\ \hline 0 & 0 & 1 & 0 & 0 \\ 0 & 0 & 1 & 1 & 1 \\ \hline 0 & 0 & 0 & 0 & 1 \end{array} \right).$$

Equivalently,

$$x'_1 = x_1,$$

$$x'_2 = x_1 + x_2 + 1,$$

$$x'_3 = x_3,$$

$$x'_4 = x_3 + x_4 + 1.$$

Let us take as new origin of the affine frame any point Q with coordinates in $\mathcal{R}$ given by

$$Q = (-1, q_2, -1, q_4).$$

All these points are fixed points.

Then, with respect to $\mathcal{R}' = \{Q; (e_1, e_2, e_3, e_4)\}$, we have

$$M(f, \mathcal{R}') = \left(\begin{array}{cc|cc|c} 1 & 0 & 0 & 0 & 0 \\ 1 & 1 & 0 & 0 & 0 \\ \hline 0 & 0 & 1 & 0 & 0 \\ 0 & 0 & 1 & 1 & 0 \\ \hline 0 & 0 & 0 & 0 & 1 \end{array}\right).$$

Thus, we have changed the affine frame in such a way that the Jordan matrix has not changed (the basis of the vector space has not changed), and the vector $\overrightarrow{Qf(Q)}$ has components $(0, 0, 0, 0)$ in this new basis, since Q is a fixed point. The origin has changed.

Example 4.12 (Coexistence)

Let f be an affinity of an affine space of dimension 4. Let us assume that in some affine frame

$$\mathcal{R} = \{P; (e_1, e_2, e_3, e_4)\}$$

we have

$$M(f, \mathcal{R}) = \left(\begin{array}{cc|cc|c} 1 & 0 & 0 & 0 & \alpha \\ 1 & 1 & 0 & 0 & \beta \\ \hline 0 & 0 & a & 0 & 1 \\ 0 & 0 & 0 & b & 3 \\ \hline 0 & 0 & 0 & 0 & 1 \end{array}\right),$$

with $a, b \neq 1$ (the eigenvalue 1 coexists with eigenvalues different from 1). Equivalently,

$$x_1' = x_1 + \alpha,$$

$$x_2' = x_1 + x_2 + \beta,$$

$$x_3' = ax_3 + 1,$$

$$x_4' = bx_4 + 3.$$

As the new origin of the affine frame we take any point Q with coordinates in $\mathcal{R}$ given by

$$Q = \left(q_1, q_2, \frac{1}{1-a}, \frac{3}{1-b}\right).$$

Then, with respect to $\mathcal{R}' = \{Q; (e_1, e_2, e_3, e_4)\}$, we have

$$M(f, \mathcal{R}') = \left(\begin{array}{cccc|c} 1 & 0 & 0 & 0 & \alpha \\ 1 & 1 & 0 & 0 & q_1 + \beta \\ 0 & 0 & a & 0 & 0 \\ 0 & 0 & 0 & b & 0 \\ 0 & 0 & 0 & 0 & 1 \end{array} \right).$$

Thus, we have changed the affine frame in such a way that the Jordan matrix has not changed (the basis in the vector space has not changed), and the vector $\overrightarrow{Qf(Q)}$ has components $(*, *, 0, 0)$ in this basis. The origin has changed.

This is what we wanted to show, but we can go further and simplify the expression of the matrix associated to f by taking $q_1 = -\beta$. If $\alpha = 0$, we have

$$M(f, \mathcal{R}') = \left(\begin{array}{cccc|c} 1 & 0 & 0 & 0 & 0 \\ 1 & 1 & 0 & 0 & 0 \\ 0 & 0 & a & 0 & 0 \\ 0 & 0 & 0 & b & 0 \\ 0 & 0 & 0 & 0 & 1 \end{array} \right),$$

and if $\alpha \neq 0$, we make the change of basis $e_1' = \alpha e_1$, $e_2' = \alpha e_2$, $e_3' = e_3$, $e_4' = e_4$ and we have

$$M(f, \mathcal{R}'') = \left(\begin{array}{cccc|c} 1 & 0 & 0 & 0 & 1 \\ 1 & 1 & 0 & 0 & 0 \\ 0 & 0 & a & 0 & 0 \\ 0 & 0 & 0 & b & 0 \\ 0 & 0 & 0 & 0 & 1 \end{array} \right),$$

where $\mathcal{R}' = \{Q; (e_1', e_2', e_3', e_4')\}$.

4.5 Classification of Affinities in Arbitrary Dimension

Let f be an affinity of an affine space $\mathbb{A}$, of dimension n, and let

$$E = E_1 \oplus E_2$$

be the canonical decomposition induced by f. Notice that

$$(\tilde{f} - \mathrm{id})^s_{|E_1} = 0,$$

where s is the multiplicity of the root 1 in the characteristic polynomial $p_c(\tilde{f})$ of the associated endomorphism $\tilde{f}$, and that

$$(\tilde{f} - \mathrm{id})_{|E_2}$$

is an isomorphism.

We know that there exists a basis $\mathcal{B}_1$ of E_1 and a basis $\mathcal{B}_2$ of E_2 such that the matrix of f in an affine frame $\mathcal{R} = \{P; \mathcal{B}\}$, with $\mathcal{B} = (\mathcal{B}_1, \mathcal{B}_2)$, is

$$M(f, \mathcal{R}) = \begin{pmatrix} J & 0 & b \\ 0 & K & c \\ 0 & 0 & 1 \end{pmatrix}, \tag{4.8}$$

where J is a Jordan matrix with 1s on the diagonal, K is a certain matrix, and b and c are column matrices. That is, $\mathcal{B}$ is a *Jordan basis adapted to* $\tilde{f}$. In this case we also say that $\mathcal{R} = \{P; \mathcal{B}\}$ is a *Jordan affine frame adapted to* f.

Concretely,

$$J = M(\tilde{f}_{|E_1}, \mathcal{B}_1) \quad (\text{expression (4.2)}),$$

$$K = M(\tilde{f}_{|E_2}, \mathcal{B}_2),$$

$$\overrightarrow{Pf(P)} = u + v, \quad u \in E_1, \ v \in E_2, \tag{4.9}$$

$$b = C(u, \mathcal{B}_1),$$

$$c = C(v, \mathcal{B}_2).$$

Here, $C(u, \mathcal{B}_1)$ is the column matrix formed by the components of the vector u in the basis $\mathcal{B}_1$.

The matrix K does not play a special role in the following sections but, as we have remarked in Observation 4.1, we can think of it as a Jordan matrix without the eigenvalue 1, or as a matrix in rational form. We cannot be more precise because in this chapter k is an arbitrary field.

Since the basis $\mathcal{B}_1$ is written as in expression (4.3), page 131, we shall write the matrix b in the form

$$b = \begin{pmatrix} b_1 \\ b_{12} \\ \vdots \\ b_{1s_1} \\ \hline b_2 \\ b_{22} \\ \vdots \\ b_{2s_2} \\ \hline \vdots \\ \hline b_r \\ b_{r2} \\ \vdots \\ b_{rs_r} \end{pmatrix}. \tag{4.10}$$

We remark that b_i is the component of the vector $\overrightarrow{Pf(P)}$ with respect to the i-th generator.

Next we shall see that, by simply changing the origin of $\mathcal{R}$, we can show that the column matrix c in expression (4.8) is zero.

Proposition 4.13

Let f be an affinity of an affine space $\mathbb{A}$. Then there exists a Jordan affine frame adapted to f, $\mathcal{R} = \{P; \mathcal{B}\}$, such that

$$M(f, \mathcal{R}) = \begin{pmatrix} J & 0 & b \\ 0 & K & 0 \\ 0 & 0 & 1 \end{pmatrix},$$

where J, K and b are as in (4.9).

Proof

We know that there exists a Jordan affine frame adapted to f, $\mathcal{R}_0 = \{P_0; \mathcal{B}\}$, with $\mathcal{B} = (\mathcal{B}_1, \mathcal{B}_2)$, such that

$$M(f, \mathcal{R}_0) = \begin{pmatrix} J & 0 & b \\ 0 & K & c \\ 0 & 0 & 1 \end{pmatrix}.$$

We shall see that there exists a point $P \in \mathbb{A}$ such that $\overrightarrow{Pf(P)} \in E_1$.

Hence, the components of this vector in $\mathcal{B}_2$ are zero. We shall take this point P as the origin of the new affine frame and, in this way, we shall obtain the desired result.

Inspired by Example 4.12, we construct this point P in the following way.

First, we compute the vector $\overrightarrow{P_0 f(P_0)}$ and we decompose it in the direct sum $E = E_1 \oplus E_2$ as

$$\overrightarrow{P_0 f(P_0)} = u + v, \quad u \in E_1, \ v \in E_2.$$

Then, we let

$$P = P_0 - (\tilde{f} - \mathrm{id})^{-1}_{|E_2}(v),$$

so that

$$\overrightarrow{PP_0} = (\tilde{f} - \mathrm{id})^{-1}_{|E_2}(v).$$

This is the central point where we have used the fact that $(\tilde{f} - \mathrm{id})$ is an isomorphism on E_2.

Then

$$\overrightarrow{Pf(P)} = \overrightarrow{PP_0} + \overrightarrow{P_0 f(P_0)} + \overrightarrow{f(P_0)f(P)}$$

$$= -(\tilde{f} - \mathrm{id})(\overrightarrow{PP_0}) + u + v$$

$$= u \in E_1,$$

and hence, in the affine frame $\mathcal{R} = \{P; \mathcal{B}\}$, we have

$$M(f, \mathcal{R}) = \begin{pmatrix} J & 0 & b \\ 0 & K & 0 \\ 0 & 0 & 1 \end{pmatrix},$$

and this completes the proof. $\square$

Proposition 4.14

Let f be an affinity of an affine space $\mathbb{A}$ of dimension n. Then there exists a Jordan affine frame adapted to f, $\mathcal{R} = \{P; \mathcal{B}\}$, such that

$$M(f, \mathcal{R}) = \begin{pmatrix} J & 0 & b \\ 0 & K & 0 \\ 0 & 0 & 1 \end{pmatrix},$$

with

$$b = \begin{pmatrix} b_1 \\ 0 \\ \vdots \\ 0 \\ \hline b_2 \\ 0 \\ \vdots \\ 0 \\ \hline \vdots \\ \hline b_r \\ 0 \\ \vdots \\ 0 \end{pmatrix}, \quad b_i \in \{0,1\}, \ i = 1,\ldots,r,$$

where J and K are as in (4.9).

Proof

We must show that there exists a Jordan affine frame adapted to f, $\mathcal{R} = \{P; \mathcal{B}\}$, such that $\overrightarrow{Pf(P)}$ is equal to a sum of generators.

Let $\mathcal{R}_0 = \{P_0; \mathcal{B}\}$ be a Jordan affine frame adapted to f, with $\mathcal{B} = (\mathcal{B}_1, \mathcal{B}_2)$, and write $\mathcal{B}_1$ as in (4.3), page 131. By Proposition 4.13 we can assume that

$$M(f,\mathcal{R}_0) = \begin{pmatrix} J & 0 & b \\ 0 & K & 0 \\ 0 & 0 & 1 \end{pmatrix},$$

with b as in (4.10).

We want to find a point $P \in \mathbb{A}$ (which will be the new origin) such that

$$\overrightarrow{Pf(P)} = \sum_{i=1}^{r} b_i e_i,$$

that is, with the same coefficient in e_i as that of the vector $\overrightarrow{P_0 f(P_0)}$, and with the remaining coefficients zero.

Once we have this expression, we shall modify the basis in order to transform the non-zero b_i of this expression into 1.

To find P we first write the vector $\overrightarrow{P_0P}$ with respect to $\mathcal{R}_0$ as:

$$\overrightarrow{P_0P} = \sum_i p_i e_i + \sum p_{ij} e_{ij},$$

where p_i, p_{ij}, for $i = 1, \ldots, r$, $j = 2, \ldots, s_i$, are coefficients to be determined. Note that in this expression we are assuming that $\overrightarrow{P_0P} \in E_1$.

Since the components of $\overrightarrow{Pf(P)}$ are obtained by subtracting the coordinates of P from the corresponding coordinates of $P' = f(P)$, and the equations of f (in $\mathcal{B}_1$) are

$$x_1' = x_1 + b_1,$$

$$x_{12}' = x_1 + x_{12} + b_{12},$$

$$x_{13}' = x_{12} + x_{13} + b_{13},$$

$$\vdots$$

$$x_2' = x_2 + b_2,$$

$$x_{22}' = x_2 + x_{22} + b_{22},$$

$$x_{23}' = x_{22} + x_{23} + b_{23},$$

$$\vdots$$

we have,

$$\overrightarrow{Pf(P)} = \begin{pmatrix} b_1 \\ p_1 + b_{12} \\ \vdots \\ p_{1s_1-1} + b_{1s_1} \\ \hline b_2 \\ p_2 + b_{22} \\ \vdots \\ p_{2s_2-1} + b_{2s_2} \\ \vdots \\ \hline b_r \\ p_r + b_{r2} \\ \vdots \\ p_{rs_r-1} + b_{rs_r} \end{pmatrix}.$$

Hence, we can choose

$$p_i = -b_{i2}, \quad i = 1, \ldots, r,$$
$$p_{i(j-1)} = -b_{ij}, \quad i = 1, \ldots, r; \ j = 3, \ldots, s_i,$$

(4.11)

and we have

$$\overrightarrow{Pf(P)} = \begin{pmatrix} b_1 \\ 0 \\ \vdots \\ 0 \\ \hline b_2 \\ 0 \\ \vdots \\ 0 \\ \hline \vdots \\ \hline b_r \\ 0 \\ \vdots \\ 0 \end{pmatrix}.$$

Equivalently,

$$\overrightarrow{Pf(P)} = \sum_{i=1}^{r} b_i e_i.$$

We shall take the point

$$P = (\overbrace{-b_{12}, -b_{13}, \ldots, -b_{1s_1}, *}, \ldots, \overbrace{-b_{r2}, -b_{r3}, \ldots, -b_{rs_r}, *})$$

thus obtained as the new origin of the affine frame.

Note that in expression (4.11) the terms p_{is_i} (the last in each Jordan box) are indeterminate. The asterisks in the above expression indicate arbitrary values.

Finally, for all $t = 1, \ldots, r$ such that $b_t \neq 0$, we replace the basis in the corresponding box $C_t = \langle e_t, e_{t2}, \ldots, e_{ts_t} \rangle$ by

$$e'_t = b_t e_t,$$
$$e'_{t2} = b_t e_{t2},$$
$$\vdots$$
$$e'_{ts_t} = b_t e_{ts_t},$$

and we obtain the result. That is, the affine frame we were looking for is the affine frame $\mathcal{R} = \{P; \mathcal{B}\}$ where this basis $\mathcal{B}$ is the initial basis of $\mathcal{R}_0$, but with these small modifications in some of the Jordan boxes.   □

Observation 4.15

In order to avoid having to write the large matrices appearing in the above proof, we can proceed as follows.

For every pair of points P_0, P we verify that

$$\overrightarrow{Pf(P)} = (\tilde{f} - \mathrm{id})\overrightarrow{P_0P} + \overrightarrow{P_0f(P_0)}.$$

Applying this equality to the setting described above, and using the rules given in table (4.5), page 131, the calculation of $\overrightarrow{Pf(P)}$ made on page 147 can also be performed as follows:

$$\overrightarrow{Pf(P)} = (\tilde{f} - \mathrm{id})\left(\sum_i p_i e_i + \sum_{ij} p_{ij} e_{ij}\right) + \sum_i b_i e_i + \sum_{ij} b_{ij} e_{ij}$$

$$= \sum_{i,j} p_{ij} e_{i(j+1)} + \sum_i p_i e_{i2} + \sum_i b_i e_i + \sum_{ij} b_{ij} e_{ij}$$

$$= \sum_i b_i e_i + \sum_{i,j \geq 3} (p_{i(j-1)} + b_{ij}) e_{ij} + \sum_i (p_i + b_{i2}) e_{i2}.$$

In all these summands, $i = 1, \ldots, r$, and j (which depends on the i to which it is attached) has the range $j = 2, \ldots, s_i$, except in the penultimate summand, where it is specified that the sum begins at $j = 3$.

From this relation we directly deduce (4.11).

Theorem 4.16 (Classification)

Let f be an affinity of an affine space $\mathbb{A}$ of dimension n. Then there exists a Jordan affine frame adapted to f, $\mathcal{R} = \{P; \mathcal{B}\}$, such that

$$M(f, \mathcal{R}) = \begin{pmatrix} J & 0 & b \\ 0 & K & 0 \\ 0 & 0 & 1 \end{pmatrix},$$

with

$$b = \begin{pmatrix} \varepsilon \\ 0 \\ \vdots \\ 0 \end{pmatrix}, \quad \varepsilon \in \{0, 1\},$$

where J and K are as in (4.9). We say that $M(f, \mathcal{R})$ is a canonical expression of f.

Proof

Let $\mathcal{R}' = \{P; \mathcal{B}'\}$ be the Jordan basis adapted to $\tilde{f}$ given in the above proposition. Then

$$M(f, \mathcal{R}') = \begin{pmatrix} J' & 0 & b \\ 0 & K & 0 \\ 0 & 0 & 1 \end{pmatrix}$$

and

$$\overrightarrow{Pf(P)} = \sum_i b_i e_i, \quad b_i \in \{0, 1\}.$$

Let t be such that $b_t = 1$ and such that $s_t \geq s_{t'}$, for all t' with $b_{t'} = 1$. That is, the vector e_t appears in the expression of $\overrightarrow{Pf(P)}$, but the generators corresponding to boxes of higher dimension than that of the box C_t do not.

We make the following change of basis: the vectors of $\mathcal{B}$ remain unchanged, except those of the box C_t, which are modified as follows:

$$e'_t = \overrightarrow{Pf(P)} = \sum_i b_i e_i, \quad b_i \in \{0, 1\},$$

$$e'_{t2} = (\tilde{f} - \mathrm{id})e'_t,$$

$$e'_{t3} = (\tilde{f} - \mathrm{id})^2 e'_t,$$

$$\vdots$$

$$e'_{ts_t} = (\tilde{f} - \mathrm{id})^{s_t - 1} e'_t.$$

These vectors are linearly independent. In fact, using the properties of the Jordan basis collected in (4.5), page 131, we see that the matrix with columns formed by the components, in the Jordan basis, of the s_t vectors $e'_t, e'_{t2}, \ldots, e'_{ts_t}$ is

$$\begin{pmatrix}
\vdots & \vdots & \vdots & & \vdots \\
b_t & 0 & 0 & \cdots & 0 \\
0 & b_t & 0 & \cdots & 0 \\
0 & 0 & b_t & \cdots & 0 \\
\vdots & \vdots & \vdots & & \vdots \\
0 & 0 & 0 & \cdots & b_t \\
\vdots & \vdots & \vdots & & \vdots
\end{pmatrix}.$$

Since $b_t = 1$, this matrix has rank s_t, and hence the vectors $e'_t, e'_{t2}, \ldots, e'_{ts_t}$ are linearly independent.

If we replace the vectors $e_t, e_{t2}, \ldots, e_{ts_t}$ of the initial Jordan basis by $e'_t, e'_{t2}, \ldots, e'_{ts_t}$, we obtain a new basis $\mathcal{B}''$ (see [8], page 250). Let

$$C'_t = \langle e'_t, e'_{t2}, \ldots, e'_{ts_t} \rangle.$$

In general, we have $C_t \neq C'_t$.

However, the most important thing is that the direct sum decomposition

$$E = C_1 \oplus \cdots \oplus C'_t \oplus \cdots \oplus C_r$$

is still invariant under $\tilde{f}$.

To prove that $\tilde{f}(C'_t) \subset C'_t$, it is sufficient to prove that $(\tilde{f} - \mathrm{id})e'_{ts_t} = \vec{0}$. In fact,

$$(\tilde{f} - \mathrm{id})(e'_{ts_t}) = (\tilde{f} - \mathrm{id})^{s_t}(e'_t)$$

$$= (\tilde{f} - \mathrm{id})^{s_t}\left(\sum_i b_i e_i \right)$$

$$= 0,$$

since if $b_{t'} \neq 1$, then $s_t \geq s_{t'}$ and

$$(\tilde{f} - \mathrm{id})^{s_t}(e_{t'}) = \vec{0}.$$

Note that this decomposition coincides, replacing C_t by C'_t, with decomposition (4.4).

Moreover, this new basis is a Jordan basis because

$$M(\tilde{f}_{|E_1}; \mathcal{B}') = M(\tilde{f}_{|E_1}; \mathcal{B}'').$$

Finally, in order to obtain the affine frame we are looking for it remains only to permute the vectors of $\mathcal{B}''$ in order to put in the first positions the vectors of the basis of C'_t. That is, we adapt the basis to the decomposition

$$E = C'_t \oplus C_1 \oplus \cdots \oplus C_r.$$

This is the basis $\mathcal{B}$ that we were looking for.   □

Observation 4.17

Notice that if f has a fixed point, we can assume, taking this point as origin, that $\varepsilon = 0$.

Proposition 4.18

Let f be an affinity without fixed points. Then, the dimension $\rho(f)$ of the invariant variety of smaller dimension is equal to the order of the first Jordan box in the canonical expression of f.

Proof

Let

$$M(f, \mathcal{R}) = \begin{pmatrix} J & 0 & b \\ 0 & K & 0 \\ 0 & 0 & 1 \end{pmatrix},$$

with

$$b = \begin{pmatrix} \varepsilon \\ 0 \\ \vdots \\ 0 \end{pmatrix}, \quad \varepsilon \in \{0, 1\},$$

be the canonical expression of f.

Since we are assuming that there are no fixed points, we must have $\varepsilon = 1$. Indeed, if $\varepsilon = 0$, the origin P of the affine frame $\mathcal{R}$ is a fixed point.

Let C_t be the first Jordan box of J. Then, since $\overrightarrow{Pf(P)} = e_t$, the linear variety $L = P + \langle e_t, e_{t2}, \ldots, e_{ts_t} \rangle$ is invariant, and has dimension s_t.

We must prove that there are no invariant linear varieties of lower dimension.

Let us assume $L' = Q + [H]$ is invariant. Then $u_t = \overrightarrow{Qf(Q)} \in H$.

The first s_t equations of f in $\mathcal{R}$ are

$$x_1' = x_1 + 1,$$

$$x_{12}' = x_1 + x_{12},$$

$$x_{13}' = x_{12} + x_{13},$$

$$\vdots$$

$$x_{1s_t}' = x_{1s_t-1} + x_{1s_t}.$$

Since the components of the vector $\overrightarrow{Qf(Q)}$ are obtained by subtracting the coordinates of Q from the corresponding coordinates of $Q' = f(Q)$, we have

$$u_t = \overrightarrow{Qf(Q)} = (\overbrace{1, q_1, \ldots, q_{1s_t-1}}^{s_t}, *, \ldots, *).$$

We take

$$u_{t2} = (\tilde{f} - \mathrm{id})u_t,$$

$$u_{t3} = (\tilde{f} - \mathrm{id})^2 u_t,$$

$$\vdots$$

$$u_{ts_t} = (\tilde{f} - \mathrm{id})^{s_t - 1} u_t.$$

These vectors are linearly independent. In fact, the matrix with columns the components in the initial Jordan basis of the s_t vectors $u_t, u_{t2}, \ldots, u_{ts_t}$ is

$$
\begin{pmatrix}
1 & 0 & 0 & \cdots & 0 \\
q_1 & 1 & 0 & \cdots & 0 \\
q_{12} & q_1 & 1 & \cdots & 0 \\
\vdots & \vdots & \vdots & & \vdots \\
q_{1s_t-1} & q_{1s_t-2} & q_{1s_t-3} & \cdots & 1 \\
* & * & * & \cdots & * \\
\vdots & \vdots & \vdots & & \vdots
\end{pmatrix},
$$

and this matrix has rank s_t.

Since all of these vectors are in H, it follows that $\dim H \geq s_t$, and this completes the proof.   □

Theorem 4.19 (Characterization)

Let f and g be affinities of an affine space $\mathbb{A}$ of dimension n. Then, f is similar to g if and only if $\tilde{f}$ is similar to $\tilde{g}$ and $\rho(f) = \rho(g)$.

Proof

We have seen, in Propositions 3.5 and 3.15 respectively, that if f is similar to g then $\tilde{f}$ is similar to $\tilde{g}$ and $\rho(f) = \rho(g)$.

Conversely, let us assume that we have two affinities, f and g, such that $\tilde{f}$ is similar to $\tilde{g}$ and $\rho(f) = \rho(g)$.

First case: $\rho(f) = \rho(g) = 0$.

It follows, from Theorem 4.16 and Observation 4.17, that there exists a Jordan affine frame adapted to f, $\mathcal{R} = \{P; \mathcal{B}\}$, such that

$$
M(f, \mathcal{R}) =
\begin{pmatrix}
J & 0 & 0 \\
0 & K & 0 \\
0 & 0 & 1
\end{pmatrix},
$$

and there exists a Jordan affine frame adapted to g, $\mathcal{R}' = \{P'; \mathcal{B}'\}$, such that

$$M(g, \mathcal{R}') = \begin{pmatrix} J' & 0 & 0 \\ 0 & K' & 0 \\ 0 & 0 & 1 \end{pmatrix}.$$

Since $\tilde{f}$ is similar to $\tilde{g}$, there exists a matrix $C \in \mathcal{M}_{n \times n}(k)$ such that

$$C \begin{pmatrix} J & 0 \\ 0 & K \end{pmatrix} = \begin{pmatrix} J' & 0 \\ 0 & K' \end{pmatrix} C.$$

Hence,

$$\begin{pmatrix} C & 0 \\ 0 & 1 \end{pmatrix} M(f, \mathcal{R}) = M(g, \mathcal{R}') \begin{pmatrix} C & 0 \\ 0 & 1 \end{pmatrix}.$$

But this implies, by part (ii) of Proposition 3.8, that f is similar to g, and this completes the proof.

Second case: $\rho(f) = \rho(g) \neq 0$.

By Proposition 4.18 there exists a Jordan affine frame adapted to f, $\mathcal{R} = \{P; \mathcal{B}\}$, such that

$$M(f, \mathcal{R}) = \begin{pmatrix} J_t & 0 & 0 & b \\ 0 & J_0 & 0 & 0 \\ 0 & 0 & K & 0 \\ 0 & 0 & 0 & 1 \end{pmatrix},$$

with

$$b = \begin{pmatrix} 1 \\ 0 \\ \vdots \\ 0 \end{pmatrix}, \qquad J_t = \begin{pmatrix} 1 & 0 & \cdots & \cdots & 0 \\ 1 & 1 & \ddots & & \vdots \\ 0 & \ddots & \ddots & \ddots & \vdots \\ \vdots & \ddots & \ddots & 1 & 0 \\ 0 & \cdots & 0 & 1 & 1 \end{pmatrix}.$$

This Jordan box J_t has order $\rho(f)$. J_0 is the matrix formed by the remaining Jordan boxes (once we remove J_t) of the canonical expression of f.

Analogously, there exists a Jordan affine frame adapted to g, $\mathcal{R}' = \{P'; \mathcal{B}'\}$, such that

$$M(g, \mathcal{R}') = \begin{pmatrix} J_t & 0 & 0 & b \\ 0 & J_0' & 0 & 0 \\ 0 & 0 & K' & 0 \\ 0 & 0 & 0 & 1 \end{pmatrix},$$

with the same J_t and the same b that appear in $M(f, \mathcal{R})$. J_0' is the matrix formed by the remaining Jordan boxes (once we remove J_t) of the canonical expression of g.

To simplify the notation, we shall write

$$
M(f, \mathcal{R}) = \begin{pmatrix} J & 0 & c \\ 0 & K & 0 \\ 0 & 0 & 1 \end{pmatrix},
$$

with

$$
J = \begin{pmatrix} J_t & 0 \\ 0 & J_0 \end{pmatrix},
$$

and

$$
M(g, \mathcal{R}') = \begin{pmatrix} J' & 0 & c \\ 0 & K' & 0 \\ 0 & 0 & 1 \end{pmatrix},
$$

with

$$
J' = \begin{pmatrix} J_t & 0 \\ 0 & J_0' \end{pmatrix}.
$$

We have put

$$
c = \begin{pmatrix} 1 \\ 0 \\ \vdots \\ 0 \end{pmatrix},
$$

a matrix of one column and as many rows as the dimension of J. It coincides with b when the Jordan matrix is reduced to the first box.

Since $\tilde{f}$ is similar to $\tilde{g}$ and we are using adapted bases there exists, by Proposition 4.4, an invertible matrix C of the form

$$
C = \begin{pmatrix} P_1 & P_2 \\ 0 & P_4 \end{pmatrix}
$$

such that

$$
CM(\tilde{f}, \mathcal{B}) = M(\tilde{g}, \mathcal{B})C.
$$

Therefore, we have

$$
\begin{pmatrix} P_1 & P_2 \\ 0 & P_4 \end{pmatrix} \begin{pmatrix} J & 0 \\ 0 & K \end{pmatrix} = \begin{pmatrix} J' & 0 \\ 0 & K' \end{pmatrix} \begin{pmatrix} P_1 & P_2 \\ 0 & P_4 \end{pmatrix}.
$$

In particular, $P_1 J = J' P_1$ and $P_4 K = K' P_4$. We want to replace the matrix P_1 with another matrix P_1' such that the equality $P_1' J = J' P_1'$ still holds, but with its first column having a 1 in the first position, and zeros elsewhere.

Note first that, since C is invertible, P_1 and P_4 are also invertible. In particular, J and J' are conjugate. Hence, they have the same number of Jordan boxes, perhaps in a different order.

In particular, J_0 and J_0' also have the same number of Jordan boxes, perhaps in a different order, and hence they are conjugate. Reordering boxes is simply a change of basis.

Therefore, there exists an invertible Q such that

$$Q J_0 = J_0' Q.$$

But then, the matrix

$$P_1' = \begin{pmatrix} I_t & 0 \\ 0 & Q \end{pmatrix},$$

with I_t the $\rho(f) \times \rho(f)$ identity matrix, is invertible and it satisfies

$$\begin{pmatrix} I_t & 0 \\ 0 & Q \end{pmatrix} \begin{pmatrix} J_t & 0 \\ 0 & J_0 \end{pmatrix} = \begin{pmatrix} J_t & 0 \\ 0 & J_0' \end{pmatrix} \begin{pmatrix} I_t & 0 \\ 0 & Q \end{pmatrix}.$$

That is, we have found an invertible matrix, P_1', such that

$$P_1' J = J' P_1',$$

and such that *its first column is equal to c, i.e., it has a 1 in the first position, and zeros elsewhere.*

Finally, again using the block product of matrices, we see that

$$\left(\begin{array}{cc|c} P_1' & 0 & 0 \\ 0 & P_4 & 0 \\ \hline 0 & 0 & 1 \end{array}\right) \left(\begin{array}{cc|c} J & 0 & c \\ 0 & K & 0 \\ \hline 0 & 0 & 1 \end{array}\right) = \left(\begin{array}{cc|c} J' & 0 & c \\ 0 & K' & 0 \\ \hline 0 & 0 & 1 \end{array}\right) \left(\begin{array}{cc|c} P_1' & 0 & 0 \\ 0 & P_4 & 0 \\ \hline 0 & 0 & 1 \end{array}\right).$$

In fact, this equality is equivalent to

$$P_1' J = J' P_1',$$

$$P_1' c = c,$$

$$P_4 K = K' P_4.$$

The first and third equalities have already been verified, and the second equality simply says that the first column of P_1' is c.

But this implies, by part (ii) of Proposition 3.8, that f is similar to g, and this completes the proof. □

EXERCISES

4.1. Study the affinities of a real (or complex) affine space of dimension 3.

4.2. Rewrite the classification of affinities in dimension two using the general classification theorem of affinities, Theorem 4.16.

4.3. Classify the affinity of an affine space of dimension 3 given by

$$x_1' = 3x_1 + x_2 - x_3 + 1,$$

$$x_2' = x_1 + 8,$$

$$x_3' = x_1 + x_2 + x_3.$$

4.4. Classify the affinity of an affine space of dimension 4 given by

$$M(f, \mathcal{R}) = \left(\begin{array}{cccc|c} 1 & 0 & 0 & 0 & 2 \\ 1 & 1 & 0 & 0 & 3 \\ \hline 0 & 0 & 2 & 0 & 1 \\ 0 & 0 & 0 & 3 & 3 \\ \hline 0 & 0 & 0 & 0 & 1 \end{array} \right).$$

4.5. Classify the affinity of an affine space of dimension 4 given by

$$M(f, \mathcal{R}) = \left(\begin{array}{cc|cc|c} -1 & -4 & 0 & 0 & 1 \\ 1 & 3 & 0 & 0 & 2 \\ \hline 0 & 0 & 1 & 0 & 3 \\ 0 & 0 & 1 & 1 & 4 \\ \hline 0 & 0 & 0 & 0 & 1 \end{array} \right).$$

4.6. Classify the affinity of an affine space of dimension 4 given by

$$M(f, \mathcal{R}) = \left(\begin{array}{cc|c|c} -1 & -4 & 0 & 1 \\ 1 & 3 & 0 & 2 \\ \hline 0 & 0 & 2 & 3 \\ \hline 0 & 0 & 0 & 1 \end{array} \right).$$

5

Euclidean Affine Spaces

5.1 Introduction

In this chapter we introduce the concept of distance in affine spaces. For this
we shall "copy" the fundamental properties of the Euclidean distance of $\mathbb{R}^n$,
the distance given by Pythagoras' Theorem.

5.2 Definition of Euclidean Affine Space. Pythagoras' Theorem

Definition 5.1

A *Euclidean affine space* is an affine space $\mathbb{A}$ such that the associated vector
space E is a Euclidean vector space.

Recall that a *Euclidean vector space* is an $\mathbb{R}$-vector space E on which a scalar
product is defined. A *scalar product* is a bilinear, positive definite, symmetric
map $\phi : E \times E \longrightarrow \mathbb{R}$, see Definition A.8, page 326. The scalar product of two
vectors $u, v \in E$ is denoted by

$$\phi(u,v) = \langle u,v \rangle, \quad \text{for all } u,v \in E,$$

and the *modulus* of a vector $v \in E$ is defined by $|v| = \sqrt{\langle v,v \rangle}$.

A. Reventós Tarrida, *Affine Maps, Euclidean Motions and Quadrics*,
Springer Undergraduate Mathematics Series,
DOI 10.1007/978-0-85729-710-5_5, © Springer-Verlag London Limited 2011

The most important properties of these vector spaces can be found in Appendix A, page 319.

Example 5.2

The paradigmatic example is the affine space $\mathbb{A} = \mathbb{R}^n$, modeled on the vector space $E = \mathbb{R}^n$, Example 2 on page 3. When we consider on E the standard scalar product

$$(x_1, \ldots, x_n) \cdot (y_1, \ldots, y_n) = \sum_{i=1}^{n} x_i y_i,$$

$\mathbb{A}$ is a Euclidean affine space. In this case, we simply refer to *the Euclidean affine space* $\mathbb{R}^n$ or, when $n = 2$, *the Euclidean affine plane* $\mathbb{R}^2$.

Example 5.3

Every linear variety $L = P + [F]$ of a Euclidean affine space $\mathbb{A}$, modeled on E, is a Euclidean affine space. Indeed, we know that L is an affine space with associated vector space F, and F is a Euclidean vector space with scalar product induced by the scalar product on E.

Definition 5.4

Let $\mathbb{A}$ be a Euclidean affine space. The *distance* $d(P,Q)$ between the points $P, Q \in \mathbb{A}$ is given by

$$d(P,Q) = |\overrightarrow{PQ}|.$$

Therefore, we have a map

$$d : \mathbb{A} \times \mathbb{A} \longrightarrow \mathbb{R}$$
$$P, Q \mapsto d(P,Q)$$

The most important properties of this map are the following.

Proposition 5.5

For all $P, Q, R \in \mathbb{A}$ we have:
(1) $d(P,Q) \geq 0$;
(2) $d(P,Q) = 0$ if and only if $P = Q$;

(3) $d(P,Q) = d(Q,P)$; and

(4) $d(P,Q) \leq d(P,R) + d(R,Q)$.

Proof

The first three properties are a direct consequence of the definition. The fourth, known as the *triangle inequality*, is a consequence of Corollary A.10, page 327. □

Theorem 5.6 (Pythagoras' Theorem)

If the triangle $\triangle PQR$ is right angled at P, then

$$d(Q,R)^2 = d(P,Q)^2 + d(P,R)^2.$$

Proof

That the triangle is right angled at P means, by definition, that the vectors $\overrightarrow{PQ}$ and $\overrightarrow{PR}$ are orthogonal, that is,

$$\langle \overrightarrow{PQ}, \overrightarrow{PR} \rangle = 0.$$

Thus,

$$\begin{aligned}
d(Q,R)^2 &= \langle \overrightarrow{QR}, \overrightarrow{QR} \rangle \\
&= \langle \overrightarrow{QP} + \overrightarrow{PR}, \overrightarrow{QP} + \overrightarrow{PR} \rangle \\
&= d(P,Q)^2 + d(P,R)^2,
\end{aligned}$$

and this completes the proof.   □

Definition 5.7

Two linear varieties $L_1 = P + [F]$ and $L_2 = Q + [G]$ of a Euclidean affine space $\mathbb{A}$ are said to be *orthogonal* if the vector subspaces F and G are orthogonal.

Let us recall that two vector subspaces F and G of a Euclidean vector space E are said to be orthogonal if $\langle u,v \rangle = 0$, for all $u \in F$, $v \in G$ or, equivalently, when $F \subset G^\perp$ where

$$G^\perp = \{ u \in E : \langle u,v \rangle = 0 \text{ for all } v \in G \}.$$

For the most important properties of orthogonal subspaces, see Section A.5, page 331. Notice that $F \subset G^\perp$ implies $F \cap G = \{\vec{0}\}$.

5.3 The Distance Between Two Varieties

It is natural to define the distance between two subsets as the infimum of the distances between their respective points. Concretely:

Definition 5.8

The distance between two linear varieties $L_1 = P + [F]$ and $L_2 = Q + [G]$ of a Euclidean affine space $\mathbb{A}$ is defined as

$$d(L_1, L_2) = \inf_{X \in L_1, Y \in L_2} d(X, Y).$$

To compute this distance we proceed as follows:

(1) First we compute the vector space $F + G$ and its orthogonal $(F + G)^\perp$. We have

$$E = (F + G) \oplus (F + G)^\perp.$$

(2) Then we decompose the vector $\overrightarrow{PQ}$ in this direct sum, yielding

$$\overrightarrow{PQ} = u + v, \quad u \in (F + G), \; v \in (F + G)^\perp.$$

(3) The distance is then given by

$$d(L_1, L_2) = |v|.$$

Justification of the Method   Every point $X \in L_1$ can be written as $X = P + u_1$ with $u_1 \in F$ and every point $Y \in L_2$ can be written as $Y = Q + u_2$ with $u_2 \in G$. Hence,

$$\inf_{X \in L_1, Y \in L_2} d(X, Y) = \inf_{u_1 \in F, u_2 \in G} d(P + u_1, Q + u_2)$$

$$= \inf_{u_1 \in F, u_2 \in G} |\overrightarrow{PQ} + u_2 - u_1|$$

$$= \inf_{u_1 \in F, u_2 \in G} |u + v + u_2 - u_1|$$

$$= \inf_{u_1 \in F, u_2 \in G} |(u + u_2 - u_1) + v|$$

$$= \inf_{u_1 \in F, u_2 \in G} \sqrt{|(u + u_2 - u_1)|^2 + |v|^2}$$

$$= |v|,$$

and this completes the proof.

5.4 Common Perpendicular

The method described in the previous section allows us to compute the distance between two varieties. Sometimes, however, we need to find a pair of points $X \in L_1$ and $Y \in L_2$ such that

$$d(X, Y) = d(L_1, L_2),$$

that is, points attaining the infimum of the distances between the points of the two varieties.

Proposition 5.9

Let $L_1 = P + [F]$ and $L_2 = Q + [G]$ be linear varieties of an affine space $\mathbb{A}$. Then there are points $X \in L_1$ and $Y \in L_2$ such that

$$d(L_1, L_2) = d(X, Y).$$

If $F \cap G = \{\vec{0}\}$, these points are unique.

Proof

We decompose $\overrightarrow{PQ}$ in the direct sum

$$E = (F + G) \oplus (F + G)^{\perp}.$$

We then have

$$\overrightarrow{PQ} = u + v, \quad u \in (F + G), \ v \in (F + G)^{\perp},$$

and we know that

$$d(L_1, L_2) = |v|.$$

The vectors u and v are uniquely determined. Now we decompose u as a sum of one element of F and one of G, see Figure 5.1 (this decomposition is unique if and only if $F \cap G = \{\vec{0}\}$).

We have

$$u = u_1 + u_2, \quad u_1 \in F, \ u_2 \in G.$$

Hence,

$$\overrightarrow{PQ} = u_1 + u_2 + v, \quad u_1 \in F, \ u_2 \in G, \ v \in (F + G)^{\perp}.$$

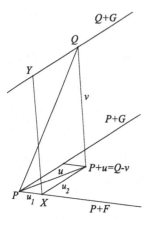

Figure 5.1. Decomposition of $\overrightarrow{PQ}$

Then the equality

$$Q = P + \overrightarrow{PQ} = P + u_1 + u_2 + v$$

can be written as

$$Q - u_2 = P + u_1 + v,$$

which, setting

$$X = P + u_1 \in L_1,$$

$$Y = Q - u_2 \in L_2,$$

is equivalent to $Y = X + v$, and hence

$$\overrightarrow{XY} = v.$$

In particular, we have found points $X \in L_1$ and $Y \in L_2$ such that

$$d(L_1, L_2) = |v| = d(X, Y).$$

Observe that the straight line $X + \langle v \rangle$ is a common perpendicular to L_1 and L_2.

These points X, Y are not, in general, unique. Indeed, note that, for any $w \in F \cap G$, the points

$$X' = X + w,$$

$$Y' = Y + w$$

satisfy

$$X' \in L_1, \qquad Y' \in L_2, \qquad \overrightarrow{XY} = \overrightarrow{X'Y'}$$

and, hence,

$$d(L_1, L_2) = d(X', Y').$$

The points X and Y are uniquely determined only when $F \cap G = \{\vec{0}\}$, i.e. when the sum of F and G is a direct sum. In fact, in this case, there are unique vectors $u_1 \in F$, $u_2 \in G$, $v \in (F + G)^\perp$ such that

$$\overrightarrow{PQ} = u_1 + u_2 + v,$$

and we know that the points $X = P + u_1 \in L_1$, $Y = Q - u_2 \in L_2$ satisfy

$$\overrightarrow{XY} = v.$$

Let us suppose that there are other points $X' = P + u_1' \in L_1$, $Y' = Q + u_2' \in L_2$, such that $|\overrightarrow{X'Y'}| = d(L_1, L_2)$. Then

$$\overrightarrow{X'Y'} = \overrightarrow{PQ} + u_2' - u_1' = (u_1 - u_1') + (u_2 + u_2') + v.$$

By Pythagoras' Theorem and the equality $|\overrightarrow{X'Y'}| = |v|$ we have

$$u_1 - u_1' + u_2 + u_2' = \vec{0}.$$

Hence, $u_1 - u_1' = -u_2 - u_2'$, and since the first term of this equality belongs to F and the second term belongs to G, both must be zero, and we have $u_1 = u_1'$ and $u_2 = -u_2'$. Hence, $X = X'$ and $Y = Y'$, and this completes the proof. $\square$

At the other extreme to the case where F and G form a direct sum is the case $F = G$. Observe that, in this case, that is, if $L_1 = P + [F]$ and $L_2 = Q + [F]$, in order to find points where the distance is attained it is sufficient to decompose

$$\overrightarrow{PQ} = u + v \in F \oplus F^\perp,$$

and then, according to the calculations made in the above proposition, any pair of points of the form

$$X = P + w,$$

$$Y = P + w + v = Q - u + w, \quad \text{for all } w \in F,$$

is such that $d(L_1 L_2) = d(X, Y)$.

Proposition 5.10

Let $L_1 = P + [F]$ and $L_2 = Q + [G]$ be linear varieties with $F \cap G = H \neq \{\vec{0}\}$. Then there are parallel linear varieties $H_1 \subset L_1$ and $H_2 \subset L_2$, of the same dimension, such that

$$d(H_1, H_2) = d(L_1, L_2).$$

Proof

We know that there are points $X \in L_1$, $Y \in L_2$ such that the straight line they determine is perpendicular to L_1 and L_2 and such that $d(X, Y) = d(L_1, L_2)$. Consider the linear varieties $H_1 = X + [H]$ and $H_2 = Y + [H]$. It is clear that $H_1 \subset L_1$, $H_2 \subset L_2$, and that they are parallel, because they have the same direction.

To find the distance between H_1 and H_2 we must decompose the vector $\overrightarrow{XY}$ in the direct sum $E = H \oplus H^\perp$ and take the modulus of the second component. But $\overrightarrow{XY} \in (F + G)^\perp \subset H^\perp$, and hence

$$d(H_1, H_2) = |\overrightarrow{XY}| = d(X, Y) = d(L_1, L_2).$$

$\square$

Observe that every pair of points of the form

$$P' = P + w,$$

$$Q' = Q + w, \quad w \in H,$$

attains the minimum distance.

Example 5.11

Find, in the Euclidean affine space $\mathbb{R}^3$, the distance and the common perpendicular between the straight lines $L_1 = (1, 2, 3) + \langle (1, 0, 0) \rangle$ and $L_2 = (0, 0, 0) + \langle (0, 2, 4) \rangle$.

Solution

Let $P = (1, 2, 3)$, $F = \langle (1, 0, 0) \rangle$, $Q = (0, 0, 0)$, $G = \langle (0, 2, 4) \rangle$, so that $L_1 = P + [F]$ and $L_2 = Q + [G]$. Then

$$F + G = \langle (1, 0, 0), (0, 2, 4) \rangle.$$

Since we are in $\mathbb{R}^3$, we can compute $(F + G)^\perp$ using the vector product of $(1, 0, 0)$ and $(0, 2, 4)$. Another method is to write the vectors of $(F + G)^\perp$ as

(α, β, γ), and then impose the condition that they are perpendicular to $(1,0,0)$ and $(0,2,4)$. We have

$$(F+G)^{\perp} = \langle(0,4,-2)\rangle.$$

Now we write

$$\overrightarrow{PQ} = u + v, \quad u \in (F+G), \quad v \in (F+G)^{\perp}.$$

Concretely,

$$\overrightarrow{PQ} = (-1,-2,-3)$$
$$= \lambda(1,0,0) + \mu(0,2,4) + \nu(0,4,-2) \in (F+G) \oplus (F+G)^{\perp}.$$

We deduce that $\lambda = -1$, $\mu = -\frac{4}{5}$, $\nu = -\frac{1}{10}$. In particular,

$$d(L_1, L_2) = \left| -\frac{1}{10}(0,4,-2) \right| = \frac{\sqrt{5}}{5}.$$

To find points X, Y attaining this distance, we recall that $X = P + u_1$ and $Y = Q - u_2$ (see Figure 5.2) with $u_1 \in F$, $u_2 \in G$ and $v \in (F+G)^{\perp}$, such that

$$\overrightarrow{PQ} = u_1 + u_2 + v.$$

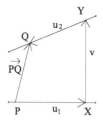

Figure 5.2. Construction of points attaining the distance

Therefore, in our case, $u_1 = \lambda(1,0,0) = (-1,0,0)$ and $u_2 = \mu(0,2,4) = (0, -\frac{8}{5}, \frac{-16}{5})$. Hence,

$$X = (1,2,3) + (-1,0,0) = (0,2,3),$$

$$Y = (0,0,0) - \left(0, -\frac{8}{5}, \frac{-16}{5}\right) = \left(0, \frac{8}{5}, \frac{16}{5}\right).$$

Clearly, $d(X, Y) = \frac{\sqrt{5}}{5}$. □

Example 5.12

Find the distance and the common perpendicular between the planes L_1 and L_2 of the Euclidean affine space $\mathbb{R}^4$ given by

$$L_1 = (0,1,2,3) + \langle(1,0,0,0),(2,6,-1,0)\rangle,$$

$$L_2 = (0,0,0,0) + \langle(0,2,4,1),(2,6,-1,0)\rangle.$$

Solution

Let $P = (0,1,2,3)$, $Q = (0,0,0,0)$, $F = \langle(1,0,0,0),(2,6,-1,0)\rangle$, $G = \langle(0,2,4,1),$
$(2,6,-1,0)\rangle$.
   Thus $L_1 = P + [F]$, $L_2 = Q + [G]$,

$$F + G = \langle(1,0,0,0),(2,6,-1,0),(0,2,4,1)\rangle,$$

and

$$(F+G)^\perp = \langle(0,1,6,-26)\rangle.$$

   Now we write

$$\overrightarrow{PQ} = u + v, \quad u \in (F+G), \ v \in (F+G)^\perp.$$

Concretely,

$$\overrightarrow{PQ} = (0,-1,-2,-3)$$

$$= \lambda(1,0,0,0) + \mu(0,2,4,1)$$

$$+ \nu(2,6,-1,0) + \delta(0,1,6,-26). \tag{5.1}$$

We find $\lambda = -40/713$, $\mu = -449/713$, $\nu = 20/713$, $\delta = 65/713$.
In particular,

$$d(L_1,L_2) = \left| \frac{65}{713}(0,1,6,-26) \right| = \frac{65}{\sqrt{713}}.$$

   To find points X,Y attaining this distance, we recall that $X = P + u_1$, $Y = Q - u_2$ with $u_1 \in F$, $u_2 \in G$ and $v \in (F+G)^\perp$, such that

$$\overrightarrow{PQ} = u_1 + u_2 + v.$$

   One possibility, among others since $F \cap G \neq \{\vec{0}\}$, is to take, comparing with formula (5.1),

$$u_1 = \lambda(1,0,0,0) + \mu(0,2,4,1) = \frac{1}{713}(-40,-898,-1796,-449)$$

$$u_2 = \nu(2,6,-1,0) = \frac{1}{713}(40,120,-20,0).$$

Thus,

$$X = P + u_1 = \frac{1}{713}(-40, -185, -370, 1690),$$

$$Y = Q - u_2 = \frac{1}{713}(-40, -120, 20, 0).$$

Clearly, $d(X,Y) = \frac{65}{\sqrt{713}}$. $\qquad\qquad\qquad\qquad\qquad\qquad\square$

EXERCISES

5.1. Let L_1 and L_2 be linear varieties of a Euclidean affine space $\mathbb{A}$. Prove that if $X \in L_1$ and $Y \in L_2$ are such that $d(L_1, L_2) = d(X,Y)$, then the straight line determined by the points X and Y is a common perpendicular to L_1 and L_2.

5.2. Let A, B, C be three collinear points in a Euclidean affine space $\mathbb{A}$. Prove that

$$|(A, B, C)| = \frac{d(A,B)}{d(A,C)}.$$

Does this mean that the simple ratio depends on the distance?

5.3. Let

$$L_1: \begin{cases} x + y + z = 2, \\ x - y + z = 1, \end{cases} \qquad L_2: \begin{cases} x = 2\lambda + \mu, \\ y = 3 + 2\lambda + 2\mu, \\ z = 4 + 3\lambda + 2\mu \end{cases}$$

be two linear varieties of the Euclidean affine space $\mathbb{R}^3$.
 (a) Find the distance from $P = (0,0,0)$ to L_1 and L_2.
 (b) Find the straight line through P intersecting L_1 and the straight line $x = z = 0$.

5.4. Let L be the linear variety of the Euclidean affine space $\mathbb{R}^4$ given by

$$\begin{cases} x - y = 0, \\ x + t = 3. \end{cases}$$

Find the plane Π orthogonal to L such that $L \cap \Pi = \{(1,1,1,1)\}$.

5.5. Let $\mathbb{A}$ be the affine space $\mathbb{R}^3$ considered as a Euclidean affine space with the scalar product given, in the canonical basis of the vector space $\mathbb{R}^3$, by the matrix

$$\begin{pmatrix} 5 & 1 & 0 \\ 1 & 1 & 0 \\ 0 & 0 & 3 \end{pmatrix}.$$

Find the distance from the point $P = (-1, 1, -2)$ to the plane through the points $A = (1, -1, 1)$, $B = (-2, 1, 3)$ and $C = (4, -5, -2)$.

5.6. Let A be the affine space $\mathbb{R}^4$ considered as a Euclidean affine space with the scalar product given, in the canonical basis of the vector space $\mathbb{R}^4$, by the matrix

$$\begin{pmatrix} 5 & 1 & 0 & 1 \\ 1 & 1 & 0 & 0 \\ 0 & 0 & 3 & 0 \\ 1 & 0 & 0 & 1 \end{pmatrix}.$$

Find the distance between the following linear varieties of A:

(a) The plane determined by $(0, 1, 1, 0)$, $(1, 1, 1, 1)$ and $(-1, 0, 0, 1)$ and the straight line

$$\begin{cases} x + y - z = 1, \\ 2x + t = 2, \\ 2x + y - z + t = 0. \end{cases}$$

(b) The planes

$$\begin{cases} x + y - z = 0, \\ 2x + t = 3, \end{cases}$$

and

$$\begin{cases} x + y + z + t = 2, \\ 4x + 2y + 2t = 0. \end{cases}$$

(c) The straight lines $(1, -1, 0, 0) + \langle (0, 1, 0, 1) \rangle$ and

$$\begin{cases} x + y - 2z = 2, \\ y + z + t = 3, \\ x + z = -1. \end{cases}$$

5.7. Find, in the Euclidean affine space $\mathbb{R}^4$, the distance between the planes

$$\Pi_1: \quad \begin{cases} x - t - 2 = 0, \\ y - z + 2 = 0. \end{cases}$$

$$\Pi_2: \quad \begin{cases} x - z - 2 = 0, \\ y - t + 1 = 0. \end{cases}$$

Find points $P \in \Pi_1$ and $Q \in \Pi_2$ such that $d(P,Q) = d(\Pi_1, \Pi_2)$. Repeat the exercise considering in $\mathbb{R}^4$ the scalar product given in Exercise 5.6.

5.8. Find, in the Euclidean affine space $\mathbb{R}^4$, the distance from the plane

$$\Pi: \quad \begin{cases} x - t - 2 = 0, \\ y - z + 2 = 0 \end{cases}$$

to the straight line

$$r: \quad \begin{cases} x - z - 2 = 0, \\ y - t + 1 = 0, \\ y = 0. \end{cases}$$

Find points $P \in \Pi$ and $Q \in r$ such that $d(P,Q) = d(\Pi, r)$. Repeat the exercise considering in $\mathbb{R}^4$ the scalar product given in Exercise 5.6.

5.9. Find, in the Euclidean affine space $\mathbb{R}^4$, the distance from the plane

$$\Pi: \quad \begin{cases} 2y - 3z - 2 = 0, \\ y - 3t - 1 = 0 \end{cases}$$

to the straight line

$$r: \quad (1, 2, 3, 1) + \lambda(0, 3, 2, 1).$$

Find points $P \in \Pi$ and $Q \in r$ such that $d(P,Q) = d(\Pi, r)$.

5.10. Let

$$r: \quad x - 1 = -y = z, \qquad s: \quad x = y = z,$$

be two straight lines in the Euclidean affine space $\mathbb{R}^3$.
(a) Find the common perpendicular to r and s.
(b) Find the distance $d(r,s)$. Find points $P \in r$ and $Q \in s$ such that $d(P,Q) = d(r,s)$. Repeat the exercise considering in $\mathbb{R}^3$ the scalar product given in Exercise 5.5.

5.11. Consider a triangle in a Euclidean affine plane. Prove the following statements:
(a) The three perpendicular bisectors intersect in a single point (the *circumcenter*).
(b) The three angle bisectors intersect in a single point (the *incenter*).
(c) The three altitudes intersect in a single point (the *orthocenter*).

5.12. Consider a triangle in a Euclidean affine plane. Prove that the product of the lengths of the two line segments determined by the orthocenter on each altitude is constant.

5.13. Suppose that the interior and exterior bisectors of the angle C of a triangle $\triangle ABC$ meet the straight line AB in points C_i and C_e respectively, see Figure 5.3. Prove that

$$(C_e, A, B) = -(C_i, A, B).$$

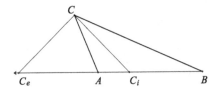

Figure 5.3. Metric property of the bisectors

5.14. Let O, A, B be points in the Euclidean affine plane $\mathbb{R}^2$ and suppose that

$$d(O, A) = 3, \qquad d(O, B) = 5, \qquad \angle OAB = \frac{\pi}{3}.$$

(a) Find the matrix of the standard scalar product on the vector space $\mathbb{R}^2$ with respect to the basis $(\overrightarrow{OA}, \overrightarrow{OB})$.

(b) Find $d(A, B)$ and find the distance from O to the straight line AB.

5.15. Find, in the Euclidean affine space $\mathbb{R}^3$, the distance between the linear varieties L_1 and L_2, given by

$$L_1: \begin{cases} x = 1 + \lambda, \\ y = 2, \\ z = 3. \end{cases} \qquad L_2: \begin{cases} x = 0, \\ z = 2y. \end{cases}$$

Find points $P \in L_1$ and $Q \in L_2$ such that $d(P, Q) = d(L_1, L_2)$. Repeat the exercise considering on $\mathbb{R}^3$ the scalar product given in Exercise 5.5.

5.16. Find, in the Euclidean affine space $\mathbb{R}^4$, the distance between the planes

$$\Pi_1: \begin{cases} 3t - y = 0, \\ 3z - 3x + 2y = 0. \end{cases}$$

$$\Pi_2: \quad (1, 0, 0, 1) + \langle (3, 3, 1, 1), (5, 3, 3, 1) \rangle.$$

Find points $P \in \Pi_1$ and $Q \in \Pi_2$ such that $d(P,Q) = d(\Pi_1, \Pi_2)$. Repeat the exercise considering on $\mathbb{R}^4$ the scalar product given in Exercise 5.6.

5.17. Find, in the Euclidean affine space $\mathbb{R}^4$, the distance between the linear varieties

$$L_1 = (0,0,0,0) + \langle (1,0,1,0), (0,1,2,3), (1,0,0,1) \rangle,$$

$$L_2 = (1,1,1,1) + \langle (2,1,3,4) \rangle.$$

Find points $P \in L_1$ and $Q \in L_2$ such that $d(P,Q) = d(L_1, L_2)$.

5.18. Let $\triangle ABC$ be a triangle in a Euclidean affine plane $\mathbb{A}$ and let P be any point of $\mathbb{A}$.

Prove that the three straight lines through P perpendicular, respectively, to the straight lines AB, AC and the median corresponding to the vertex A (the straight line determined by A and the midpoint M of BC) cut the altitude through vertex A in equal line segments (see Figure 5.4).

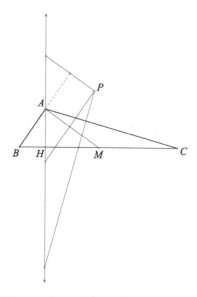

Figure 5.4. Altitude and median

5.19. Consider the set

$$\mathbb{A} = \{(x,y,z,t) \in \mathbb{R}^4 : x+y+z-2t = -1\}$$

and the vector space

$$E = \{(u_1, u_2, u_3, u_4) \in \mathbb{R}^4 : u_1 + u_2 + u_3 - 2u_4 = 0\}.$$

Prove that $(\mathbb{A}, E)$, with the action $\varphi : \mathbb{A} \times E \longrightarrow \mathbb{A}$ given by

$$\varphi((x, y, z, t), (u_1, u_2, u_3, u_4)) = (x + u_1, y + u_2, z + u_3, t + u_4),$$

is an affine space.

Observe that the vector space E admits two scalar products: one, denoted by $\langle \cdot, \cdot \rangle_1$, induced by the standard scalar product on $\mathbb{R}^4$, and the other, denoted by $\langle \cdot, \cdot \rangle_2$, given in Exercise A.9 of Appendix A, page 345. Hence, we have two Euclidean affine spaces: $(\mathbb{A}, E, \langle \cdot, \cdot \rangle_1)$ and $(\mathbb{A}, E, \langle \cdot, \cdot \rangle_2)$.

Find, for each of these two Euclidean affine spaces:

(a) The distance between the point $P = (1, 1, 1, 2)$ and the point $Q = (0, 0, 0, 1/2)$.

(b) The angles of the triangle $\triangle PQR$, with $R = (1, 0, 0, 1)$, and their sum.

(c) The distance from the point P to the linear variety

$$L = \{(x, y, z, t) \in \mathbb{A}; x + y + z + t = 0\}.$$

Note: The hyperplane $\mathbb{A}$ is the hyperplane tangent to the hyperboloid of two sheets $x^2 + y^2 + z^2 - t^2 = -1$ at the point $(1, 1, 1, 2)$. This hyperboloid, with the metric induced by the Lorentz metric of $\mathbb{R}^4$, is a model of three-dimensional Hyperbolic Geometry. See Exercise A.9 of Appendix A, page 345. Although the Lorentz metric is not positive definite, its restriction to this hyperplane is positive definite. This exercise was suggested by Carmen Safont.

6
Euclidean Motions

6.1 Introduction

In this chapter we will use some results on isometries, essentially Theorems A.21 and A.22, page 333, in order to study a special class of affinities: those preserving distance. These distance preserving affinities are called *Euclidean motions*.

We give the classification of Euclidean motions in arbitrary dimension, see Theorem 6.21. Readers only interested in low dimensions can go directly to Chapter 7, page 197.

We recall that if $(E, \langle \cdot, \cdot \rangle)$ is a Euclidean vector space, an *isometry* is an endomorphism $\tilde{f} : E \longrightarrow E$ such that

$$\langle \tilde{f}(u), \tilde{f}(v) \rangle = \langle u, v \rangle, \quad \text{for all } u, v \in E.$$

$\tilde{f}$ is also said to *preserve the scalar product* or be *scalar product preserving*.

Among the most important properties of isometries we wish to emphasize that their eigenvalues have modulus 1 and that eigenvectors corresponding to different eigenvalues, one equal to 1 and the other equal to -1, are orthogonal.

We also recall that two isometries $\tilde{f}$ and $\tilde{g}$ are *isometrically similar* if and only if there is an isometry $\tilde{h}$ conjugating them, that is, such that $\tilde{f} = \tilde{h}^{-1} \circ \tilde{g} \circ \tilde{h}$; see Definition A.36, page 343.

A. Reventós Tarrida, *Affine Maps, Euclidean Motions and Quadrics*, Springer Undergraduate Mathematics Series, DOI 10.1007/978-0-85729-710-5_6, © Springer-Verlag London Limited 2011

6.2 Definition of Euclidean Motion

Definition 6.1

Let $\mathbb{A}$ be a Euclidean affine space. A map $f : \mathbb{A} \to \mathbb{A}$ is called a *Euclidean motion* if

$$d(f(P), f(Q)) = d(P, Q), \quad \text{for all } P, Q \in \mathbb{A}.$$

In other words, the Euclidean motions are the *distance-preserving maps*.

Proposition 6.2

Let $\mathbb{A}$ be a Euclidean affine space. A map $f : \mathbb{A} \longrightarrow \mathbb{A}$ is a *Euclidean motion* if and only if it is an affinity and its associated linear map $\tilde{f}$ is an isometry.

Proof

Let us suppose that f is an affinity and that $\tilde{f}$ is an isometry. Then

$$d(f(P), f(Q)) = |\overrightarrow{f(P)f(Q)}| = |\tilde{f}(\overrightarrow{PQ})| = |\overrightarrow{PQ}| = d(P, Q).$$

Conversely, suppose that f preserves the distance. Fix $P \in \mathbb{A}$ and consider the map $\tilde{f}_P$. Recall that, by definition, $\tilde{f}_P(\overrightarrow{PQ}) = \overrightarrow{f(P)f(Q)}$. We must prove that $\tilde{f}_P$ is linear and preserves the scalar product. By Theorem A.21, page 333, it is sufficient to prove that it preserves the scalar product.

Let $u, v \in E$ be two vectors and set $u = \overrightarrow{PQ}$, $v = \overrightarrow{PR}$. We have

$$d(Q, R)^2 = |\overrightarrow{QR}|^2 = |\overrightarrow{QP} + \overrightarrow{PR}|^2$$
$$= |\overrightarrow{QP}|^2 + |\overrightarrow{PR}|^2 + 2\langle \overrightarrow{QP}, \overrightarrow{PR} \rangle,$$

and

$$d(f(Q), f(R))^2 = |\overrightarrow{f(Q)f(R)}|^2$$
$$= |\overrightarrow{f(Q)f(P)} + \overrightarrow{f(P)f(R)}|^2$$
$$= |\overrightarrow{QP}|^2 + |\overrightarrow{PR}|^2 + 2\langle \overrightarrow{f(Q)f(P)}, \overrightarrow{f(P)f(R)} \rangle.$$

Equating the distances,

$$\langle u, v \rangle = \langle \overrightarrow{PQ}, \overrightarrow{PR} \rangle = \langle \tilde{f}_P(\overrightarrow{PQ}), \tilde{f}_P(\overrightarrow{PR}) \rangle = \langle \tilde{f}_P(u), \tilde{f}_P(v) \rangle.$$

Thus, $\tilde{f}_P$ preserves the scalar product, and in particular, it is linear. $\square$

6.3 Examples of Euclidean Motions

Let us review the examples of affinities given in Section 2.8, page 66, and determine whether the affinities considered there are, or are not, Euclidean motions, when considered in the context of Euclidean affine spaces.

6.3.1 Translations

The simplest examples of Euclidean motions are the translations, since if f is a translation, then $\tilde{f} = \mathrm{id}$, and the identity is, obviously, an isometry.

6.3.2 Homotheties

The homotheties, with similitude ratio $\lambda \neq -1$, are *not* Euclidean motions, since if f is a homothety, then $\tilde{f} = \lambda \, \mathrm{id}$, with $\lambda \neq 1$, and hence $\tilde{f}$ is not an isometry, because $\langle \tilde{f}(u), \tilde{f}(v) \rangle = \lambda^2 \langle u, v \rangle \neq \langle u, v \rangle$.

6.3.3 Orthogonal Symmetries

Every linear variety $L = P + [F]$ of a Euclidean affine space $\mathbb{A}$ gives rise to a symmetry $s_L : \mathbb{A} \longrightarrow \mathbb{A}$ defined by

$$s_L(P + v) = P + v_1 - v_2, \quad \text{for all } v = v_1 + v_2 \in E, \text{ with } v_1 \in F, v_2 \in F^{\perp}.$$

This symmetry is called the *orthogonal symmetry with respect to L*, or simply the *symmetry with respect to L*.

It is easy to find, proceeding as in page 71, an orthonormal basis $\mathcal{B}$ (Definition A.11, page 327) such that if $\tilde{f}$ is the linear map associated to an orthogonal symmetry f, then

$$M(\tilde{f}, \mathcal{B}) = \begin{pmatrix} 1 & & & & & \\ & \ddots & & & & \\ & & 1 & & & \\ & & & -1 & & \\ & & & & \ddots & \\ & & & & & -1 \end{pmatrix}.$$

Here, there are as many 1s on the diagonal as the dimension of F and as many -1s as the dimension of $F^{\perp}$, all other entries being equal to zero.

Since this matrix is orthogonal it follows from Corollary A.25, page 335, that $\tilde{f}$ is an isometry. Hence, *orthogonal symmetries are Euclidean motions*. See also Exercise 6.1, page 192.

In Observation 2.37, page 72, we studied symmetries in affine spaces without a Euclidean structure. In that setting one cannot attach a meaning to $F^\perp$, and thus in order to give a symmetry with respect to F we had to specify a decomposition of E as a direct sum of two subspaces $E = F \oplus G$, and thus obtain the *symmetry with respect to F in the direction G*. This kind of symmetry on a Euclidean affine space, where F and G are not orthogonal to each other, is not a Euclidean motion.

6.3.4 Orthogonal Projections

Every linear variety $L = P + [F]$ of a Euclidean affine space $\mathbb{A}$ gives rise to a projection $p_L : \mathbb{A} \longrightarrow \mathbb{A}$ defined by

$$p_L(P + v) = P + v_1, \quad \text{for all } v = v_1 + v_2 \in E, \text{ with } v_1 \in F, v_2 \in F^\perp.$$

This projection is called the *orthogonal projection on L*. It is easy to find, proceeding as on page 73, an orthonormal basis $\mathcal{B}$ such that, if $\tilde{f}$ is the linear map associated to an orthogonal projection f, then

$$M(\tilde{f}, \mathcal{B}) = \begin{pmatrix} 1 & & & & & \\ & \ddots & & & & \\ & & 1 & & & \\ & & & 0 & & \\ & & & & \ddots & \\ & & & & & 0 \end{pmatrix}.$$

Here, there are as many 1s on the diagonal as the dimension of F, and zeros elsewhere.

Since this matrix is not orthogonal, we conclude that *orthogonal projections are not Euclidean motions*. In Observation 2.39, page 74, we studied projections in affine spaces without a Euclidean structure. In that setting one cannot define the subspace $F^\perp$, and thus in order to give a projection on F we had to specify a decomposition of E as a direct sum of two subspaces $E = F \oplus G$, and we obtained the *projection on F in the direction G*.

6.4 Similar Euclidean Motions

Definition 6.3

We say that two Euclidean motions $f, g : \mathbb{A} \longrightarrow \mathbb{A}$ of a Euclidean affine space $\mathbb{A}$ are *similar as Euclidean motions* if and only if there is a Euclidean motion $h : \mathbb{A} \longrightarrow \mathbb{A}$ such that $f = h^{-1} \circ g \circ h$.

Notice that it is possible that f and g can be similar as affinities, in the sense of Definition 3.4, page 92, but not similar as Euclidean motions. We shall see an example of this shortly, but first we study the relationship between similar Euclidean motions and similar isometries; see Definition A.36, page 343.

Proposition 6.4

If two Euclidean motions are similar as Euclidean motions, then the associated linear maps are isometrically similar.

Proof

From

$$f = h^{-1} \circ g \circ h,$$

we deduce

$$\tilde{f} = \tilde{h}^{-1} \circ \tilde{g} \circ \tilde{h},$$

and hence $\tilde{f}$ is isometrically similar to $\tilde{g}$, because $\tilde{f}$ and $\tilde{g}$ are conjugated by an isometry. □

The converse, however, is not true. For instance, all translations have the same associated linear map (the identity), but not all translations are *similar as Euclidean motions* to each other. Concretely, we have the following.

Proposition 6.5

Let T_u and T_v be two translations of a Euclidean affine space $\mathbb{A}$. Then T_u is similar as a Euclidean motion to T_v if and only if $|u| = |v|$.

Proof

Let us suppose $|u| = |v|$. Take any isometry φ such that $\varphi(u) = v$. The existence of φ is easy to prove, see Exercise A.1 of Appendix A, page 344. Then any affinity h such that $\tilde{h} = \varphi$ conjugates T_u and T_v. Indeed, for all points $P \in \mathbb{A}$ we have

$$(h \circ T_u)(P) = h(P + u) = h(P) + \varphi(u) = h(P) + v = (T_v \circ h)(P).$$

Conversely, if there is an isometry h such that $h \circ T_u = T_v \circ h$, then, for every point $P \in \mathbb{A}$, we have

$$h(P + u) = h(P) + \tilde{h}(u) = h(P) + v,$$

and hence $\tilde{h}(u) = v$. Since $\tilde{h}$ is an isometry, $|u| = |v|$. □

In particular, if u and v are non-zero vectors with $|u| \neq |v|$, then T_u is similar to T_v as affinities, but T_u is not similar to T_v as Euclidean motions.

6.5 Calculations in Coordinates

In arbitrary vector spaces there are no privileged bases. Nevertheless, on vector spaces endowed with a scalar product, orthonormal bases (see Definition A.11, page 327) hold a certain privileged status: computations are easier when working with these bases. A similar situation occurs in the context of Euclidean affine spaces.

Definition 6.6

An affine frame $\mathcal{R} = \{P; \mathcal{B}\}$ of a Euclidean affine space $\mathbb{A}$ is called *orthonormal* if $\mathcal{B}$ is an orthonormal basis of E.

If $f : \mathbb{A} \longrightarrow \mathbb{A}$ is a Euclidean motion of a Euclidean affine space $\mathbb{A}$ of dimension n, and $\mathcal{R}$ is an orthonormal affine frame, we have

$$M(f, \mathcal{R}) = \begin{pmatrix} A & a \\ 0 & 1 \end{pmatrix},$$

where $A = M(\tilde{f}, \mathcal{B})$. This matrix satisfies

$$AA^\mathsf{T} = I_n,$$

since it is the matrix associated to an isometry in an orthonormal basis, see Corollary A.25, page 335. Matrices satisfying the condition $AA^\mathsf{T} = I_n$ are called *orthogonal matrices*, and they satisfy $\det A = \pm 1$. Orthogonal matrices form a group, see Section A.4, page 329.

Conversely, every matrix of the form $\left(\begin{smallmatrix} A & a \\ 0 & 1 \end{smallmatrix}\right)$, with A orthogonal, can be interpreted via an orthonormal affine frame as a Euclidean motion.

For convenience, we define matrices *similar as matrices of Euclidean motions*.

Definition 6.7

We say that the matrices $\left(\begin{smallmatrix} A & a \\ 0 & 1 \end{smallmatrix}\right)$ and $\left(\begin{smallmatrix} B & b \\ 0 & 1 \end{smallmatrix}\right)$ with $AA^\mathsf{T} = BB^\mathsf{T} = I_n$, are *similar as matrices of Euclidean motions* if there is an invertible matrix $\left(\begin{smallmatrix} C & c \\ 0 & 1 \end{smallmatrix}\right)$, with $CC^\mathsf{T} = I_n$, such that

$$\begin{pmatrix} A & a \\ 0 & 1 \end{pmatrix} = \begin{pmatrix} C & c \\ 0 & 1 \end{pmatrix}^{-1} \begin{pmatrix} B & b \\ 0 & 1 \end{pmatrix} \begin{pmatrix} C & c \\ 0 & 1 \end{pmatrix}.$$

Proposition 6.8

Let f and g be Euclidean motions of a Euclidean affine space $\mathbb{A}$. Then
   (i) The matrices $M(f, \mathcal{R}_1)$ and $M(f, \mathcal{R}_2)$ of f in two different orthonormal affine frames $\mathcal{R}_1$ and $\mathcal{R}_2$ are similar as matrices of Euclidean motions.
  (ii) f and g are similar as Euclidean motions if and only if the matrices $M(f, \mathcal{R})$ and $M(g, \mathcal{R})$, where $\mathcal{R}$ is an orthonormal affine frame, are similar as matrices of Euclidean motions.
 (iii) f and g are similar as Euclidean motions if and only if there are orthonormal affine frames $\mathcal{R}_1$ and $\mathcal{R}_2$ of $\mathbb{A}$ such that

$$M(f, \mathcal{R}_1) = M(g, \mathcal{R}_2).$$

Proof

The proof is an adaptation of the proof of Proposition 3.8.   □

Corollary 6.9

Let $\mathcal{R} = \{P; \mathcal{B}\}$ be an orthonormal affine frame of a Euclidean affine space $\mathbb{A}$ of dimension n. Let f, g be Euclidean motions and set

$$M(f, \mathcal{R}) = \begin{pmatrix} A & a \\ 0 & 1 \end{pmatrix}, \qquad M(g, \mathcal{R}) = \begin{pmatrix} B & b \\ 0 & 1 \end{pmatrix}.$$

Then f is similar to g as Euclidean motions if and only if there is an orthogonal matrix $C \in \mathcal{M}_{n \times n}(\mathbb{R})$ and a column matrix $c \in \mathcal{M}_{n \times 1}(\mathbb{R})$ such that

$$\begin{cases} A = C^{\mathsf{T}} B C, \\ (B - I_n)c = Ca - b. \end{cases}$$

Proof

The proof is an adaptation of the proof of Proposition 3.13. $\square$

Observe that Proposition 6.5 is also a corollary of Proposition 6.8. In fact we have the following.

Corollary 6.10

Let T_u and T_v be two translations of a Euclidean affine space $\mathbb{A}$ of dimension n. Then T_u is similar as a Euclidean motion to T_v if and only if $|u| = |v|$.

Proof

Complete the unit vector $\frac{u}{|u|}$ to an orthonormal basis $\mathcal{B} = (\frac{u}{|u|}, e_2, \ldots, e_n)$ of the vector space E. In the orthonormal affine frame $\mathcal{R} = \{P; \mathcal{B}\}$, with P an arbitrary point, the equations of T_u are (by the same argument as on page 67),

$$\begin{cases} x_1' = x_1 + |u|, \\ x_2' = x_2, \\ \quad \vdots \\ x_n' = x_n. \end{cases}$$

Analogously, we construct an orthonormal affine frame $\mathcal{R}' = \{P'; \mathcal{B}'\}$ such that the equations of T_v in $\mathcal{R}'$ are

$$\begin{cases} x_1' = x_1 + |v|, \\ x_2' = x_2, \\ \quad \vdots \\ x_n' = x_n. \end{cases}$$

Hence, if $|u| = |v|$, it follows from part (iii) of Proposition 6.8 that T_u is similar as a Euclidean motion to T_v.

Conversely, if T_u is similar as a Euclidean motion to T_v, we set

$$M(T_u, \mathcal{R}) = \begin{pmatrix} I_n & a \\ 0 & 1 \end{pmatrix}, \qquad M(T_v, \mathcal{R}) = \begin{pmatrix} I_n & b \\ 0 & 1 \end{pmatrix},$$

where $\mathcal{R}$ is an orthonormal affine frame, and the above Corollary 6.9 says that there is an orthogonal matrix C such that $Ca = b$.

But a and b represent, respectively, the components of u and v in the given orthonormal basis. Hence, $|v|^2 = b^{\mathsf{T}}b = a^{\mathsf{T}}C^{\mathsf{T}}Ca = a^{\mathsf{T}}a = |u|^2$. $\qquad\square$

6.6 Glide Vector

Euclidean motions without fixed points have associated to them, in a natural way, an eigenvector of the associated endomorphism with eigenvalue 1. This eigenvector plays an important role in the study and classification of Euclidean motions. Before giving a precise definition, let us recall the following result.

Proposition 6.11

Let f be a Euclidean motion in a Euclidean affine space $\mathbb{A}$, with associated linear map $\tilde{f}$. Let

$$E = \ker(\tilde{f} - \mathrm{id}) \oplus \ker(\tilde{f} - \mathrm{id})^{\perp}.$$

Let $P \in \mathbb{A}$ and set

$$\overrightarrow{Pf(P)} = u_f + v; \quad u_f \in \ker(\tilde{f} - \mathrm{id}), \ v \in \ker(\tilde{f} - \mathrm{id})^{\perp}.$$

Then the vector u_f does not depend on the selected point P.

Proof

In order to prove that u_f does not depend on the selected point P, we take another point $Q \in \mathbb{A}$ and observe that

$$\overrightarrow{Pf(P)} = (\tilde{f} - \mathrm{id})\overrightarrow{QP} + \overrightarrow{Qf(Q)}.$$

But, by Proposition A.30, page 338 (or Observation 6.13), we have

$$(\tilde{f} - \mathrm{id})\overrightarrow{QP} \in \mathrm{Im}(\tilde{f} - \mathrm{id}) = \ker(\tilde{f} - \mathrm{id})^{\perp},$$

and hence the components in $\ker(\tilde{f} - \mathrm{id})$ that we obtain by decomposing $\overrightarrow{Pf(P)}$ and $\overrightarrow{Qf(Q)}$ in the direct sum $E = \ker(\tilde{f} - \mathrm{id}) \oplus \ker(\tilde{f} - \mathrm{id})^\perp$ are equal. $\square$

This proposition is of interest only when 1 is an eigenvalue of $\tilde{f}$, since, otherwise, $\ker(\tilde{f} - \mathrm{id}) = \{\vec{0}\}$ and $u_f = \vec{0}$. Recall that $\tilde{f}$ does not have eigenvalue 1 if and only if f has a unique fixed point (see Proposition 2.28, page 65).

Definition 6.12

Let f be a Euclidean motion of a Euclidean affine space $\mathbb{A}$, with associated linear map $\tilde{f}$. Let

$$E = \ker(\tilde{f} - \mathrm{id}) \oplus \ker(\tilde{f} - \mathrm{id})^\perp.$$

Let $P \in \mathbb{A}$ and set

$$\overrightarrow{Pf(P)} = u_f + v; \quad u_f \in \ker(\tilde{f} - \mathrm{id}), \ v \in \ker(\tilde{f} - \mathrm{id})^\perp.$$

We say that u_f is the *glide vector* of f, and its modulus $\tau(f) = |u_f|$ is called the *glide modulus* of f.

Notice that the glide vector of f is an eigenvector of $\tilde{f}$ with eigenvalue 1. It is denoted by u_f to emphasize that it is determined by f, in the sense that it does not depend on the point P selected to compute $\overrightarrow{Pf(P)}$. The importance of the glide modulus is that it is the same for all Euclidean motions that are *similar as Euclidean motions* (the glide vector, however, can change). See Propositions 6.16 and 6.22.

Observation 6.13

The equality

$$\mathrm{Im}(\tilde{f} - \mathrm{id}) = \ker(\tilde{f} - \mathrm{id})^\perp,$$

proved in Proposition A.30, page 338, can be proved directly in the following way: Let $w \in \ker(\tilde{f} - \mathrm{id})$. Then,

$$\langle (\tilde{f} - \mathrm{id})(v), w \rangle = \langle \tilde{f}(v), w \rangle - \langle v, w \rangle = \langle \tilde{f}(v), \tilde{f}(w) \rangle - \langle v, w \rangle = 0.$$

This proves that $\mathrm{Im}(\tilde{f} - \mathrm{id}) \subset \ker(\tilde{f} - \mathrm{id})^\perp$. But we know, by the Isomorphism Theorem ([8], page 284), that these subspaces have the same dimension and, hence, they are equal.

Observation 6.14

There are some cases in which the calculation of the glide modulus $\tau(f)$ is especially simple. For instance:

(1) When f has fixed points, because if P is a fixed point of f then $\overrightarrow{Pf(P)} = \vec{0}$, and hence $\tau(f) = 0$.

(2) When f has invariant straight lines, or invariant linear varieties, directed by eigenvectors with eigenvalue 1, because in this case we can take a point P on an invariant straight line or on an invariant linear variety and we will have $\overrightarrow{Pf(P)} \in \ker(\tilde{f} - \mathrm{id})$, and hence $\tau(f) = |\overrightarrow{Pf(P)}|$.

(3) When f is a translation T_u, because then $\tau(f) = |u|$, see Proposition 6.16.

Observation 6.15 (Method to find the glide vector)

The easiest way to find u_f and $\tau(f)$ is to take an orthonormal basis $(e_1, \ldots, e_k)$ of $\ker(\tilde{f} - \mathrm{id})$ and an orthonormal basis $(e_{k+1}, \ldots, e_n)$ of $\ker(\tilde{f} - \mathrm{id})^\perp$. Then we put

$$\overrightarrow{Pf(P)} = a_1 e_1 + \cdots + a_k e_k + a_{k+1} e_{k+1} + \cdots + a_n e_n,$$

and we have

$$u_f = a_1 e_1 + \cdots + a_k e_k,$$

$$\tau(f) = \sqrt{a_1^2 + \cdots + a_k^2}.$$

As particular cases we have:

Computation of $\tau(f)$ in dimension two. Let f be a Euclidean motion with associated linear map $\tilde{f}$ in a Euclidean affine space $\mathbb{A}$ of dimension two. Then

$$\tau(f) = \begin{cases} 0, & \text{if } \dim \ker(\tilde{f} - \mathrm{id}) = 0, \\ |\langle \overrightarrow{Pf(P)}, e \rangle|, & \text{if } \dim \ker(\tilde{f} - \mathrm{id}) = 1, \\ |\overrightarrow{Pf(P)}|, & \text{if } \dim \ker(\tilde{f} - \mathrm{id}) = 2 \ (\tilde{f} = \mathrm{id}), \end{cases}$$

where P is an arbitrary point and e is a unit eigenvector of $\tilde{f}$ with eigenvalue 1.

Computation of $\tau(f)$ in dimension three. Let f be a Euclidean motion with associated linear map $\tilde{f}$ in a Euclidean affine space $\mathbb{A}$ of dimension three. Then

$$\tau(f) = \begin{cases} 0, & \text{if } \dim \ker(\tilde{f} - \mathrm{id}) = 0, \\ |\langle \overrightarrow{Pf(P)}, e_1 \rangle|, & \text{if } \dim \ker(\tilde{f} - \mathrm{id}) = 1, \\ \sqrt{\langle \overrightarrow{Pf(P)}, e_1 \rangle^2 + \langle \overrightarrow{Pf(P)}, e_2 \rangle^2}, & \text{if } \dim \ker(\tilde{f} - \mathrm{id}) = 2, \\ |\overrightarrow{Pf(P)}|, & \text{if } \dim \ker(\tilde{f} - \mathrm{id}) = 3, \end{cases}$$

where P is, in all cases, an arbitrary point; in the second case e_1 is a unit eigenvector of $\tilde{f}$ with eigenvalue 1; and in the third case e_1, e_2 are unit orthogonal eigenvectors of $\tilde{f}$ with eigenvalue 1.

Proposition 6.16

The glide vector of a translation by vector u, T_u, is u. In particular, $\tau(T_u) = |u|$.

Proof

Since $\widetilde{T_u} = \mathrm{id}$,

$$E = \ker(\widetilde{T_u} - \mathrm{id}) \oplus \ker(\widetilde{T_u} - \mathrm{id})^\perp = \ker(\widetilde{T_u} - \mathrm{id}),$$

and, hence, the glide vector u_{T_u} coincides with the vector $\overrightarrow{PT_u(P)}$, for every point P. But we know that $\overrightarrow{PT_u(P)} = u$, and hence $u_{T_u} = u$, and this completes the proof. □

Proposition 6.17

Let f be a Euclidean motion of a Euclidean affine space $\mathbb{A}$. Then $\tau(f) = 0$ if and only if f has a fixed point.

Proof

Let us suppose that f has a fixed point P. Applying the definition of $\tau(f)$ to the vector $\overrightarrow{Pf(P)}$ we obtain $\tau(f) = 0$.

Conversely, suppose $\tau(f) = 0$. Fix a point $P \in \mathbb{A}$ and set

$$\overrightarrow{Pf(P)} = u_f + v; \quad u_f \in \ker(\tilde{f} - \mathrm{id}), \ v \in \ker(\tilde{f} - \mathrm{id})^\perp.$$

Since $\tau(f) = 0$, $u_f = \vec{0}$, and hence

$$\overrightarrow{Pf(P)} \in \ker(\tilde{f} - \mathrm{id})^\perp.$$

But we have seen in Proposition A.30, page 338 (and in Observation 6.13) that $\ker(\tilde{f} - \mathrm{id})^\perp = \mathrm{Im}(\tilde{f} - \mathrm{id})$, so that

$$\overrightarrow{Pf(P)} \in \mathrm{Im}(\tilde{f} - \mathrm{id}).$$

Equivalently, there is a $v \in E$ such that

$$\overrightarrow{Pf(P)} = (\tilde{f} - \mathrm{id})(v). \tag{6.1}$$

From this equality we can prove that f has a fixed point. Indeed, finding a fixed point is equivalent to finding a vector $w \in E$ such that the point $Q = P + w$ is fixed. That is,

$$P + w = Q = f(Q) = f(P + w) = f(P) + \tilde{f}(w),$$

or, equivalently,

$$\overrightarrow{Pf(P)} = w - \tilde{f}(w) = -(\tilde{f} - \mathrm{id})(w).$$

By (6.1), it is sufficient to take $w = -v$ and the point $Q = P - v$ is fixed. $\square$

6.7 Classification of Euclidean Motions

Theorem 6.18

Let f be a Euclidean motion without fixed points in a Euclidean affine space $\mathbb{A}$. Then the Euclidean motion $g = f \circ T_{-u_f}$ has a fixed point. In fact, the glide vector u_f is the only eigenvector with eigenvalue 1 satisfying this property.

Proof

Fix a point $P \in \mathbb{A}$ and set

$$\overrightarrow{Pf(P)} = u_f + v; \quad u \in \ker(\tilde{f} - \mathrm{id}),\ v \in \ker(\tilde{f} - \mathrm{id})^{\perp} = \mathrm{Im}(\tilde{f} - \mathrm{id}).$$

In particular, there is a $w \in E$ such that $v = (\tilde{f} - \mathrm{id})(w)$.

Inspired by the proof of the above proposition, we take $Q = P - w$ as a candidate for the fixed point. Let us check it:

$$\begin{aligned}
g(Q) &= f \circ T_{-u_f}(Q) \\
&= f(Q - u_f) \\
&= f(P) - \tilde{f}(w) - u_f \\
&= P + u_f + \tilde{f}(w) - w - \tilde{f}(w) - u_f \\
&= P - w \\
&= Q.
\end{aligned}$$

To prove the uniqueness of u_f, let us assume that there is an eigenvector u' with eigenvalue 1 such that $f \circ T_{-u'}$ has a fixed point Q'. Then

$$Q' = f \circ T_{-u'}(Q') = f(Q' - u') = f(Q') - \tilde{f}(u') = f(Q') - u',$$

and, hence,

$$\overrightarrow{Q'f(Q')} = u'.$$

But, since u' is an eigenvector with eigenvalue 1, this equality can be written as

$$\overrightarrow{Q'f(Q')} = u' + \vec{0} \in \ker(\tilde{f} - \mathrm{id}) \oplus \ker(\tilde{f} - \mathrm{id})^{\perp}.$$

Thus, by Proposition 6.11, we have $u_f = u'$, and this completes the proof. $\square$

Thus $\tau(f)$ can be interpreted as the modulus of the translation that we must perform in order to transform a Euclidean motion without fixed points into a Euclidean motion with a fixed point.

Observe that $f \circ T_{-u_f} = T_{-u_f} \circ f$. In fact, $w \in E$ is an eigenvector of $\tilde{f}$ with eigenvalue 1 if and only if $f \circ T_w = T_w \circ f$.

Corollary 6.19

Every Euclidean motion f of a Euclidean affine space $\mathbb{A}$ is the composition of a translation with a Euclidean motion that has a fixed point.

Proof

If f has a fixed point, we take as the translation the identity. Otherwise we know that there is a Euclidean motion g with a fixed point such that $g = f \circ T_{-u_f}$. Hence, $f = g \circ T_{u_f}$. $\square$

Due to this result, the classification of Euclidean motions is practically reduced to the classification of Euclidean motions with fixed points.

Observation 6.20

The fixed points of $g = f \circ T_{-u_f}$ belong to an invariant straight line of f of direction u_f. Indeed, if $g(P) = P$, we have

$$g(P) = f(P - u_f) = f(P) - u_f = P;$$

hence,

$$\overrightarrow{Pf(P)} = u_f,$$

and the straight line through P with direction vector u_f (an eigenvector with eigenvalue 1) is invariant.

In particular, *a Euclidean motion has either a fixed point or an invariant straight line.* Hence, its invariance level ρ is 0 or 1.

Theorem 6.21 (Classification)

Let f be a Euclidean motion of a Euclidean affine space $\mathbb{A}$ of dimension n. Then there is an orthonormal affine frame $\mathcal{R}$ of $\mathbb{A}$ such that

$$M(f,\mathcal{R}) = \left(\begin{array}{c|c} M & c \\ \hline 0 & 1 \end{array}\right),$$

where the matrix M is the matrix given on page 340, and

$$c = \begin{pmatrix} \tau(f) \\ 0 \\ \vdots \\ 0 \end{pmatrix}.$$

Proof

Since $\tilde{f}$ is an isometry, it follows from Theorem A.33, page 340, that there is an orthonormal basis $\mathcal{B}$ such that

$$M(\tilde{f},\mathcal{B}) = M,$$

where M is the matrix given on page 340. Notice that the number of 1s, -1s and rotation boxes appearing in M, as well as the angles of these rotations, are completely determined by the characteristic polynomial of $\tilde{f}$.

If f has a fixed point P, $\tau(f) = 0$, and in the affine frame $\mathcal{R} = \{P; \mathcal{B}\}$ we have

$$M(f,\mathcal{R}) = \left(\begin{array}{c|c} M & 0 \\ \hline 0 & 1 \end{array}\right),$$

which completes the proof in this case.

We now assume that f does not have fixed points.

We know, by Theorem 6.18, that the Euclidean motion $g = f \circ T_{-u_f} = T_{-u_f} \circ f$, where u_f is the glide vector of f, has a fixed point. Notice that

$$\tilde{g} = \tilde{f} \circ \widetilde{T_{-u_f}} = \tilde{f} \circ \mathrm{id} = \tilde{f},$$

and hence, u_f is also an eigenvector of $\tilde{g}$ with eigenvalue 1.

Let P be a fixed point of g. Since $\tilde{f} = \tilde{g}$, we have

$$M(g, \mathcal{R}) = \left(\begin{array}{c|c} M & 0 \\ \hline 0 & 1 \end{array} \right).$$

The basis $\mathcal{B}$ has been explicitly constructed in the proof of Theorem A.33, page 340, and it is adapted to the decomposition of the vector space E as an orthogonal direct sum of the form

$$E = \ker(\tilde{f} - \mathrm{id}) \oplus \cdots,$$

given explicitly in equality (A.4), page 340. In particular, the first vectors are unit eigenvectors of $\tilde{f}$ with eigenvalue 1, the choice of which are restricted only by the condition that they form an orthonormal basis of $\ker(\tilde{f} - \mathrm{id})$.

We modify $\mathcal{B}$, taking a basis of $\ker(\tilde{f} - \mathrm{id})$ with first vector $e_1 = \frac{1}{|u_f|} u_f = \frac{1}{\tau(f)} u_f$. The basis of $\ker(\tilde{f} - \mathrm{id})^{\perp}$ remains unchanged. Denote by $\mathcal{B}_1$ this new basis and let $\mathcal{R}_1 = \{P; \mathcal{B}_1\}$. Observe that $M(g, \mathcal{R}) = M(g, \mathcal{R}_1)$.

Since $f = T_{u_f} \circ g$, we have:

$$M(f, \mathcal{R}_1) = M(T_{u_f}, \mathcal{R}_1) M(g, \mathcal{R}_1)$$

$$= \left(\begin{array}{cc} I_n & c \\ 0 & 1 \end{array} \right) \left(\begin{array}{c|c} M & 0 \\ \hline 0 & 1 \end{array} \right)$$

$$= \left(\begin{array}{c|c} M & c \\ \hline 0 & 1 \end{array} \right),$$

where

$$c = \left(\begin{array}{c} \tau(f) \\ 0 \\ \vdots \\ 0 \end{array} \right),$$

and this completes the proof. □

6.8 Invariance of the Glide Modulus

Let us prove that $\tau(f)$ is invariant within each equivalence class of similar Euclidean motions.

Proposition 6.22

Let f and g be two Euclidean motions of a Euclidean affine space $\mathbb{A}$, similar as Euclidean motions, then $\tau(f) = \tau(g)$.

Proof

Let h be an invertible Euclidean motion such that $h \circ f = g \circ h$. Then the isometries $\tilde{f}$ and $\tilde{g}$ are conjugated by $\tilde{h}$, and hence

$$\tilde{h}(\ker(\tilde{f} - \mathrm{id})) = \ker(\tilde{g} - \mathrm{id}).$$

Since $\tilde{h}$ preserves orthogonality,

$$\tilde{h}(\ker(\tilde{f} - \mathrm{id})^{\perp}) = \ker(\tilde{g} - \mathrm{id})^{\perp}.$$

Let us take an arbitrary point P and set

$$\overrightarrow{Pf(P)} = u_f + v, \quad u_f \in \ker(\tilde{f} - \mathrm{id}), \ v \in \ker(\tilde{f} - \mathrm{id})^{\perp}.$$

Then $\tau(f) = |u_f|$.

On the other hand, since the definition of $\tau(g)$ does not depend on the selected point Q used to compute $\overrightarrow{Qg(Q)}$, we apply it to the point $h(P)$, and we have

$$\overrightarrow{h(P)gh(P)} = \tilde{h}(\overrightarrow{Pf(P)})$$

$$= \tilde{h}(u_f + v)$$

$$= \tilde{h}(u_f) + \tilde{h}(v).$$

Since $\tilde{h}(u_f) \in \ker(\tilde{g} - \mathrm{id})$ and $\tilde{h}(v) \in \ker(\tilde{g} - \mathrm{id})^{\perp}$, this last equality is the decomposition of $\overrightarrow{h(P)gh(P)}$ in the direct sum

$$E = \ker(\tilde{g} - \mathrm{id}) \oplus \ker(\tilde{g} - \mathrm{id})^{\perp}.$$

In particular, $\tilde{h}(u_f)$ is the glide vector of g and, hence, $\tau(g) = |\tilde{h}(u_f)| = |u_f| = \tau(f)$, and this completes the proof. □

Theorem 6.23 (Characterization)

Two Euclidean motions f and g of a Euclidean affine space $\mathbb{A}$ are similar as Euclidean motions if and only if they have the same glide modulus and the associated endomorphisms $\tilde{f}$ and $\tilde{g}$ are isometrically similar.

Proof

We know that if f is similar to g as Euclidean motions then $\tilde{f}$ is similar to $\tilde{g}$ as isometries and $\tau(f) = \tau(g)$, see Propositions 6.4 and 6.22, respectively.

To prove the converse, let us recall that if $\tilde{f}$ is isometrically similar to $\tilde{g}$, then $\tilde{f}$ and $\tilde{g}$ have the same characteristic polynomial; see Theorem A.38, page 343. Hence, there are bases $\mathcal{B}$ and $\mathcal{B}'$ such that $M(\tilde{f}, \mathcal{B}) = M(\tilde{g}, \mathcal{B}') = M$, where M is the matrix on page 340.

By the Classification Theorem 6.21, we know that there are orthonormal affine frames $\mathcal{R}$ and $\mathcal{R}'$ such that

$$M(f, \mathcal{R}) = \begin{pmatrix} M & c \\ 0 & 1 \end{pmatrix}, \qquad M(g, \mathcal{R}') = \begin{pmatrix} M & c' \\ 0 & 1 \end{pmatrix},$$

with $c^{\mathsf{T}} = (\tau(f), 0, \ldots, 0)$ and $c'^{\mathsf{T}} = (\tau(g), 0, \ldots, 0)$. Hence, $c = c'$ and the above matrices are equal. Hence, f is similar as a Euclidean motion to g, and this completes the proof. $\qquad\square$

The same theorem can also be stated as follows:

Theorem 6.24 (Characterization)

Two Euclidean motions of a Euclidean affine space $\mathbb{A}$ are similar as Euclidean motions if and only if they have the same glide modulus and their associated endomorphisms have the same cha racteristic polynomial.

Proof

Simply recall that two isometries are isometrically similar if and only if they have the same characteristic polynomial, see Theorem A.38, page 343. $\qquad\square$

EXERCISES

6.1. Prove, directly from Definition 6.1 (without using Corollary A.25, page 335), that orthogonal symmetries are Euclidean motions.

6.2. Find, in the Euclidean affine space $\mathbb{R}^3$, the equations of the straight line r' symmetric to the straight line

$$r: \quad \frac{x-1}{2} = \frac{-y+3}{1} = \frac{z-2}{3}$$

with respect to the plane $\Pi: 2x - y + z - 1 = 0$.

6.3. Let $\mathcal{B} = (u, v)$, with $u = (1, -1)$ and $v = (1, 1)$, be a basis of the Euclidean affine plane $\mathbb{R}^2$. Let $T : \mathbb{R}^2 \longrightarrow \mathbb{R}^2$ be the affinity given in the affine frame $\{(0, 0); \mathcal{B}\}$ by the equation:

$$T \begin{pmatrix} x \\ y \end{pmatrix} = \frac{1}{2} \begin{pmatrix} -\sqrt{3} & -1 \\ -1 & \sqrt{3} \end{pmatrix} \begin{pmatrix} x \\ y \end{pmatrix} + \begin{pmatrix} 1 \\ 2 \end{pmatrix}.$$

Prove that T is a Euclidean motion.

6.4. Let $T : \mathbb{R}^3 \longrightarrow \mathbb{R}^3$ be the affinity of the Euclidean affine space $\mathbb{R}^3$ given in the canonical affine frame by

$$T(x, y, z) = \frac{1}{4} \begin{pmatrix} 2+\sqrt{3} & -\sqrt{2} & -2+\sqrt{3} \\ \sqrt{2} & 2\sqrt{3} & \sqrt{2} \\ -2+\sqrt{3} & -\sqrt{2} & 2+\sqrt{3} \end{pmatrix} \begin{pmatrix} x \\ y \\ z \end{pmatrix} + \begin{pmatrix} 1 \\ 0 \\ 1 \end{pmatrix}.$$

Prove that T is a Euclidean motion.

6.5. Let

$$L_1: \quad \begin{cases} x + y + z = 2, \\ x - y + z = 1. \end{cases}$$

$$L_2: \quad \begin{cases} x = 2\lambda + \mu, \\ y = 3 + 2\lambda + 2\mu, \\ z = 4 + 3\lambda + 2\mu \end{cases}$$

be two linear varieties of the Euclidean affine space $\mathbb{R}^3$.
  (a) Find the orthogonal projection of the point $P = (2, 1, 3)$ on L_1 and on L_2.
  (b) Find the symmetry of P with respect to L_2.

6.6. Find, in the Euclidean affine space $\mathbb{R}^3$, the symmetry of the point $(-1, 1, 3)$ with respect to each of the following linear varieties:

$$L_1: \quad \begin{cases} x - y - z = 1, \\ y - 2z = 0. \end{cases}$$

$$L_2: \quad \begin{cases} \dfrac{x-1}{1} = \dfrac{y}{-1} = \dfrac{z+1}{2}. \end{cases}$$

6.7. Let A be the affine space $\mathbb{R}^2$ considered as a Euclidean affine space with the scalar product given, in the canonical basis of the vector space $\mathbb{R}^2$, by

$$\begin{pmatrix} 4 & 1 \\ 1 & 1 \end{pmatrix}.$$

Find the symmetry of the point $P = (2,3)$ with respect to the straight line $x + y + 1 = 0$.

6.8. Find, in the Euclidean affine space $\mathbb{R}^3$, the straight line r' symmetric to the straight line

$$r: \quad \frac{x-1}{2} = \frac{-y+3}{1} = \frac{z-2}{3}$$

with respect to the plane $\Pi: 2x - y + z = 1$.

6.9. Find, with respect to the canonical affine frame of the Euclidean affine plane $\mathbb{R}^2$, the equations of the axial symmetry f given by

$$M(f, \mathcal{R}) = \begin{pmatrix} 1 & 0 & 0 \\ 0 & -1 & 0 \\ 0 & 0 & 1 \end{pmatrix},$$

where $\mathcal{R}$ is the affine frame

$$\mathcal{R} = \{(0,0); ((1/2, \sqrt{3}/2), (\sqrt{3}/2, -1/2))\}.$$

Repeat the exercise when the Euclidean structure (the scalar product) on the affine space $\mathbb{R}^2$ is given, in the canonical affine frame, by the matrix

$$\begin{pmatrix} 3 & 2 \\ 2 & 2 \end{pmatrix}.$$

6.10. Let

$$\Pi: \quad -x + 2y + z - 2 = 0, \qquad r: \quad \frac{x+2}{3} = \frac{y-1}{1} = \frac{z-3}{2},$$

be a plane and a straight line of the Euclidean affine space $\mathbb{R}^3$.

(a) Find the equations of the straight line r' symmetric to r with respect to Π.

(b) Find the equations of the plane Π' symmetric to Π with respect to r.

Repeat the exercise when the Euclidean structure on the affine space $\mathbb{R}^3$ is that given in Exercise 5.5 of Chapter 5, page 169.

6.11. Suppose that an affinity of a Euclidean affine space is given, in some affine frame $\mathcal{R}$, by the equations

$$x_1' = x_1,$$

$$x_2' = x_1 + x_2.$$

Is it a Euclidean motion? And if $\mathcal{R}$ is orthonormal?

6.12. Suppose that an affinity of a Euclidean affine space is given, in some affine frame $\mathcal{R}$, by the equations

$$\begin{pmatrix} x_1' \\ x_2' \end{pmatrix} = \begin{pmatrix} 9 & 20 \\ -4 & -9 \end{pmatrix} \begin{pmatrix} x_1 \\ x_2 \end{pmatrix} + \begin{pmatrix} 1 \\ 5 \end{pmatrix}.$$

Is it a Euclidean motion? And if $\mathcal{R}$ is orthonormal?

6.13. Let f be an affinity of a Euclidean affine space $\mathbb{A}$. Prove that

$$\tau(f) = \inf_{P \in \mathbb{A}} d(P, f(\overset{\cdot}{P})).$$

Euclidean Motions of the Line, the Plane and of Space

7.1 Introduction

The matrix M appearing in the Classification Theorem of Euclidean motions (Theorem 6.21) is made up of 1s and -1s on the diagonal and boxes of sines and cosines centered on the diagonal. As the dimension increases, this leads to many possibilities, the number of combinations of diagonal entries and boxes rapidly growing. However, in low dimensions the number of possibilities is relatively small.

In this chapter we study the cases of dimension 1, 2 and 3. Thus we shall have, for these dimensions, a more explicit version of the Classification Theorem. We present the results without appealing to this theorem, so that those readers only interested in low dimensions may omit the previous chapter.

7.2 Classification of Euclidean Motions of the Line

Theorem 7.1 (Classification Theorem)

Let f be a Euclidean motion of a Euclidean affine space $\mathbb{A}$ of dimension 1. Then there is an orthonormal affine frame $\mathcal{R} = \{P; e\}$ such that the equation

A. Reventós Tarrida, *Affine Maps, Euclidean Motions and Quadrics*,
Springer Undergraduate Mathematics Series,
DOI 10.1007/978-0-85729-710-5_7, © Springer-Verlag London Limited 2011

of f in $\mathcal{R}$ is one and only one of the following:

$$x' = -x \qquad\qquad \text{Symmetry,}$$
$$x' = x + d, \quad d \geq 0 \quad \text{Translation (identity if } d = 0).$$

Proof

By Theorem A.33, page 340, there is an orthonormal basis $\mathcal{B} = (e_1)$ such that

$$M(\tilde{f}, \mathcal{B}) = (1), \quad \text{or}$$

$$M(\tilde{f}, \mathcal{B}) = (-1).$$

If f has a fixed point P, in the affine frame $\mathcal{R} = \{P; \mathcal{B}\}$, we have

$$M(f, \mathcal{R}) = \left(\begin{array}{c|c} 1 & 0 \\ \hline 0 & 1 \end{array} \right), \quad \text{or}$$

$$M(f, \mathcal{R}) = \left(\begin{array}{c|c} -1 & 0 \\ \hline 0 & 1 \end{array} \right).$$

In the first case, f is the identity, and the equation is $x' = x$. In the second case, f is a symmetry, and the equation is $x' = -x$.

If f does not have fixed points, we know, by Proposition 2.28, page 65, that $\tilde{f}$ has at least one eigenvector v with eigenvalue 1. Since we are in dimension one, $E = \langle v \rangle = \ker(\tilde{f} - \mathrm{id})$, and hence $\tilde{f} = \mathrm{id}$, that is, f is a translation. We can take v with $|v| = 1$, so that $\mathcal{B} = (v)$ is an orthonormal basis of E.

Since for all $P \in \mathbb{A}$ we have $\overrightarrow{Pf(P)} = \tau(f)v$, in the affine frame $\mathcal{R} = \{P; \mathcal{B}\}$ the matrix of f is

$$M(f, \mathcal{R}) = \left(\begin{array}{c|c} 1 & \tau(f) \\ \hline 0 & 1 \end{array} \right),$$

and, setting $d = \tau(f) > 0$, the equation of f is $x' = x + d$, $d > 0$. $\qquad\square$

Thus, *every Euclidean motion of the straight line different from the identity is similar as a Euclidean motion to a symmetry or to a translation by a vector of modulus $d > 0$.*

Notice that translations of the form $x' = x - d$, with $d > 0$, do not appear in this list, because changing the affine frame (by simply replacing v with $-v$) this translation can be written as $x' = x + d$, with $d > 0$.

Notice also that we do not separate translations $T(d)$ with $d > 0$ from the identity $(d = 0)$, as we did in the affine case, because now we have as many equivalence classes of translations as real numbers $d \geq 0$. However, in the affine case, there are only two equivalence classes: one formed only by the identity and the other formed by all translations different from the identity. Roughly speaking, the assertion *the identity is a translation by vector zero* has more sense in the context of Euclidean motions than in the context of affinities.

This result is summarized in Table 7.1.

	Characteristic of $\tilde{f}$	$\tau(f)$
Translation	$x - 1$	$d \geq 0$
Symmetry	$x + 1$	0

Table 7.1. Euclidean motions of the line

7.3 Classification of Euclidean Motions of the Plane

The aim of this section is to study the Euclidean motions of a Euclidean affine space of dimension two.

We know that given an isometry $\tilde{f}$ of a Euclidean vector space E of dimension two, there is an orthonormal basis $\mathcal{B}$ of E such that $M(\tilde{f}, \mathcal{B})$ is equal to one and only one of the matrices of the following list (see Theorem A.34, page 342).

7.3.1 List of Canonical Expressions of Isometries

$$\tilde{R}(\alpha) = \begin{pmatrix} \cos\alpha & -\sin\alpha \\ \sin\alpha & \cos\alpha \end{pmatrix}, \quad 0 < \alpha \leq \pi,$$

$$\tilde{S} = \begin{pmatrix} 1 & 0 \\ 0 & -1 \end{pmatrix},$$

$$\tilde{\mathrm{id}} = \begin{pmatrix} 1 & 0 \\ 0 & 1 \end{pmatrix} = I_2.$$

The notation $\tilde{R}(\alpha)$ is chosen to remind us that this type of isometry is a *rotation by angle* α. Observe that $\tilde{R}(0) = \tilde{\mathrm{id}}$, however when we write $\tilde{R}(\alpha)$ we assume that $\alpha \neq 0$, as we want to separate the cases depending on whether these matrices have the eigenvalue 1 or not. The notation $\tilde{S}$ recalls that this type of isometry is a *symmetry*.

Inspired by the above list, we now give a list of matrices that will help us to formulate the Classification Theorem more easily.

7.3.2 List of Canonical Expressions of Euclidean Motions

$$R(\alpha) = \left(\begin{array}{cc|c} \cos\alpha & -\sin\alpha & 0 \\ \sin\alpha & \cos\alpha & 0 \\ \hline 0 & 0 & 1 \end{array} \right) = \left(\begin{array}{c|c} \tilde{R}(\alpha) & 0 \\ & 0 \\ \hline 0 \quad 0 & 1 \end{array} \right), \quad 0 < \alpha \leq \pi,$$

$$Gl(d) = \left(\begin{array}{cc|c} 1 & 0 & d \\ 0 & -1 & 0 \\ \hline 0 & 0 & 1 \end{array} \right) = \left(\begin{array}{c|c} \tilde{S} & d \\ & 0 \\ \hline 0 \quad 0 & 1 \end{array} \right), \quad d \geq 0,$$

$$T(d) = \left(\begin{array}{cc|c} 1 & 0 & d \\ 0 & 1 & 0 \\ \hline 0 & 0 & 1 \end{array} \right) = \left(\begin{array}{c|c} I_2 & d \\ & 0 \\ \hline 0 \quad 0 & 1 \end{array} \right), \quad d \geq 0.$$

The notation is again chosen to remind us that these motions are *rotations* by angle α, *glide reflections* of modulus d, and *translations* by a vector of modulus d.

In order to classify a Euclidean motion we must study its associated endomorphism and its glide modulus.

Theorem 7.2 (Classification Theorem)

Let f be a Euclidean motion of a Euclidean affine space of dimension two. Then there is an orthonormal affine frame $\mathcal{R}$ such that the matrix $M(f, \mathcal{R})$ is equal to one of the matrices in the list of canonical expressions.

Proof

Since $\tilde{f}$ is an isometry, it follows from Theorem A.34, page 342, that there is an orthonormal basis $\mathcal{B} = (e_1, e_2)$ of E such that

$$M(\tilde{f}, \mathcal{B}) = \tilde{R}(\alpha), \quad \text{or}$$

$$M(\tilde{f}, \mathcal{B}) = \tilde{S}, \quad \text{or}$$

$$M(\tilde{f}, \mathcal{B}) = \tilde{\mathrm{id}}.$$

If f has a fixed point P, $\tau(f) = 0$, and in the affine frame $\mathcal{R} = \{P; \mathcal{B}\}$ we have, respectively,

$$M(f, \mathcal{R}) = R(\alpha), \quad \text{or}$$

$$M(f, \mathcal{R}) = Gl(0), \quad \text{or}$$

$$M(f, \mathcal{R}) = T(0),$$

each of which is a matrix in the list of canonical expressions.

We now assume that f does not have fixed points.

We know, by Theorem 6.18, that the Euclidean motion $g = f \circ T_{-u_f} = T_{-u_f} \circ f$, where $u_f \neq 0$ is the glide vector of f, has a fixed point. Since $\tilde{g} = \tilde{f}$, u_f is also an eigenvector of $\tilde{g}$ with eigenvalue 1. Set $d = \tau(f) = |u_f| > 0$.

Let P be a fixed point of g. We know, as before, that there is an orthonormal affine frame $\mathcal{R} = \{P; \mathcal{B}\}$ such that

$$M(g, \mathcal{R}) = \left(\begin{array}{c|c} M & 0 \\ \hline 0 & 1 \end{array} \right),$$

where, by Proposition 2.28, $M = \tilde{S}$ or $M = \tilde{\mathrm{id}}$.

In both cases, the first vector of the basis, e_1, is an eigenvector of $\tilde{g} = \tilde{f}$ with eigenvalue 1, which can be taken (modifying if necessary the initial basis) as the normalized glide vector, that is, we can always select the above basis $\mathcal{B}$ so that $u_f = de_1$.

Since $f = T_{u_f} \circ g$, in the first case ($M = \tilde{S}$) we have:

$$M(f, \mathcal{R}) = M(T_{u_f}, \mathcal{R}) M(g, \mathcal{R})$$

$$= \left(\begin{array}{cc|c} 1 & 0 & d \\ 0 & 1 & 0 \\ \hline 0 & 0 & 1 \end{array} \right) \left(\begin{array}{cc|c} 1 & 0 & 0 \\ 0 & -1 & 0 \\ \hline 0 & 0 & 1 \end{array} \right)$$

$$= \left(\begin{array}{cc|c} 1 & 0 & d \\ 0 & -1 & 0 \\ \hline 0 & 0 & 1 \end{array} \right) = Gl(d),$$

and in the second case ($M = \tilde{\mathrm{id}}$) we have:

$$M(f, \mathcal{R}) = M(T_{u_f}, \mathcal{R}) M(g, \mathcal{R})$$

$$= \left(\begin{array}{cc|c} 1 & 0 & d \\ 0 & 1 & 0 \\ \hline 0 & 0 & 1 \end{array} \right) \left(\begin{array}{cc|c} 1 & 0 & 0 \\ 0 & 1 & 0 \\ \hline 0 & 0 & 1 \end{array} \right)$$

$$= \begin{pmatrix} 1 & 0 & d \\ 0 & 1 & 0 \\ \hline 0 & 0 & 1 \end{pmatrix} = T(d).$$

In summary, there always exists an orthonormal affine frame in which the matrix of f coincides with one of the matrices in the list of canonical expressions. This completes the proof. $\qquad\square$

Thus, *every Euclidean motion of a Euclidean affine plane is similar as a Euclidean motion to a rotation, a glide reflection, or a translation.*

This result is summarized in Table 7.2.

	Characteristic of $\tilde{f}$	$\tau(f)$
Rotation	$x^2 - (2\cos\alpha)x + 1$	0
Glide reflection	$(x-1)(x+1)$	$d \geq 0$
Translation	$(x-1)^2$	$d \geq 0$

Table 7.2. Euclidean motions of the plane

The characteristic polynomial of the rotations have complex roots, except if $\alpha = 0$, in which case the polynomial is $(x-1)^2$, or when $\alpha = \pi$, in which case the polynomial is $(x+1)^2$.

7.4 Geometrical Interpretation

If $M(f, \mathcal{R}) = R(\alpha)$, then f is a *rotation* with center the origin of $\mathcal{R}$. Since

$$\begin{pmatrix} \cos\alpha & -\sin\alpha \\ \sin\alpha & \cos\alpha \end{pmatrix} = \begin{pmatrix} \cos\alpha & \sin\alpha \\ \sin\alpha & -\cos\alpha \end{pmatrix} \begin{pmatrix} 1 & 0 \\ 0 & -1 \end{pmatrix},$$

it follows that *a rotation by angle α is the composition of two symmetries with respect to axes that meet in the center of the rotation, with an angle equal to $\alpha/2$.* See Exercises 7.1 and 7.4, page 217. In the particular case $\alpha = \pi$, the rotation is called a *half turn* or *central symmetry*, because it is effectively a central symmetry according to the definition given on page 72.

If $M(f, \mathcal{R}) = Gl(0)$, then f is a *symmetry* with respect to the straight line through the origin with direction vector the second vector of the basis. This straight line is comprised of fixed points, and all straight lines perpendicular to it are invariant. f is also called an *axial symmetry*, since it is effectively an axial symmetry according to the definition given on page 72.

If $M(f, \mathcal{R}) = Gl(d)$, $d > 0$, then f is a *glide reflection*, the composition of a symmetry with a translation in the direction of the axis of symmetry. The translation vector has modulus d. There is an invariant straight line and no fixed points.

If $M(f, \mathcal{R}) = T(d)$, then f is a *translation* with translation vector of modulus d. Every straight line with this direction vector is invariant. If $d = 0$, f is the *identity*.

In summary, every Euclidean motion of a Euclidean affine space of dimension two is equivalent to one and only one of the Euclidean motions of Table 7.3.

Euclidean motion	Name	Equation	Affinity
$R(\alpha)$	*Rotation* One fixed point. $\tau = 0$.	$x' = x \cos \alpha - y \sin \alpha,$ $y' = x \sin \alpha + y \cos \alpha,$ $0 < \alpha \le \pi.$	*Elliptic* $\rho = 0$
$Gl(d)$	*Glide reflection* No fixed points. One inv. str. line. $\tau = d$.	$x' = x + d,$ $y' = -y, \quad d > 0.$	$T \circ h_g(-1)$ $\rho = 1$
	Symmetry Fixed str. line. Infinite inv. str. lines. $\tau = 0$.	$x' = x,$ $y' = -y, \quad d = 0.$	*Symmetry* $\rho = 0$
$T(d)$	*Translation* No fixed points. Infinite inv. str. lines. $\tau = d$	$x' = x + d, \quad d > 0,$ $y' = y.$	*Translation* $\rho = 1$
	Identity $\tau = 0$.	$x' = x,$ $y' = y, \quad d = 0.$	*Identity* $\rho = 0$

Table 7.3. Euclidean motions of the plane

Recall that the notation $T \circ h_g(-1)$, used in the second row of the table, denotes the composition of a general homology of ratio -1 followed by a translation.

Note that, for the same reasons as given on page 198, in this table we neither separate translations from the identity nor symmetries from glide reflections.

7.5 Classification of Euclidean Motions of Space

The aim of this section is to study the Euclidean motions of a Euclidean affine space of dimension three.

We know that given an isometry $\tilde{f}$ of a Euclidean vector space E of dimension three, there is an orthonormal basis $\mathcal{B}$ of E such that $M(\tilde{f}, \mathcal{B})$ is equal to one and only one of the following matrices (see Theorem A.35, page 342).

7.5.1 List of Canonical Expressions of Isometries

$$\tilde{R}(\alpha) = \begin{pmatrix} 1 & 0 & 0 \\ 0 & \cos\alpha & -\sin\alpha \\ 0 & \sin\alpha & \cos\alpha \end{pmatrix}, \quad 0 \le \alpha \le \pi,$$

$$\widetilde{SR}(\alpha) = \begin{pmatrix} -1 & 0 & 0 \\ 0 & \cos\alpha & -\sin\alpha \\ 0 & \sin\alpha & \cos\alpha \end{pmatrix}, \quad 0 < \alpha \le \pi,$$

$$\tilde{S} = \begin{pmatrix} 1 & 0 & 0 \\ 0 & 1 & 0 \\ 0 & 0 & -1 \end{pmatrix}.$$

The notation $\tilde{R}(\alpha)$ reminds us that we are talking about a *rotation by angle* α *about an axis*. The axis is the vector subspace generated by the first vector of the basis $\mathcal{B}$. Likewise, the $\tilde{S}$ denotes a *symmetry* and the $\widetilde{SR}(\alpha)$ denotes a rotation composed with a symmetry.

We write $\tilde{S}$ instead of $\widetilde{SG}(0)$ because we wish to separate the cases depending on whether these matrices admit the eigenvalue 1 or not.

Next we give a list of matrices which will help us to formulate the Classification Theorem more easily.

7.5.2 List of Canonical Expressions of Euclidean Motions

$$R(\alpha, d) = \left(\begin{array}{ccc|c} 1 & 0 & 0 & cd \\ 0 & \cos\alpha & -\sin\alpha & c0 \\ 0 & \sin\alpha & \cos\alpha & c0 \\ \hline 0 & 0 & 0 & c1 \end{array} \right), \quad 0 \le \alpha \le \pi, \ d \ge 0,$$

$$SR(\alpha) = \left(\begin{array}{ccc|c} -1 & 0 & 0 & 0 \\ 0 & \cos\alpha & -\sin\alpha & 0 \\ 0 & \sin\alpha & \cos\alpha & 0 \\ \hline 0 & 0 & 0 & 1 \end{array}\right), \quad 0 < \alpha \le \pi,$$

$$Gl(d) = \left(\begin{array}{ccc|c} 1 & 0 & 0 & d \\ 0 & 1 & 0 & 0 \\ 0 & 0 & -1 & 0 \\ \hline 0 & 0 & 0 & 1 \end{array}\right), \quad d \ge 0.$$

The matrices of this list are called *canonical matrices*.

The notation is chosen to remind us that we are speaking, respectively, of rotations by angle α composed with translations by a vector of modulus d, $R(\alpha, d)$ (*helicoidal* Euclidean motions); rotations by angle α composed with symmetries, $SR(\alpha)$ (*anti-rotations*); and *glide reflections* of modulus d, $Gl(d)$.

In order to classify a Euclidean motion, we must study its associated endomorphism and its glide modulus.

Theorem 7.3 (Classification Theorem)

Let f be a Euclidean motion of a Euclidean affine space of dimension three. Then, there is an orthonormal affine frame $\mathcal{R}$ such that $M(f, \mathcal{R})$ is a canonical matrix.

Proof

Since $\tilde{f}$ is an isometry, it follows from Theorem A.35, page 342, that there is an orthonormal basis $\mathcal{B} = (e_1, e_2, e_3)$ of E such that

$$M(\tilde{f}, \mathcal{B}) = \tilde{R}(\alpha), \quad \text{or}$$

$$M(\tilde{f}, \mathcal{B}) = \widetilde{SR}(\alpha), \quad \text{or}$$

$$M(\tilde{f}, \mathcal{B}) = \tilde{S}.$$

If f has a fixed point P, $\tau(f) = 0$, and in the affine frame $\mathcal{R} = \{P; \mathcal{B}\}$ we have, respectively,

$$M(f, \mathcal{R}) = R(\alpha, 0), \quad \text{or}$$

$$M(f, \mathcal{R}) = SR(\alpha), \quad \text{or}$$

$$M(f, \mathcal{R}) = Gl(0),$$

each of which is a canonical matrix.

We now assume that f does not have fixed points.

We know, by Theorem 6.18, that the Euclidean motion $g = f \circ T_{-u_f} = T_{-u_f} \circ f$, where $u_f \neq 0$ is the glide vector of f, has a fixed point. Since $\tilde{g} = \tilde{f}$, u_f is also an eigenvector of $\tilde{g}$ with eigenvalue 1. Put $d = \tau(f) = |u_f| > 0$.

Let P be a fixed point of g. Then, there is an orthonormal affine frame $\mathcal{R} = \{P; \mathcal{B}\}$, with $\mathcal{B} = (e_1, e_2, e_3)$, such that

$$M(g, \mathcal{R}) = \left(\begin{array}{c|c} M & 0 \\ \hline 0 & 1 \end{array} \right),$$

where, by Proposition 2.28, page 65, $M = \tilde{R}(\alpha)$ or $M = \tilde{S}$. In both cases, the first vector of the basis, e_1, is an eigenvector of $\tilde{g} = \tilde{f}$ with eigenvalue 1, which can be considered (modifying if necessary the initial basis) as the normalized glide vector, that is, we can always select the above basis $\mathcal{B}$ in such a way that $u_f = de_1$.

Since $f = T_{u_f} \circ g$, in the first case ($M = \tilde{R}(\alpha)$) we have:

$$M(f, \mathcal{R}) = M(T_{u_f}, \mathcal{R}) M(g, \mathcal{R})$$

$$= \left(\begin{array}{ccc|c} 1 & 0 & 0 & d \\ 0 & 1 & 0 & 0 \\ 0 & 0 & 1 & 0 \\ \hline 0 & 0 & 0 & 1 \end{array} \right) \left(\begin{array}{ccc|c} 1 & 0 & 0 & 0 \\ 0 & \cos\alpha & -\sin\alpha & 0 \\ 0 & \sin\alpha & \cos\alpha & 0 \\ \hline 0 & 0 & 0 & 1 \end{array} \right) = R(\alpha, d),$$

and in the second case ($M = \tilde{S}$) we have:

$$M(f, \mathcal{R}) = M(T_{u_f}, \mathcal{R}) M(g, \mathcal{R})$$

$$= \left(\begin{array}{ccc|c} 1 & 0 & 0 & d \\ 0 & 1 & 0 & 0 \\ 0 & 0 & 1 & 0 \\ \hline 0 & 0 & 0 & 1 \end{array} \right) \left(\begin{array}{ccc|c} 1 & 0 & 0 & 0 \\ 0 & 1 & 0 & 0 \\ 0 & 0 & -1 & 0 \\ \hline 0 & 0 & 0 & 1 \end{array} \right) = Gl(d).$$

Summarizing, there always exists an orthonormal affine frame $\mathcal{R}$ such that $M(f, \mathcal{R})$ is a canonical matrix. This completes the proof. $\qquad \square$

Thus, *every Euclidean motion of a Euclidean affine space of dimension three is similar as a Euclidean motion to a helicoidal Euclidean motion (including rotations, translations and the identity), an anti-rotation, or a glide reflection (including symmetries).*

This result is summarized in Table 7.4.

The characteristic polynomial of a helicoidal Euclidean motion has complex roots, except if $\alpha = 0$, in which case the characteristic polynomial is $(x - 1)^3$, or if $\alpha = \pi$, in which case the characteristic polynomial is $(x - 1)(x + 1)^2$.

	Characteristic of $\tilde{f}$	$\tau(f)$
Helicoidal	$(x-1)(x^2 - 2\cos(\alpha)x + 1)$	$d \geq 0$
Anti-rotation	$(x+1)(x^2 - 2\cos(\alpha)x + 1)$	0
Glide Reflection	$(x-1)^2(x+1)$	$d \geq 0$

Table 7.4. Euclidean motions of the space

The characteristic polynomial of an anti-rotation has complex roots, except if $\alpha = 0$, in which case the polynomial is $(x+1)(x-1)^2$, or if $\alpha = \pi$, in which case the polynomial is $(x+1)^3$.

7.6 Geometrical Interpretation

If $M(f, \mathcal{R}) = R(\alpha, 0)$, $\alpha \neq 0$, then the Euclidean motion f reduces to a *rotation around an axis*. It has a fixed straight line (the axis) and planes perpendicular to the axis are invariant. The glide modulus is $\tau = 0$. In the particular case $\alpha = \pi$, this Euclidean motion is called a *half turn*, or *axial symmetry*.

If $M(f, \mathcal{R}) = R(\alpha, d)$, $\alpha \neq 0$, $d > 0$, then f is a rotation around an axis followed by a translation in the direction of this axis. f is called a *helicoidal Euclidean motion*. There is an invariant straight line and there are no fixed points. The glide modulus is $\tau = d$.

If $M(f, \mathcal{R}) = R(0, d)$, then the helicoidal Euclidean motion f reduces to a *translation*. If $d = 0$, it is the *identity*.

If $M(f, \mathcal{R}) = SR(\alpha)$, $\alpha \neq 0$, then f is the composition of a symmetry with respect to a plane with a rotation around an axis perpendicular to this plane. The axis of the rotation is the unique invariant straight line, and $\tau = 0$. We say that f is an *anti-rotation*. If $\alpha = \pi$, f is called a *central symmetry*.

If $M(f, \mathcal{R}) = Gl(d)$, $d > 0$, then f is the composition of a symmetry with respect to a plane with a translation by a vector in the direction of this plane. It has infinitely many invariant straight lines and $\tau = d$. By analogy with the comparable case in dimension two, we say that f is a *glide reflection*.

If $M(f, \mathcal{R}) = Gl(0)$, then f is a symmetry with respect to a plane and $\tau = 0$. f is said to be a *mirror symmetry*.

Summing up, every Euclidean motion of a Euclidean affine space of dimension three is equivalent to one and only one of the Euclidean motions in Table 7.5.

Euclidean motion	Name and properties	Equation
$R(\alpha, d)$	*Helicoidal Euclidean motion* One inv. straight line. No fixed points. $\tau = d$.	$x' = x + d,$ $y' = y \cos\alpha - z \sin\alpha,$ $z' = y \sin\alpha + z \cos\alpha,$ $\alpha \neq 0, \quad d > 0.$
	Rotation $(\alpha \neq 0, d = 0)$ One fixed str. line. Invariant $\perp$ plane. $\tau = 0$. *Axial symmetry if $\alpha = \pi$.*	$x' = x,$ $y' = y \cos\alpha - z \sin\alpha,$ $z' = y \sin\alpha + z \cos\alpha.$
	Translation $(\alpha = 0, d > 0)$ No fixed points. $\tau = d$.	$x' = x + d, \quad d > 0,$ $y' = y,$ $z' = z.$
	Identity $(\alpha = 0, d = 0)$ $\tau = 0$.	$x' = x,$ $y' = y,$ $z' = z.$
$SR(\alpha)$	*Anti-rotation* $R \circ S$ $S =$ Mirror symmetry. $R =$ Rotation $\perp$ plane. One inv. str. line $\perp$ plane. $\tau = 0$.	$x' = -x,$ $y' = y \cos\alpha - z \sin\alpha,$ $z' = y \sin\alpha + z \cos\alpha,$ $\alpha \neq 0.$
$Gl(d)$	*Glide* $T \circ S$ $S =$ Mirror symmetry. $T =$ Translation $\parallel$ plane. ∞ inv. str. lines. $\tau = d$.	$x' = x + d, \quad d > 0,$ $y' = y,$ $z' = -z.$
	Mirror symmetry One fixed plane. $\tau = 0$.	$x' = x,$ $y' = y,$ $z' = -z.$

Table 7.5. Euclidean motions of the space

Example 7.4

Classify the Euclidean motion f of the Euclidean affine space $\mathbb{R}^3$ given, in the canonical affine frame $\mathcal{C}$, by

$$M(f,\mathcal{C}) = \frac{1}{16} \begin{pmatrix} \sqrt{3}+2 & 11\sqrt{2} & \sqrt{3}-2 & 16\sqrt{2} \\ -7\sqrt{2}-2\sqrt{6} & 2\sqrt{3} & -7\sqrt{2}+2\sqrt{6} & -32 \\ -3\sqrt{3}+10 & -\sqrt{2} & -3\sqrt{3}-10 & 16\sqrt{2} \\ 0 & 0 & 0 & 16 \end{pmatrix}.$$

Find the fixed points, the invariant straight lines, the glide vector and an orthonormal affine frame $\mathcal{R}$ such that $M(f,\mathcal{R})$ is a canonical matrix.

Solution

It is advisable to use Maple or some similar program, because the calculation is a little long. The characteristic polynomial of the associated endomorphism $\tilde{f}$ is

$$x^3 + \frac{1}{2}x^2 + \frac{1}{2}x + 1,$$

which can be written as

$$(x+1)\left(x^2 - \frac{1}{2}x + 1\right).$$

Looking at Table 7.4 on page 207 we directly see that f is an *anti-rotation* by angle α with $\cos(\alpha) = \frac{1}{4}$.

We also arrive at the same conclusion by noting that $\det \tilde{f} = -1$ and that 1 is not an eigenvalue of $\tilde{f}$. This implies that f is an anti-rotation. Then, we have $2\cos(\alpha) = \text{trace}(M(f,\mathcal{C}))$, see Exercise 7.7, page 218. Since this trace is equal to $\frac{1}{2}$, we again obtain $\cos(\alpha) = \frac{1}{4}$.

To find the fixed points, we solve the system (with Maple or similar),

$$(M(f,\mathcal{C}) - I_4) \begin{pmatrix} x \\ y \\ z \\ 1 \end{pmatrix} = \begin{pmatrix} 0 \\ 0 \\ 0 \\ 0 \end{pmatrix},$$

and take the solutions with fourth coordinate 1. We find that the *unique fixed point* is the point

$$P = \left(-\frac{\sqrt{2}}{2}, -2, \frac{\sqrt{2}}{2}\right).$$

In particular, the *glide vector* vanishes.

The axis of the anti-rotation (the unique invariant straight line) is a straight line through P directed by the eigenvector of $\tilde{f}$ with eigenvalue -1.
To find this vector we must solve the system

$$(M(\tilde{f},\mathcal{C}) + I_3) \begin{pmatrix} x \\ y \\ z \end{pmatrix} = \begin{pmatrix} 0 \\ 0 \\ 0 \end{pmatrix}.$$

We get the eigenvector

$$v = (1, -\sqrt{2}, -2\sqrt{3} - 5),$$

or, of course, any multiple of it. Therefore the unique *invariant straight line* is $l: P + \langle v \rangle$.
To construct an affine frame $\mathcal{R}$ such that

$$M(f,\mathcal{R}) = \left(\begin{array}{ccc|c} -1 & 0 & 0 & 0 \\ 0 & \cos\alpha & -\sin\alpha & 0 \\ 0 & \sin\alpha & \cos\alpha & 0 \\ \hline 0 & 0 & 0 & 1 \end{array} \right),$$

we must complete $e_1 = v/|v|$ to an orthonormal basis (e_1, e_2, e_3), such that the scalar product $\langle e_3, \tilde{f}(e_2) \rangle$ is positive, and take P as the origin. The reason, as can be seen by inspecting the matrix, is that we must have $\langle e_3, \tilde{f}(e_2) \rangle = \sin\alpha$, and we are assuming $0 \leq \alpha \leq \pi$, that is, $\sin\alpha \geq 0$. Since $\cos\alpha = \frac{1}{4}$, we have $\sin\alpha = \frac{\sqrt{15}}{4}$.
For instance, we can take $e_2 = \frac{1}{\sqrt{3}}(\sqrt{2}, 1, 0)$ (a unit vector orthogonal to e_1), and e_3 is the vector product $e_1 \wedge e_2$ (see page 215). Thus, $\langle e_3, \tilde{f}(e_2) \rangle = 15/4\sqrt{2}(\sqrt{3} + 1) > 0$. If this value is negative, we replace e_3 by $-e_3$. □

Example 7.5

Classify the Euclidean motion f of the Euclidean affine space $\mathbb{R}^3$ given, in the canonical affine frame $\mathcal{C}$, by

$$M(f,\mathcal{C}) = \left(\begin{array}{ccc|c} (\sqrt{2}+2)/4 & (-\sqrt{2}+2)/4 & 1/2 & 0 \\ (-\sqrt{2}+2)/4 & (\sqrt{2}+2)/4 & -1/2 & 4 \\ 1/2 & -1/2 & -1/\sqrt{2} & 2\sqrt{2} \\ \hline 0 & 0 & 0 & 1 \end{array} \right).$$

Find the fixed points, the invariant straight lines, the glide vector and an orthonormal affine frame $\mathcal{R}$ such that the $M(f,\mathcal{R})$ is a canonical matrix.

Solution

The characteristic polynomial of the associated endomorphism $\tilde{f}$ is

$$x^3 - x^2 - x + 1,$$

which can be written as

$$(x+1)(x-1)^2.$$

Looking at Table 7.4 on page 207 we see directly that f *is a glide reflection* (or a mirror symmetry).

To find the fixed points, we solve the system (with Maple or similar)

$$(M(f,\mathcal{C}) - I_4) \begin{pmatrix} x \\ y \\ z \\ 1 \end{pmatrix} = \begin{pmatrix} 0 \\ 0 \\ 0 \\ 0 \end{pmatrix},$$

and take the solutions with fourth coordinate 1. Or, equivalently, we study the system

$$(M(\tilde{f},\mathcal{C}) - I_3) \begin{pmatrix} x \\ y \\ z \end{pmatrix} = \begin{pmatrix} 0 \\ -4 \\ -2\sqrt{2} \end{pmatrix}.$$

This system has no solutions, and hence, *there are no fixed points.* Thus, f is a glide reflection with glide modulus different from zero; in particular, it is not a mirror symmetry.

To find the glide vector we find an orthonormal basis adapted to the decomposition

$$\mathbb{R}^3 = \ker(\tilde{f} - \mathrm{id}) \oplus \ker(\tilde{f} - \mathrm{id})^{\perp}.$$

In our case this decomposition coincides with

$$\mathbb{R}^3 = \ker(\tilde{f} - \mathrm{id}) \oplus \ker(\tilde{f} + \mathrm{id}),$$

because eigenvectors of different eigenvalues are orthogonal; see Theorem A.22, page 333, and Theorem A.32, page 340.

From this it is easy to see that the *glide vector* u_f is given by

$$u_f = \frac{1}{2}(\tilde{f} + \mathrm{id})\overrightarrow{Pf(P)},$$

where P is any point; see Exercise 7.13, page 219. To simplify our calculations, let us take $P = (0,0,0)$; we get

$$\overrightarrow{Pf(P)} = (0, 4, 2\sqrt{2}),$$

and

$$u_f = (1, 3, -2 + \sqrt{2}).$$

Therefore, the glide modulus is

$$\tau(f) = 2\sqrt{4 - \sqrt{2}}.$$

We know that the Euclidean motion $g = f \circ T_{-u_f}$ has a fixed point. In fact, since $\tilde{f} = \tilde{g}$, g is a mirror symmetry.

To find its plane of symmetry, that is, the set of fixed points, we observe that

$$M(g, \mathcal{C}) = M(f, \mathcal{C}) M(T_{-u_f}, \mathcal{C})$$

$$= \left(\begin{array}{ccc|c} (2+\sqrt{2})/4 & (2-\sqrt{2})/4 & 1/2 & -1 \\ (2-\sqrt{2})/4 & (2+\sqrt{2})/4 & -1/2 & 1 \\ 1/2 & -1/2 & -\sqrt{2}/2 & 2+\sqrt{2} \\ \hline 0 & 0 & 0 & 1 \end{array} \right).$$

Thus, we must solve the system

$$(M(\tilde{g}, \mathcal{C}) - I_3) \begin{pmatrix} x \\ y \\ z \end{pmatrix} = \begin{pmatrix} 1 \\ -1 \\ -2 - \sqrt{2} \end{pmatrix}.$$

The solution of this system is the plane

$$\Pi: \quad x - y - (2 + \sqrt{2})z = -2\sqrt{2} - 4.$$

Equivalently,

$$\Pi: \quad (0, 2\sqrt{2} + 4, 0) + \langle (1, 1, 0), (1, 3, -2 + \sqrt{2}) \rangle.$$

Recall that the linear variety of fixed points is directed by $\ker(\tilde{f} - \mathrm{id})$, and hence we know directly that u_f belongs to it.

Now we look for an affine frame in which the matrix of f is a canonical matrix. For this we shall take the origin in Π, and an orthonormal basis (e_1, e_2, e_3), with e_1 the normalized glide vector, e_2 in the direction of Π, and e_3 normal to Π. Concretely, we take $\mathcal{R} = \{P; (e_1, e_2, e_3)\}$ with

$$P = (0, 2\sqrt{2} + 4, 0),$$

$$e_1 = \frac{1}{2\sqrt{4 - \sqrt{2}}}(1, 3, -2 + \sqrt{2}) \quad (e_1 = u_f / |u_f|),$$

$$e_2 = \frac{1}{2\sqrt{7}\sqrt{6-2\sqrt{2}}}(10-\sqrt{2}, 2-3\sqrt{2}, 6-2\sqrt{2}),$$

$$e_3 = \frac{1}{2\sqrt{2+\sqrt{2}}}(1,-1,-2-\sqrt{2}).$$

Thus we have

$$M(g,\mathcal{R}) = \left(\begin{array}{ccc|c} 1 & 0 & 0 & 0 \\ 0 & 1 & 0 & 0 \\ 0 & 0 & -1 & 0 \\ \hline 0 & 0 & 0 & 1 \end{array}\right).$$

Therefore g is a mirror symmetry with respect to the plane Π. Since $f = g \circ T_{u_f} = T_{u_f} \circ g$, we see that f *is the glide reflection obtained as the composition of the symmetry with respect to Π with a translation by the vector u_f in the direction of Π.*

Matricially,

$$M(f,\mathcal{R}) = M(g,\mathcal{R})M(T_{u_f},\mathcal{R})$$

$$= \left(\begin{array}{ccc|c} 1 & 0 & 0 & 0 \\ 0 & 1 & 0 & 0 \\ 0 & 0 & -1 & 0 \\ \hline 0 & 0 & 0 & 1 \end{array}\right) \left(\begin{array}{ccc|c} 1 & 0 & 0 & 2\sqrt{4-\sqrt{2}} \\ 0 & 1 & 0 & 0 \\ 0 & 0 & 1 & 0 \\ \hline 0 & 0 & 0 & 1 \end{array}\right)$$

$$= \left(\begin{array}{ccc|c} 1 & 0 & 0 & 2\sqrt{4-\sqrt{2}} \\ 0 & 1 & 0 & 0 \\ 0 & 0 & -1 & 0 \\ \hline 0 & 0 & 0 & 1 \end{array}\right).$$

Finally, we observe that any straight line contained in Π and with direction u_f is an *invariant straight line.* □

7.7 Composition of Rotations in Dimension Three

A rotation by angle α, $0 \leq \alpha \leq \pi$, around a straight line l is not determined by the data l, α. We also need to know the sense of this rotation.

However, if we assume that the given Euclidean affine space of dimension three is oriented (that is, a basis $\mathcal{B}$ is given on the associated vector space E), and that the given straight line l is also oriented (that is, we choose one of the two unit vectors determining its direction), then this line and angle do determine a unique rotation.

Concretely, given a unit vector $u \in E$, a straight line l: $P + \langle u \rangle$, and an angle α, we define the *rotation by angle α around the axis l* as the Euclidean motion given by the matrix $R(\alpha, 0)$ in the orthonormal affine frame $\mathcal{R}' = \{P; \mathcal{B}'\}$, where $\mathcal{B}' = (u, v, w)$, with v any vector orthogonal to u, and w the vector determined by the condition that the basis $\mathcal{B}'$ is orthonormal and positive, that is, $\det M(\mathcal{B}', \mathcal{B}) > 0$. We denote this rotation by $R(l, u, \alpha)$ and we say that it is the *rotation by angle α around the straight line l, oriented by u*. In Exercise 7.19 we give an explicit formula for $R(l, u, \alpha)$.

The aim of this section is to study the Euclidean motion obtained by composing two rotations $R(l, u, \alpha)$ and $R(m, v, \beta)$, of concurrent axes l, m.

Let us first see two different ways of decomposing a rotation as a product of symmetries.

7.7.1 First Decomposition of a Rotation as a Product of Symmetries

Let $\{P; (e_1, e_2, e_3)\}$ be an orthonormal affine frame. Denote by l_i the straight lines l_i: $P + \langle e_i \rangle$, $i = 1, 2$, and let l_α be the straight line

$$l_\alpha: \quad P + \left\langle \cos\left(\frac{\alpha}{2}\right) e_2 + \sin\left(\frac{\alpha}{2}\right) e_3 \right\rangle.$$

The matricial equality

$$\begin{pmatrix} 1 & 0 & 0 \\ 0 & \cos\alpha & -\sin\alpha \\ 0 & \sin\alpha & \cos\alpha \end{pmatrix} = \begin{pmatrix} -1 & 0 & 0 \\ 0 & \cos\alpha & \sin\alpha \\ 0 & \sin\alpha & -\cos\alpha \end{pmatrix} \begin{pmatrix} -1 & 0 & 0 \\ 0 & 1 & 0 \\ 0 & 0 & -1 \end{pmatrix}$$

shows that the rotation $R(l_1, e_1, \alpha)$ is equal to the composition of the *axial symmetry* S_2 (the rotation by angle π around the axis l_2) with the axial symmetry S_α of axis l_α.

That is,

$$R(l_1, e_1, \alpha) = S_\alpha \circ S_2. \tag{7.1}$$

7.7.2 Second Decomposition of a Rotation as a Product of Symmetries

Analogously, the matricial equality

$$\begin{pmatrix} 1 & 0 & 0 \\ 0 & \cos\beta & -\sin\beta \\ 0 & \sin\beta & \cos\beta \end{pmatrix} = \begin{pmatrix} -1 & 0 & 0 \\ 0 & 1 & 0 \\ 0 & 0 & -1 \end{pmatrix} \begin{pmatrix} -1 & 0 & 0 \\ 0 & \cos\beta & -\sin\beta \\ 0 & -\sin\beta & -\cos\beta \end{pmatrix}$$

shows that a rotation by angle β around the axis l_1, oriented by e_1, is equal to the composition of the *axial symmetry* $S_{-\beta}$ of axis

$$l_{-\beta}: \quad P + \left\langle \cos\left(\frac{\beta}{2}\right) e_2 - \sin\left(\frac{\beta}{2}\right) e_3 \right\rangle$$

with the axial symmetry S_2 of axis l_2.

That is,

$$R(l_1, e_1, \beta) = S_2 \circ S_{-\beta}. \tag{7.2}$$

These two ways of decomposing a rotation as a composition of symmetries, (7.1) and (7.2), allow us to study the composition of rotations. Indeed, let us consider in an oriented three-dimensional Euclidean affine space two rotations with respect to the axes l, m concurrent in a point P, $R(l, u_1, \alpha)$ and $R(m, v_1, \beta)$.

From now on we shall write $R(u_1, \alpha)$ and $R(v_1, \beta)$ instead of $R(l, u_1, \alpha)$ and $R(m, v_1, \beta)$ respectively, since, once we have fixed the point P, it is clear that the axes of rotation are, respectively, l: $P + \langle u_1 \rangle$ and m: $P + \langle v_1 \rangle$.

Let us take $e \in E$ orthogonal to v_1 and u_1, and such that the basis $\mathcal{B}' = (v_1, u_1, e)$ is positive. We have chosen v_1 as the first element of this basis because we make the rotation $R(v_1, \beta)$ first, followed by the rotation $R(u_1, \alpha)$. We find e via the formula

$$e = \frac{v_1 \wedge u_1}{|v_1 \wedge u_1|},$$

where

$$v_1 \wedge u_1 = (bc' - b'c, a'c - ac', ab' - a'b),$$

$v_1 = (a, b, c)$ and $u_1 = (a', b', c')$ being the components of v_1 and u_1 in a positive orthonormal basis. This is the great advantage of the vector product: it gives a vector orthogonal to the two multiplied vectors, in such a way that the basis $(u, v, u \wedge v)$ is positive.

We define the vectors u_2, v_2 by the condition that the bases (u_1, e, u_2) and (v_1, e, v_2) are positive and orthonormal (see Figure 7.1).

Using formulas (7.1) and (7.2) and taking into account that e is orthogonal to u_1 and v_1, one obtains

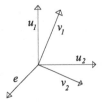

Figure 7.1. Orthonormal basis

$$R(u_1, \alpha) \circ R(v_1, \beta) = S_\alpha \circ S_e \circ S_e \circ S_{-\beta}$$

$$= S_\alpha \circ S_{-\beta},$$

where S_α is the symmetry of axis l_α, the straight line through P with direction

$$w_\alpha = \cos\left(\frac{\alpha}{2}\right) e + \sin\left(\frac{\alpha}{2}\right) u_2, \tag{7.3}$$

and $S_{-\beta}$ is the symmetry of axis $l_{-\beta}$, the straight line through P with direction

$$w_{-\beta} = \cos\left(\frac{\beta}{2}\right) e - \sin\left(\frac{\beta}{2}\right) v_2. \tag{7.4}$$

Observe that

$$\langle w_\alpha, w_{-\beta} \rangle = \cos\left(\frac{\alpha}{2}\right) \cos\left(\frac{\beta}{2}\right) - \sin\left(\frac{\alpha}{2}\right) \sin\left(\frac{\beta}{2}\right) \cos\delta,$$

where δ $(0 < \delta < \pi)$ is the angle between the vectors v_1 and u_1 (which we assume to be different and not opposite, in order to be able to define e). That is,

$$\langle v_1, u_1 \rangle = \cos\delta.$$

Let γ be the angle between the two axes of symmetry, that is, the angle between the straight lines l_α and $l_{-\beta}$. By definition of the angle between two straight lines, see Exercise 7.15, page 220, we have $0 < \gamma \le \frac{\pi}{2}$, and hence

$$\begin{cases} \cos\gamma = \langle w_\alpha, w_{-\beta} \rangle, & \text{if } \langle w_\alpha, w_{-\beta} \rangle > 0, \\ \cos\gamma = -\langle w_\alpha, w_{-\beta} \rangle, & \text{if } \langle w_\alpha, w_{-\beta} \rangle < 0. \end{cases}$$

But, again as we saw in Exercise 7.15, the composition of two symmetries of concurrent axes is a rotation around an axis perpendicular and concurrent with the other two axes. Concretely, we have

$$\boxed{R(u_1, \alpha) \circ R(v_1, \beta) = S_\alpha \circ S_{-\beta} = R(w, 2\gamma)}$$

where $R(w, 2\gamma)$ is the rotation around the third axis, oriented by the vector w, and angle 2γ, where γ is the angle between the axes of the symmetries and w

is the unit vector orthogonal to w_α and $w_{-\beta}$, given by

$$
\begin{cases}
w = \dfrac{w_{-\beta} \wedge w_\alpha}{|w_{-\beta} \wedge w_\alpha|} & \text{if } \langle w_{-\beta}, w_\alpha \rangle > 0, \\[4mm]
w = -\dfrac{w_{-\beta} \wedge w_\alpha}{|w_{-\beta} \wedge w_\alpha|} & \text{if } \langle w_{-\beta}, w_\alpha \rangle < 0,
\end{cases}
$$

with

$$
w_{-\beta} \wedge w_\alpha = \cos\left(\frac{\alpha}{2}\right)\sin\left(\frac{\beta}{2}\right) v_1 + \sin\left(\frac{\alpha}{2}\right)\cos\left(\frac{\beta}{2}\right) u_1
$$
$$
- \sin\left(\frac{\alpha}{2}\right)\sin\left(\frac{\beta}{2}\right)\sin(\delta) e.
$$

Since the initial data of the problem are the vectors v_1, u_1, it is useful to write the vectors w_α and $w_{-\beta}$ in terms of the basis (v_1, u_1, e). The expressions (7.3) and (7.4) are transformed, respectively, into

$$
w_\alpha = \cos\left(\frac{\alpha}{2}\right) e + \sin\left(\frac{\alpha}{2}\right)\frac{1}{\sin\delta} v_1 - \sin\left(\frac{\alpha}{2}\right)(\cot\delta) u_1,
$$
$$
w_{-\beta} = \cos\left(\frac{\beta}{2}\right) e - \sin\left(\frac{\beta}{2}\right)(\cot\delta) v_1 + \sin\left(\frac{\beta}{2}\right)\frac{1}{\sin\delta} u_1.
$$

EXERCISES

7.1. Let $\mathcal{R} = \{P; (e_1, e_2)\}$ be an orthonormal affine frame of a Euclidean affine space of dimension two. Prove that the equations of an orthogonal symmetry with respect to the straight line $l\colon P + \langle v \rangle$ are

$$
\begin{cases}
x' = x\cos(2\alpha) + y\sin(2\alpha), \\
y' = x\sin(2\alpha) - y\cos(2\alpha),
\end{cases}
$$

where $v = (v_1, v_2)$ is a unit vector with $v_2 \geq 0$ and $v_1 = \cos\alpha$.

7.2. In a Euclidean affine space of dimension two, what is the Euclidean motion obtained by the composition of two orthogonal symmetries, with axes that intersect with angle θ?

7.3. Prove that, in a Euclidean affine space of dimension two, every translation is a composition of symmetries with parallel axes; and every rotation is a composition of symmetries with axes that intersect at the center of the rotation. As a consequence, prove that every Euclidean motion is a composition of at most three symmetries.

7.4. Compare the decomposition of a rotation of the Euclidean affine plane as a composition of symmetries given in Section 7.4, page 202, with the decomposition of an elliptic affinity (in particular, a rotation) given in Section 3.9, page 117.

7.5. Classify the Euclidean motion of the Euclidean affine space $\mathbb{R}^2$ given, in the canonical affine frame, by

$$
\left(
\begin{array}{cc|c}
\cos(\frac{3\pi}{2}) & -\sin(\frac{3\pi}{2}) & 1 \\
\sin(\frac{3\pi}{2}) & \cos(\frac{3\pi}{2}) & 1 \\
\hline
0 & 0 & 1
\end{array}
\right).
$$

Recall that the notation $R(\alpha)$ has been introduced only for $0 < \alpha \le \pi$.

7.6. Find, in the canonical affine frame of the Euclidean affine space $\mathbb{R}^2$, the equations of:

(a) An anticlockwise rotation g of center $Q = (2,1)$ and angle $30°$ (the basis $(e_1, \tilde{g}(e_1))$ is negative with respect to the canonical basis (e_1, e_2) of the vector space $\mathbb{R}^2$, i.e., the determinant of the matrix of the change of basis is negative).

(b) An axial symmetry of axis $x + 2y - 2 = 0$.

7.7. Let f be a rotation by angle α in a Euclidean affine space of dimension two. Prove that $1 + 2\cos(\alpha) = \text{trace}(M)$, where M is the matrix of f in an affine frame.

Let f be a rotation by angle α around a given axis in a Euclidean affine space of dimension three. Prove that $2 + 2\cos(\alpha) = \text{trace}(M)$, where M is the matrix of f in an affine frame.

Let f be an anti-rotation by angle α in a Euclidean affine space of dimension three. Prove that $2\cos(\alpha) = \text{trace}(M)$, where M is the matrix of f in an affine frame.

7.8. Let S be an axial symmetry of the Euclidean affine space $\mathbb{R}^2$, with axis the straight line $L: P + \langle u \rangle$, and let v be a non-zero vector of the vector space $\mathbb{R}^2$. Prove that $v\widehat{S(v)} = 2\widehat{vu}$, where $\tilde{S}$ is the linear map associated to S. Here, the symbol "$\frown$" means "angle".

7.9. Let f be the axial symmetry of the Euclidean affine space $\mathbb{R}^2$ given by

$$
M(f, \mathcal{R}) = \left(
\begin{array}{cc|c}
1 & 0 & 0 \\
0 & -1 & 0 \\
\hline
0 & 0 & 1
\end{array}
\right),
$$

where $\mathcal{R} = \{P; (u,v)\}$, with $P = (0,0)$, $u = \frac{1}{2}(1, \sqrt{3})$ and $v = \frac{1}{2}(\sqrt{3}, -1)$. Find $M(f, \mathcal{C})$, where $\mathcal{C}$ is the canonical affine frame of the affine space $\mathbb{R}^2$.

7.10. Let $\mathcal{B} = (u, v)$ be a basis of the Euclidean vector space $\mathbb{R}^2$. Let us suppose that a rotation g of $\mathbb{R}^2$ is such that

$$M(\tilde{g}, \mathcal{B}) = \begin{pmatrix} 1 & -\frac{3}{2} \\ \frac{2}{3} & 0 \end{pmatrix},$$

where $\tilde{g}$ is the linear map associated to g. Is $\mathcal{B}$ an orthonormal basis? Find the angle of rotation of g, the angle between the vectors u, v, and the ratio of their moduli, $\frac{|u|}{|v|}$.

7.11. Find, in the canonical affine frame of the Euclidean affine space $\mathbb{R}^2$, the equations of a rotation $g : \mathbb{R}^2 \longrightarrow \mathbb{R}^2$ of angle $\pi/6$ radians in the anticlockwise sense.

(a) Let $\mathcal{C}$ be the canonical basis of the vector space $\mathbb{R}^2$. Let $\tilde{g}$ be the linear map associated to this rotation. Prove that the matrix $M(\tilde{g}, \mathcal{C})$ is an orthogonal matrix.

(b) Find $M(\tilde{g}, \mathcal{B})$, where $\mathcal{B} = (u, v)$ with

$$u = \frac{1}{\sqrt{2}}(1, 1), \qquad v = \frac{1}{\sqrt{2}}(-1, 1).$$

(c) Prove that the matrix $M(\tilde{g}, \mathcal{C}')$, where $\mathcal{C}'$ is an orthonormal basis of $\mathbb{R}^2$, can only take two values, depending on whether or not the orientation of $\mathcal{C}'$ coincides with that of the canonical basis.

7.12. Find, in the canonical affine frame of the Euclidean affine space $\mathbb{R}^3$, the equations of a rotation $g : \mathbb{R}^3 \longrightarrow \mathbb{R}^3$ around the z axis, of angle $\pi/6$ radians in the anticlockwise sense.

(a) Let $\mathcal{C}$ be the canonical basis of the vector space $\mathbb{R}^3$. Let $\tilde{g}$ be the linear map associated to this rotation. Prove that the matrix $M(\tilde{g}, \mathcal{C})$ is an orthogonal matrix.

(b) Find $M(\tilde{g}, \mathcal{B})$, where $\mathcal{B} = (u, v, w)$ with

$$u = \frac{1}{\sqrt{2}}(1, 1, 0), \qquad v = \frac{1}{\sqrt{2}}(-1, 1, 0), \qquad w = (0, 0, 1).$$

(c) Prove that, in contrast to the situation in dimension two (see the previous exercise), the matrix $M(\tilde{g}, \mathcal{C}')$, where $\mathcal{C}'$ is an orthonormal basis of $\mathbb{R}^3$, can take many values, depending on the orthonormal basis $\mathcal{C}'$ that we choose.

7.13. If f is a glide reflection of a Euclidean affine space of dimension three, prove that the glide vector is given by

$$u_f = \frac{1}{2}(\tilde{f} + \mathrm{id})\overrightarrow{Pf(P)},$$

where $\tilde{f}$ is the linear map associated to f and P is an arbitrary point.

7.14. Prove that, in a Euclidean affine space of dimension three, every translation is a composition of mirror symmetries with parallel symmetry planes; and every rotation with respect to an axis is a composition of mirror symmetries with the symmetry planes intersecting at this axis of rotation. As a consequence, prove that every Euclidean motion is a composition of at most four mirror symmetries.

7.15. Let S_u be the symmetry with respect to the straight line r: $P + \langle u \rangle$, and let S_v be the symmetry with respect to the straight line s: $P + \langle v \rangle$, in an oriented Euclidean affine space of dimension three. Assume that u and v are linearly independent unit vectors. Note that u and v are determined up to their sign. Recall that the angle between the two straight lines is, by definition, the unique angle γ, with $0 < \gamma \le \frac{\pi}{2}$, such that

$$|\langle u, v \rangle| = \cos \gamma.$$

Prove that

$$S_v \circ S_u = R(w, 2\gamma),$$

where $R(w, 2\gamma)$ is the rotation of vector w, and angle 2γ, where γ is the angle between the two straight lines r and s, and w is the unit vector orthogonal to u and v determined by the condition that (u, v, w) is either a positive basis if $\langle u, v \rangle > 0$, or negative if $\langle u, v \rangle < 0$. That is,

$$\begin{cases} w = \dfrac{u \wedge v}{|u \wedge v|} & \text{if } \langle u, v \rangle > 0, \\[2mm] w = -\dfrac{u \wedge v}{|u \wedge v|} & \text{if } \langle u, v \rangle < 0. \end{cases}$$

7.16. Classify the Euclidean motion of a Euclidean affine space $\mathbb{A}$ of dimension three given, in an orthonormal affine frame, by the following equations:

$$\begin{pmatrix} \bar{x} \\ \bar{y} \\ \bar{z} \end{pmatrix} = \begin{pmatrix} \frac{1}{3} & -\frac{2}{3} & -\frac{2}{3} \\ -\frac{2}{3} & \frac{1}{3} & -\frac{2}{3} \\ -\frac{2}{3} & -\frac{2}{3} & \frac{1}{3} \end{pmatrix} \begin{pmatrix} x \\ y \\ z \end{pmatrix} + \begin{pmatrix} 1 \\ 0 \\ -2 \end{pmatrix}.$$

Find the fixed points, the invariant straight lines, the glide vector and an orthonormal affine frame $\mathcal{R}$ such that $M(f, \mathcal{R})$ is a canonical matrix.

7.17. Let $\mathcal{R}$ be an orthonormal affine frame of a Euclidean affine space $\mathbb{A}$ of dimension three. Study the composition of the rotation

$$\begin{pmatrix} \overline{x} \\ \overline{y} \\ \overline{z} \end{pmatrix} = \begin{pmatrix} 0 & 0 & 1 \\ 1 & 0 & 0 \\ 0 & 1 & 0 \end{pmatrix} \begin{pmatrix} x \\ y \\ z \end{pmatrix} + \begin{pmatrix} 2 \\ -3 \\ 1 \end{pmatrix},$$

with the translation by the vector $v = (2, 3, -1)$.

7.18. Find, in the Euclidean affine space $\mathbb{R}^3$, the equations of a rotation by angle α around the straight line r: $(0, -3, -2) + \lambda(1, 1, 1)$. Suppose r is oriented by $v = (1, 1, 1)$.

7.19. Let $g = R(l, u, \alpha)$ be the rotation of the Euclidean affine space $\mathbb{R}^3$ around the straight line l: $P + \langle u \rangle$, oriented by the unit vector u, and angle α. Prove that

$$\tilde{g}(v) = \cos \alpha \cdot v + (1 - \cos \alpha)(\langle u, v \rangle) \cdot u + \sin \alpha \cdot (v \wedge u),$$

where $\tilde{g}$ is the linear map associated to g.

Given a point $Q \in \mathbb{R}^3$, find the coordinates of the point $Q' = g(Q)$, in the affine frame $\mathcal{R} = \{P; \mathcal{B}\}$, where $\mathcal{B} = (\overrightarrow{PQ}, u, \overrightarrow{PQ} \wedge u)$.

7.20. Prove that the affinities f of the Euclidean affine space $\mathbb{R}^2$ given in the canonical affine frame by the following matrices are Euclidean motions and classify them:

$$\begin{pmatrix} 1 & 0 & 1 \\ 0 & 1 & 0 \\ 0 & 0 & 1 \end{pmatrix}, \quad \begin{pmatrix} \sqrt{3}/2 & -1/2 & 0 \\ 1/2 & \sqrt{3}/2 & 0 \\ 0 & 0 & 1 \end{pmatrix}, \quad \begin{pmatrix} 1/2 & -\sqrt{3}/2 & 0 \\ \sqrt{3}/2 & 1/2 & 1 \\ 0 & 0 & 1 \end{pmatrix},$$

$$\begin{pmatrix} 0 & -1 & -1 \\ 1 & 0 & 0 \\ 0 & 0 & 1 \end{pmatrix}, \quad \begin{pmatrix} \frac{1+\sqrt{5}}{4} & -\sqrt{\frac{5-\sqrt{5}}{8}} & 0 \\ \sqrt{\frac{5-\sqrt{5}}{8}} & \frac{1+\sqrt{5}}{4} & 0 \\ 0 & 0 & 1 \end{pmatrix},$$

$$\begin{pmatrix} \frac{-1+\sqrt{5}}{4} & -\sqrt{\frac{5+\sqrt{5}}{8}} & 2 \\ \sqrt{\frac{5+\sqrt{5}}{8}} & \frac{-1+\sqrt{5}}{4} & 0 \\ 0 & 0 & 1 \end{pmatrix}, \quad \begin{pmatrix} \frac{1-\sqrt{5}}{4} & -\sqrt{\frac{5+\sqrt{5}}{8}} & 0 \\ \sqrt{\frac{5+\sqrt{5}}{8}} & \frac{1-\sqrt{5}}{4} & 0 \\ 0 & 0 & 1 \end{pmatrix}.$$

Find, in each case, the fixed points, the invariant straight lines, the glide vector and an orthonormal affine frame $\mathcal{R}$ such that $M(f, \mathcal{R})$ is a canonical matrix.

7.21. Prove that the affinities f of the Euclidean affine space $\mathbb{R}^3$ given in the canonical affine frame by the following matrices are Euclidean motions and classify them:

$$\left(\begin{array}{ccc|c} -1 & 0 & 0 & 1 \\ 0 & -1 & 0 & 0 \\ 0 & 0 & -1 & 0 \\ \hline 0 & 0 & 0 & 1 \end{array}\right), \qquad \left(\begin{array}{ccc|c} 1/3 & \sqrt{2}/\sqrt{3} & -\sqrt{2}/3 & 0 \\ \sqrt{2}/\sqrt{3} & 0 & 1/\sqrt{3} & 0 \\ -\sqrt{2}/3 & 1/\sqrt{3} & 2/3 & 0 \\ \hline 0 & 0 & 0 & 1 \end{array}\right),$$

$$\left(\begin{array}{ccc|c} -7/9 & 2\sqrt{2}/3\sqrt{3} & -2\sqrt{2}/9 & 0 \\ 2\sqrt{2}/3\sqrt{3} & 1/3 & -4/3\sqrt{3} & 1 \\ -2\sqrt{2}/9 & -4/3\sqrt{3} & -5/9 & 0 \\ \hline 0 & 0 & 0 & 1 \end{array}\right).$$

Find, in each case, the fixed points, the invariant straight lines, the glide vector and an orthonormal affine frame $\mathcal{R}$ such that $M(f,\mathcal{R})$ is a canonical matrix.

7.22. Classify the Euclidean motions of a Euclidean affine plane $\mathbb{A}$ given in an orthonormal affine frame by the following equations:

$$\begin{cases} x' = -2 + \dfrac{1}{2}x - \dfrac{\sqrt{3}}{2}y, \\ y' = 1 + \dfrac{\sqrt{3}}{2}x + \dfrac{1}{2}y. \end{cases} \qquad \begin{cases} x' = \dfrac{\sqrt{2}}{2}x + \dfrac{\sqrt{2}}{2}y, \\ y' = 1 + \dfrac{\sqrt{2}}{2}x - \dfrac{\sqrt{2}}{2}y. \end{cases}$$

Find, in each case, the fixed points, the invariant straight lines, the glide vector and an orthonormal affine frame $\mathcal{R}$ such that $M(f,\mathcal{R})$ is a canonical matrix.

Study the Euclidean motion obtained by the composition of the two previous Euclidean motions.

7.23. Classify the Euclidean motions of a Euclidean affine space $\mathbb{A}$ of dimension three given, in an orthonormal affine frame, by the following equations:

$$\begin{cases} x' = 1 + \dfrac{1}{2}x + \dfrac{\sqrt{2}}{2}y + \dfrac{1}{2}z, \\ y' = -1 + \dfrac{\sqrt{2}}{2}x - \dfrac{\sqrt{2}}{2}z, \\ z' = \dfrac{1}{2}x - \dfrac{\sqrt{2}}{2}y + \dfrac{1}{2}z. \end{cases} \qquad \begin{cases} x' = 1 + y, \\ y' = 1 - z, \\ z' = -x. \end{cases}$$

Find, in each case, the fixed points, the invariant straight lines, the glide vector and an orthonormal affine frame $\mathcal{R}$ such that $M(f,\mathcal{R})$ is a canonical matrix.

Study the Euclidean motion obtained by the composition of the two previous Euclidean motions.

7.24. Consider the following two affinities of the Euclidean affine plane $\mathbb{R}^2$:

$$f(x,y) = (y+1, x+1),$$

$$g(x,y) = \left(-\frac{4}{5}x + \frac{3}{5}y + 1, \frac{3}{5}x + \frac{4}{5}y + 3 \right).$$

(a) Find their fixed points and their invariant straight lines.

(b) Are there affine frames $\mathcal{R}$ and $\mathcal{R}'$ such that

$$M(f,\mathcal{R}) = M(g,\mathcal{R}')?$$

(c) Prove that f and g are Euclidean motions. Are they similar as Euclidean motions?

7.25. Consider the octahedron of the Euclidean affine space $\mathbb{R}^3$ with vertices $(1,0,0), (-1,0,0), (0,1,0), (0,-1,0), (0,0,1), (0,0,-1)$.

(a) Prove that Euclidean motions leaving the octahedron invariant form a subgroup G of the group of Euclidean motions of $\mathbb{R}^3$ having the origin $(0,0,0)$ as a fixed point.

(b) Prove that the elements of G leave the set of faces (respectively, edges and vertices) of the octahedron invariant.

(c) Prove that the subgroup of G comprising the orientation-preserving Euclidean motions is made up of twenty-four rotations. This group is called the *octahedral group*. Prove that there are twenty-four Euclidean motions in G that reverse the orientation.

(d) Find the equations of these twenty-four rotations. How many axes of rotation are there?

(e) How many symmetries with respect to a plane are there in G?

Affine Classification of Real Quadrics

8.1 Introduction

We have seen that hyperplanes of affine spaces are given by linear equations. The coordinates of their points are zeros of linear polynomials.

Quadrics, the objects that we are about to study, are zeros of quadratic polynomials, and hence they are in some sense the most natural objects to consider after hyperplanes.

Since linear varieties of any dimension can be considered as intersections of hyperplanes, it would be interesting to generalize this to the study of the intersection of quadrics, but we leave this for another occasion.

We restrict ourselves to the study of real affine spaces, that is, affine spaces modeled on $\mathbb{R}$-vector spaces. The calculations are similar, but simpler, over $\mathbb{C}$. See Exercise 8.11, page 281.

Let us begin with a short review of polynomials.

8.2 Quadratic Polynomials

A polynomial of second degree (quadratic) with n variables, and real coefficients, is an expression of the form

$$r(x_1, \ldots, x_n) = x^{\mathsf{T}} A x + B x + C,$$

A. Reventós Tarrida, *Affine Maps, Euclidean Motions and Quadrics*,
Springer Undergraduate Mathematics Series,
DOI 10.1007/978-0-85729-710-5_8, © Springer-Verlag London Limited 2011

where $A = (A_{ij}) \in \mathcal{M}_{n \times n}(\mathbb{R})$ is symmetric, $B = (B_1, \ldots, B_n) \in \mathcal{M}_{1 \times n}(\mathbb{R})$ (one row), $C \in \mathbb{R}$, and $x \in \mathcal{M}_{n \times 1}(k)$ (one column) given by

$$
x = \begin{pmatrix} x_1 \\ \vdots \\ x_n \end{pmatrix}.
$$

Thus,

$$
r(x_1, \ldots, x_n) = \begin{pmatrix} x_1 & \cdots & x_n \end{pmatrix} A \begin{pmatrix} x_1 \\ \vdots \\ x_n \end{pmatrix}
$$

$$
+ \begin{pmatrix} B_1 & \cdots & B_n \end{pmatrix} \begin{pmatrix} x_1 \\ \vdots \\ x_n \end{pmatrix} + C
$$

$$
= \sum_{i,j=1}^{n} A_{ij} x_i x_j + \sum_{i=1}^{n} B_i x_i + C. \tag{8.1}
$$

For instance, if $n = 2$, we have

$$
r(x_1, x_2) = \begin{pmatrix} x_1 & x_2 \end{pmatrix} \begin{pmatrix} A_{11} & A_{12} \\ A_{12} & A_{22} \end{pmatrix} \begin{pmatrix} x_1 \\ x_2 \end{pmatrix} + \begin{pmatrix} B_1 & B_2 \end{pmatrix} \begin{pmatrix} x_1 \\ x_2 \end{pmatrix} + C
$$

$$
= A_{11} x_1^2 + 2 A_{12} x_1 x_2 + A_{22} x_2^2 + B_1 x_1 + B_2 x_2 + C
$$

$$
= \sum_{i,j=1}^{2} A_{ij} x_i x_j + \sum_{i=1}^{2} B_i x_i + C,
$$

with $A_{ij}, B_i, C \in \mathbb{R}$ and $A_{12} = A_{21}$. We say that A is the matrix of the quadratic part and B the matrix of the linear part of $r(x)$.

Since polynomials are not usually written as in (8.1), but as

$$
r(x_1, \ldots, x_n) = \sum_{i,j=1,\ i \leq j}^{n} A_{ij} x_i x_j + \sum_{i=1}^{n} B_i x_i + C,
$$

if we want to write it matricially, we must divide by two the coefficients that are not on the diagonal of the quadratic part. For instance,

$$
3x_1^2 + x_1 x_2 + 2x_2 + 3 = \begin{pmatrix} x_1 & x_2 \end{pmatrix} \begin{pmatrix} 3 & 1/2 \\ 1/2 & 0 \end{pmatrix} \begin{pmatrix} x_1 \\ x_2 \end{pmatrix}
$$

$$
+ \begin{pmatrix} 0 & 2 \end{pmatrix} \begin{pmatrix} x_1 \\ x_2 \end{pmatrix} + 3.
$$

Finally, we say that a point $(a_1, \ldots, a_n) \in \mathbb{R}^n$ is a *zero* of the polynomial $r(x_1, \ldots, x_n)$ if $r(a_1, \ldots, a_n) = 0$.

Notation   From now on we shall simply write $r(x)$ instead of $r(x_1, \ldots, x_n)$, but note that in the expression

$$r(x) = x^{\mathsf{T}} A x + B x + C,$$

the x on the left and the x on the right are slightly different (the x on the right is a column matrix).

8.2.1 The Symmetric Matrix Associated to a Polynomial

Observe that every quadratic polynomial, $r(x) = x^{\mathsf{T}} A x + B x + C$, can be written as a product of matrices. Indeed, we have

$$r(x) = x^{\mathsf{T}} A x + B x + C = \begin{pmatrix} x^{\mathsf{T}} & 1 \end{pmatrix} \begin{pmatrix} A & B^{\mathsf{T}}/2 \\ B/2 & C \end{pmatrix} \begin{pmatrix} x \\ 1 \end{pmatrix}. \qquad (8.2)$$

To prove this equality, we need only recall the properties of the block product of matrices. Concretely, we have

$$\begin{pmatrix} x^{\mathsf{T}} & 1 \end{pmatrix} \begin{pmatrix} A & B^{\mathsf{T}}/2 \\ B/2 & C \end{pmatrix} \begin{pmatrix} x \\ 1 \end{pmatrix}$$

$$= \begin{pmatrix} x^{\mathsf{T}} & 1 \end{pmatrix} \begin{pmatrix} Ax + B^{\mathsf{T}}/2 \\ (B/2)x + C \end{pmatrix}$$

$$= x^{\mathsf{T}} A x + x^{\mathsf{T}} B^{\mathsf{T}}/2 + (B/2)x + C$$

$$= x^{\mathsf{T}} A x + B x + C.$$

For instance, if $n = 2$,

$$\begin{pmatrix} x_1 & x_2 & 1 \end{pmatrix} \begin{pmatrix} A_{11} & A_{12} & B_1/2 \\ A_{12} & A_{22} & B_2/2 \\ B_1/2 & B_2/2 & C \end{pmatrix} \begin{pmatrix} x_1 \\ x_2 \\ 1 \end{pmatrix}$$

$$= A_{11} x_1^2 + 2 A_{12} x_1 x_2 + A_{22} x_2^2 + B_1 x_1 + B_2 x_2 + C$$

$$= x^{\mathsf{T}} A x + B x + C$$

$$= r(x).$$

Thus, to pass from the polynomial to its associated matrix we must divide by two the coefficients that are not on the diagonal. For instance,

$$3x_1^2 + x_1x_2 + 2x_2 + 3 = \begin{pmatrix} x_1 & x_2 & 1 \end{pmatrix} \begin{pmatrix} 3 & 1/2 & 0 \\ 1/2 & 0 & 1 \\ 0 & 1 & 3 \end{pmatrix} \begin{pmatrix} cx_1 \\ x_2 \\ 1 \end{pmatrix}.$$

Since symmetric matrices are associated to symmetric bilinear maps, the study of quadrics reduces essentially to the study of these maps. The most important properties of symmetric bilinear maps are collected in Appendices B and C.

8.3 Definition of a Quadric

Definition 8.1

A *quadric*, in a real affine space $\mathbb{A}$ of dimension n, is the set of points of $\mathbb{A}$ such that their coordinates, in a given affine frame, are the zeros of a quadratic polynomial with n variables and real coefficients.

Hence, in order to give a quadric we must give an affine frame $\mathcal{R}$ of $\mathbb{A}$ and a quadratic polynomial $r(x)$ with n variables and real coefficients.

From now on we will consider only polynomials with real coefficients.

The quadric determined by the quadratic polynomial $r(x)$ and the affine frame $\mathcal{R}$, denoted by $\mathcal{Q}(r(x), \mathcal{R})$, is the subset of $\mathbb{A}$ given by

$$\mathcal{Q}(r(x), \mathcal{R}) = \{P \in \mathbb{A} : r(p) = 0\},$$

where $p = (p_1, \ldots, p_n)$ are the coordinates of the point P in the affine frame $\mathcal{R}$.

Once we fix an affine frame $\mathcal{R}$ of $\mathbb{A}$, we simply say that $\mathcal{Q}$ *is the quadric of the equation* $r(x) = 0$.

8.3.1 Polynomials or Zeros of Polynomials?

Is it possible that different polynomials give rise to the same quadric? That is, given an affine frame $\mathcal{R}$, can we have $\mathcal{Q}(r(x), \mathcal{R}) = \mathcal{Q}(s(x), \mathcal{R})$, with $r(x) \neq s(x)$?

It is clear that if we replace $r(x)$ by some multiple of it, $s(x) = \lambda r(x)$, we have $\mathcal{Q}(r(x), \mathcal{R}) = \mathcal{Q}(s(x), \mathcal{R})$. But, can we have this equality with $s(x) \neq \lambda r(x)$?

The answer is yes. For instance, the polynomials $r(x) = x_1^2 + x_2^2$ and $s(x) = x_1^2 + 5x_2^2$ satisfy $r(x) \neq \lambda s(x)$, for all $\lambda \in \mathbb{R}$, but the equations $x_1^2 + x_2^2 = 0$ and $x_1^2 + 5x_2^2 = 0$ have the same solution: $(x_1, x_2) = (0, 0)$.

That is, we have

$$\mathcal{Q}(r(x), \mathcal{C}) = \mathcal{Q}(s(x), \mathcal{C}), \quad \text{with } r(x) \neq \lambda s(x),$$

where $\mathcal{C}$ is the canonical affine frame of $\mathbb{R}^2$.

We must therefore be careful to avoid such ambiguous expressions as "let $r(x)$ be the polynomial defining $\mathcal{Q}$" or "assume that the polynomial defining $\mathcal{Q}$ has no linear part".

The normal course of action, and this is the path we shall follow, is to first study the polynomials $r(x)$ defining the quadrics, and not the quadrics themselves. Next, we pass from the polynomials to their zeros, that is, we shall study equations of the form $r(x) = 0$, taking into account that two quadratic polynomials with the same zeros are (almost always) proportional. This is the content of Hilbert's theorem, see Theorem D.4, page 399.

8.4 Change of Affine Frame

Let $\mathcal{R} = \{O; \mathcal{B}\}$ and $\mathcal{R}' = \{O'; \mathcal{B}'\}$ be two affine frames of a real affine space $\mathbb{A}$. We know, see (1.3) on page 19, that the points with coordinates x in $\mathcal{R}$ have coordinates x' in $\mathcal{R}'$ related by

$$x = ax' + b,$$

where $a = M(\mathcal{B}', \mathcal{B})$ is the matrix of the change of basis (the columns of a are the components, in $\mathcal{B}$, of the vectors of $\mathcal{B}'$), and $b = C(O', \mathcal{R})$ is the column matrix formed by the coordinates, in $\mathcal{R}$, of O'.

Recall that this relationship can be written as

$$\begin{pmatrix} x \\ 1 \end{pmatrix} = \begin{pmatrix} a & b \\ 0 & 1 \end{pmatrix} \begin{pmatrix} x' \\ 1 \end{pmatrix}. \tag{8.3}$$

Let $r(x)$ be a quadratic polynomial and let us consider the quadric $\mathcal{Q}(r(x), \mathcal{R})$. Take a point $P \in \mathcal{Q}(r(x), \mathcal{R})$. Its coordinates in $\mathcal{R}$, $p = (p_1, \ldots, p_n)$, satisfy the equation $r(p) = 0$. Let us find the equation satisfied by the coordinates of P in $\mathcal{R}'$, $p' = (p_1', \ldots, p_n')$.

If $r(x) = x^\mathsf{T} A x + B x + C$, the equation $r(p) = 0$ can be written as

$$r(p) = \begin{pmatrix} p^\mathsf{T} & 1 \end{pmatrix} \begin{pmatrix} A & B^\mathsf{T}/2 \\ B/2 & C \end{pmatrix} \begin{pmatrix} p \\ 1 \end{pmatrix} = 0. \tag{8.4}$$

The relationship between the coordinates of the point $P \in \mathbb{A}$ in the affine frames $\mathcal{R}$ and $\mathcal{R}'$ is given by

$$\begin{pmatrix} p \\ 1 \end{pmatrix} = \begin{pmatrix} a & b \\ 0 & 1 \end{pmatrix} \begin{pmatrix} p' \\ 1 \end{pmatrix}. \tag{8.5}$$

Substituting (8.5) in (8.4) we obtain

$$r(p) = (p'^{\mathsf{T}} \quad 1) \begin{pmatrix} a^{\mathsf{T}} & 0 \\ b^{\mathsf{T}} & 1 \end{pmatrix} \begin{pmatrix} A & B^{\mathsf{T}}/2 \\ B/2 & C \end{pmatrix} \begin{pmatrix} a & b \\ 0 & 1 \end{pmatrix} \begin{pmatrix} p' \\ 1 \end{pmatrix}$$

$$= (p'^{\mathsf{T}} \quad 1) \begin{pmatrix} a^{\mathsf{T}} & 0 \\ b^{\mathsf{T}} & 1 \end{pmatrix} \begin{pmatrix} Aa & Ab + B^{\mathsf{T}}/2 \\ (B/2)a & (B/2)b + C \end{pmatrix} \begin{pmatrix} p' \\ 1 \end{pmatrix}$$

$$= (p'^{\mathsf{T}} \quad 1) \begin{pmatrix} a^{\mathsf{T}}Aa & a^{\mathsf{T}}Ab + a^{\mathsf{T}}B^{\mathsf{T}}/2 \\ b^{\mathsf{T}}Aa + (B/2)a & b^{\mathsf{T}}Ab + b^{\mathsf{T}}B^{\mathsf{T}}/2 + (B/2)b + C \end{pmatrix} \begin{pmatrix} p' \\ 1 \end{pmatrix}$$

$$= (p'^{\mathsf{T}} \quad 1) \begin{pmatrix} A' & B'^{\mathsf{T}}/2 \\ B'/2 & C' \end{pmatrix} \begin{pmatrix} p' \\ 1 \end{pmatrix}$$

$$= p'^{\mathsf{T}}A'q' + B'q' + C' = 0,$$

with

$$A' = a^{\mathsf{T}}Aa,$$
$$B' = 2b^{\mathsf{T}}Aa + Ba, \tag{8.6}$$
$$C' = b^{\mathsf{T}}Ab + Bb + C.$$

(Notice that $Ba = a^{\mathsf{T}}B^{\mathsf{T}}$ and $b^{\mathsf{T}}Aa = a^{\mathsf{T}}Ab$.)

That is, if $p = (p_1, \ldots, p_n)$ is a zero of the polynomial

$$r(x) = x^{\mathsf{T}}Ax + Bx + C,$$

then $p' = (p'_1, \ldots, p'_n)$ is a zero of the polynomial

$$r'(x) = x^{\mathsf{T}}A'x + B'x + C',$$

with A', B', C' given by (8.6).

In summary, given two affine frames $\mathcal{R} = \{O; \mathcal{B}\}$ and $\mathcal{R}' = \{O'; \mathcal{B}'\}$, with $M(\mathcal{B}', \mathcal{B}) = a, C(O', \mathcal{R}) = b$, and two polynomials

$$r(x) = x^{\mathsf{T}}Ax + Bx + C,$$

$$r'(x) = x^{\mathsf{T}}A'x + B'x + C',$$

with A', B', C' related to A, B, C, a, b by relations (8.6), we have

$$\mathcal{Q}(r(x), \mathcal{R}) = \mathcal{Q}(r'(x), \mathcal{R}').$$

Notice that $r'(x) = r(ax + b)$, and hence we have

$$\boxed{\mathcal{Q}(r(x), \mathcal{R}) = \mathcal{Q}(r(ax + b), \mathcal{R}')} \qquad (8.7)$$

Hence, from the point of view of the study of quadrics, it is natural to say that these two polynomials $r(x)$ and $r(ax + b)$ are, in a certain sense, equivalent. We shall do this in Definition 8.2.

Since the affine frame $\mathcal{R}'$ is in fact the image of the affine frame $\mathcal{R}$ under the affinity with equations $x' = ax + b$, the results of this section can be reformulated in terms of affinities. We shall do this in the following section.

8.5 Image of a Quadric Under an Affinity

Let $\mathcal{R}$ be an affine frame of a real affine space $\mathbb{A}$ of dimension n. Let $f : \mathbb{A} \longrightarrow \mathbb{A}$ be a bijective affinity, given in the affine frame $\mathcal{R}$ by the equation

$$x' = ax + b,$$

where $a \in M_{n \times n}(\mathbb{R})$ is invertible and $b \in M_{n \times 1}(\mathbb{R})$ (one column).

Let $r(x)$ be a quadratic polynomial and let us consider the quadric $\mathcal{Q}$ of the equation $r(x) = 0$.

The question is: what is the equation satisfied by the coordinates of the points of $f^{-1}(\mathcal{Q})$?

Since a point $P \in f^{-1}(\mathcal{Q})$ if and only if $f(P) \in \mathcal{Q}$, the coordinates of $f(P)$ satisfy the equation $r(x) = 0$.

But since the relationship between the coordinates of the point P and those of the point $f(P)$ is given by $p' = ap + b$, we have $r(p') = r(ap + b) = 0$.

This proves

$$f^{-1}(\mathcal{Q}(r(x), \mathcal{R})) = \mathcal{Q}(r(ax + b), \mathcal{R}) \qquad (8.8)$$

or, equivalently,

$$\boxed{f(\mathcal{Q}(r(ax + b), \mathcal{R})) = \mathcal{Q}(r(x), \mathcal{R})} \qquad (8.9)$$

Summarizing, *the image of the quadric of the equation $r(ax + b) = 0$ under a bijective affinity with equation $x' = ax + b$ is the quadric of the equation $r(x) = 0$.* Recall that the relationship between the coefficients of the polynomials $r(x)$ and $r'(x) = r(ax + b)$ is given by (8.6).

For instance, the affinity of the affine space $\mathbb{R}^2$,

$$x'_1 = 2x_1 + 1,$$
$$x'_2 = x_2,$$

transforms the quadric of the equation $4x_1^2 + x_2^2 + 4x_1 + 1 = 0$ into the quadric of the equation $r(x) = x_1^2 + x_2^2 = 1$. In fact, the polynomial defining the first quadric is equal to $r(ax + b)$, with

$$a = \begin{pmatrix} 2 & 0 \\ 0 & 1 \end{pmatrix}, \qquad b = \begin{pmatrix} 1 \\ 0 \end{pmatrix}.$$

Again, as we mentioned in the previous section, it is natural to say that the polynomials $r(x)$ and $r(ax + b)$ are equivalent.

8.6 Equivalent Quadratic Polynomials

In this and in the following sections, until Section 8.12, we shall make no reference to affine spaces, instead confining our discussion to polynomials. Inspired by the calculations in the two previous sections, we define an equivalence relation between quadratic polynomials.

Definition 8.2

The quadratic polynomials in n variables

$$r(x) = x^{\mathsf{T}} A x + B x + C,$$

$$s(x) = x^{\mathsf{T}} A' x + B' x + C'$$

are *equivalent* if and only if there is an invertible $a \in \mathcal{M}_{n \times n}(\mathbb{R})$, $b \in \mathcal{M}_{n \times 1}(\mathbb{R})$ (one column), and a real number $\lambda \neq 0$, such that

$$r(ax + b) = \lambda s(x).$$

That is, two polynomials are equivalent if we can formally transform one of them into a multiple of the other by an affine change of variables.

Proposition 8.3

The quadratic polynomials in n variables

$$r(x) = x^{\mathsf{T}} A x + B x + C,$$

$$s(x) = x^{\mathsf{T}} A' x + B' x + C',$$

are equivalent if and only if there is an invertible $a \in \mathcal{M}_{n \times n}(\mathbb{R})$, $b \in \mathcal{M}_{n \times 1}(\mathbb{R})$ (one column), and a real number $\lambda \neq 0$, such that

$$\begin{pmatrix} a^\mathsf{T} & 0 \\ b^\mathsf{T} & 1 \end{pmatrix} \begin{pmatrix} A & B^\mathsf{T}/2 \\ B/2 & C \end{pmatrix} \begin{pmatrix} a & b \\ 0 & 1 \end{pmatrix} = \lambda \begin{pmatrix} A' & B'^\mathsf{T}/2 \\ B'/2 & C' \end{pmatrix}. \tag{8.10}$$

Proof

If $r(x)$ and $s(x)$ are equivalent, we have $r(ax + b) = \lambda s(x)$. We have seen, on page 230, that

$$r(ax + b) = \begin{pmatrix} x & 1 \end{pmatrix} \begin{pmatrix} a^\mathsf{T} & 0 \\ b^\mathsf{T} & 1 \end{pmatrix} \begin{pmatrix} A & B^\mathsf{T}/2 \\ B/2 & C \end{pmatrix} \begin{pmatrix} a & b \\ 0 & 1 \end{pmatrix} \begin{pmatrix} x \\ 1 \end{pmatrix}.$$

Since

$$s(x) = \begin{pmatrix} x & 1 \end{pmatrix} \begin{pmatrix} A' & B'^\mathsf{T}/2 \\ B'/2 & C' \end{pmatrix} \begin{pmatrix} x \\ 1 \end{pmatrix},$$

the proof is complete. □

The relation defined above is clearly an equivalence relation.

Corollary 8.4

The quadratic polynomials in n variables

$$r(x) = x^\mathsf{T} A x + B x + C,$$

$$s(x) = x^\mathsf{T} A' x + B' x + C'$$

are equivalent if and only if there is an invertible $a \in \mathcal{M}_{n \times n}(\mathbb{R})$, $b \in \mathcal{M}_{n \times 1}(\mathbb{R})$ (one column), and a real number $\lambda \neq 0$, such that

$$\lambda A' = a^\mathsf{T} A a,$$

$$\lambda B' = 2b^\mathsf{T} A a + B a,$$

$$\lambda C' = b^\mathsf{T} A b + B b + C.$$

Proof

Multiply the matrices of (8.10). □

Notice that $r(x)$ is equivalent to $\mu r(x)$, for all $\mu \in \mathbb{R}$, $\mu \neq 0$, as can be seen by taking $a = I_n$, $b = 0$, $\lambda = \frac{1}{\mu}$.

Corollary 8.5

The quadratic polynomials in n variables and without linear part

$$r(x) = x^\mathsf{T} A x + C,$$

$$s(x) = x^\mathsf{T} A' x + C'$$

are equivalent if and only if there is an invertible $a \in \mathcal{M}_{n \times n}(\mathbb{R})$ and a real number $\lambda \neq 0$ such that $\lambda C' = C$ and $\lambda A' = a^\mathsf{T} A a$.

Proof

The polynomials $r(x)$ and $s(x)$ are equivalent if and only if there is an invertible $n \times n$ real matrix a, a real column matrix b and a real number λ such that

$$\lambda A' = a^\mathsf{T} A a,$$

$$0 = 2b^\mathsf{T} A a,$$

$$\lambda C' = b^\mathsf{T} A b + C.$$

Since a is invertible, the second equation implies $b^\mathsf{T} A = 0$, and hence $\lambda C' = C$. To prove the converse, we just take $b = 0$. $\qquad\square$

In particular, polynomials of the form $r(x) = x^\mathsf{T} A x$ cannot be equivalent to polynomials of the form $r(x) = x^\mathsf{T} A' x + C$, with $C \neq 0$.

Observation 8.6

Polynomials that differ only by a permutation of variables are equivalent. For instance, if in the polynomial with n variables $r(x) = x^\mathsf{T} A x + B x + C$ we permute x_1 and x_2, we obtain the polynomial $s(x) = x^\mathsf{T} A' x + B' x + C$, with $A' = a^\mathsf{T} A a$ and $B' = B a$, where

$$a = \begin{pmatrix} I_2' & 0 \\ 0 & I_{n-2} \end{pmatrix},$$

and

$$I_2' = \begin{pmatrix} 0 & 1 \\ 1 & 0 \end{pmatrix}.$$

Hence, $r(x)$ and $s(x)$ are equivalent, since they satisfy (8.6), with $b = 0$. For instance, if $n = 2$, we have

$$a^\mathsf{T} Aa = \begin{pmatrix} 0 & 1 \\ 1 & 0 \end{pmatrix} \begin{pmatrix} a_{11} & a_{12} \\ a_{12} & a_{22} \end{pmatrix} \begin{pmatrix} 0 & 1 \\ 1 & 0 \end{pmatrix} = \begin{pmatrix} a_{22} & a_{12} \\ a_{12} & a_{11} \end{pmatrix},$$

and

$$Ba = \begin{pmatrix} b_1 & b_2 \end{pmatrix} \begin{pmatrix} 0 & 1 \\ 1 & 0 \end{pmatrix} = \begin{pmatrix} b_2 & b_1 \end{pmatrix}.$$

We have transformed the polynomial $a_{11}x_1^2 + 2a_2x_1x_2 + a_{22}x_2^2 + b_1x_1 + b_2x_2 + C$ into the polynomial $a_{11}x_2^2 + 2a_2x_1x_2 + a_{22}x_1^2 + b_1x_2 + b_2x_1 + C$.

If, instead of permuting x_1 and x_2, we permute x_i and x_j, the matrix a performing the permutation of the variables x_i and x_j is obtained by permuting rows i and j in the identity matrix I_n.

Example 8.7

The polynomials

$$r(x) = (x_1 - 1)^2 + (x_2 - 1)^2 - 4,$$

$$s(x) = x_1^2 + x_2^2 - 1,$$

are equivalent, since the change of variables

$$x_1 = 2x_1' + 1,$$

$$x_2 = 2x_2' + 1,$$

gives

$$r(x) = r(2x_1' + 1, 2x_2' + 1) = 4x_1'^2 + 4x_2'^2 - 4 = 4s(x').$$

We can also see this directly from Proposition 8.3. It is sufficient to observe that the above change of variables can be written as

$$\begin{pmatrix} x_1 \\ x_2 \end{pmatrix} = \begin{pmatrix} 2 & 0 \\ 0 & 2 \end{pmatrix} \begin{pmatrix} x_1' \\ x_2' \end{pmatrix} + \begin{pmatrix} 1 \\ 1 \end{pmatrix},$$

or, equivalently,

$$\begin{pmatrix} x_1 \\ x_2 \\ 1 \end{pmatrix} = \left(\begin{array}{cc|c} 2 & 0 & 1 \\ 0 & 2 & 1 \\ \hline 0 & 0 & 1 \end{array} \right) \begin{pmatrix} x_1' \\ x_2' \\ 1 \end{pmatrix}. \tag{8.11}$$

But, since

$$r(x) = \begin{pmatrix} x_1 & x_2 & 1 \end{pmatrix} \begin{pmatrix} 1 & 0 & -1 \\ 0 & 1 & -1 \\ -1 & -1 & -2 \end{pmatrix} \begin{pmatrix} x_1 \\ x_2 \\ 1 \end{pmatrix}, \qquad (8.12)$$

and

$$s(x) = \begin{pmatrix} x_1 & x_2 & 1 \end{pmatrix} \begin{pmatrix} 1 & 0 & 0 \\ 0 & 1 & 0 \\ 0 & 0 & -1 \end{pmatrix} \begin{pmatrix} x_1 \\ x_2 \\ 1 \end{pmatrix},$$

substituting (8.11) into (8.12) we obtain

$$\begin{pmatrix} 2 & 0 & 0 \\ 0 & 2 & 0 \\ 1 & 1 & 1 \end{pmatrix} \begin{pmatrix} 1 & 0 & -1 \\ 0 & 1 & -1 \\ -1 & -1 & -2 \end{pmatrix} \begin{pmatrix} 2 & 0 & 1 \\ 0 & 2 & 1 \\ 0 & 0 & 1 \end{pmatrix} = 4 \begin{pmatrix} 1 & 0 & 0 \\ 0 & 1 & 0 \\ 0 & 0 & -1 \end{pmatrix},$$

which is precisely condition (8.10) with $\lambda = 4$.

8.7 Invariants

We recall that the *index* of a real symmetric matrix A is the number

$$\text{index}(A) = \min\{r^+(A), r^-(A)\},$$

where $r^+(A)$ is the number of positive eigenvalues of A, and $r^-(A)$ is the number of negative eigenvalues of A. In particular, $\text{rank}(A) = r^+(A) + r^-(A)$. For the definition of the index of a symmetric bilinear map, see page 363, and for the definition of the index of a real symmetric matrix, see page 366.

By Descartes' theorem, Theorem C.12, page 389, to find the index of a real symmetric matrix A it is not necessary to compute the eigenvalues of A. We need only look at the number of sign differences between consecutive non-zero coefficients of its characteristic polynomial (Corollary C.13).

Definition 8.8

Given a polynomial $r(x) = x^T A x + B x + C$, let

$$\rho = \text{rank}(A),$$

$$i = \text{index}(A),$$

$$\tilde{\rho} = \operatorname{rank}(\tilde{A}),$$

$$\tilde{i} = \operatorname{index}(\tilde{A}),$$

where $\tilde{A} = \left(\begin{smallmatrix} A & B^{\mathsf{T}}/2 \\ B/2 & C \end{smallmatrix}\right)$. We say that ρ and i are respectively the quadratic rank and the quadratic index of $r(x)$ (because they depend only on the quadratic part), and that $\tilde{\rho}$ and $\tilde{i}$ are respectively the rank and the index of $r(x)$.

Observe that $\tilde{\rho}$ can only be equal to ρ, $\rho+1$ or $\rho+2$, and that $\tilde{i}$ can only be equal to i or $i+1$.

We say that these four numbers $\rho, i, \tilde{\rho}, \tilde{i}$ are *invariants* associated with the polynomial because, as we shall see immediately, they are the same for equivalent polynomials.

Theorem 8.9

Let us assume that the polynomials in n variables

$$r(x) = x^{\mathsf{T}} A x + B x + C,$$

$$s(x) = x^{\mathsf{T}} A' x + B' x + C',$$

are equivalent, and let

$$\tilde{A} = \begin{pmatrix} A & B^{\mathsf{T}}/2 \\ B/2 & C \end{pmatrix}, \qquad \tilde{A}' = \begin{pmatrix} A' & B'^{\mathsf{T}}/2 \\ B'/2 & C' \end{pmatrix}.$$

Then

$$\operatorname{rank}(A) = \operatorname{rank}(A'),$$

$$\operatorname{index}(A) = \operatorname{index}(A'),$$

$$\operatorname{rank}(\tilde{A}) = \operatorname{rank}(\tilde{A}'),$$

$$\operatorname{index}(\tilde{A}) = \operatorname{index}(\tilde{A}').$$

Proof

There exists an $n \times n$ invertible real matrix a and an $(n+1) \times (n+1)$ invertible real matrix c such that

$$\lambda A' = a^{\mathsf{T}} A a,$$

$$\lambda \tilde{A}' = c^{\mathsf{T}} \tilde{A} c.$$

This implies the respective equalities of ranks, because the rank of a matrix is invariant under left and right multiplication by invertible matrices, see for instance [8].

By Proposition B.21, page 366, we know that

$$r^+(\lambda A') = r^+(A),$$

$$r^+(\lambda \tilde{A}') = r^+(\tilde{A}).$$

But, for any real symmetric matrix P we have

$$r^+(\lambda P) = \begin{cases} r^+(P) & \lambda > 0, \\ r^-(P) & \lambda < 0. \end{cases}$$

Hence,

$$\text{index}(A') = \min\{r^+(A'), r^-(A')\} = \min\{r^+(A), r^-(A)\} = \text{index}(A),$$

and

$$\text{index}(\tilde{A}') = \min\{r^+(\tilde{A}'), r^-(\tilde{A}')\} = \min\{r^+(\tilde{A}), r^-(\tilde{A})\} = \text{index}(\tilde{A}),$$

and this completes the proof. $\square$

Observe that the invariant $r^+(A)$, so useful in the classification of symmetric bilinear maps, has had to be replaced by the index, as a consequence of the possible change of sign coming from the scalar λ appearing in Definition 8.2.

8.8 Canonical Representatives Without Linear Part

In this and in the following section we shall study the equivalence classes of polynomials induced by the equivalence relation introduced in Definition 8.2. We shall see that within each class there is an especially simple polynomial that will be called the *canonical representative* of the class.

Theorem 8.10

Let $r(x) = x^{\mathsf{T}} A x + B x + C$ be a quadratic polynomial with

$$\text{rank}(A) = \text{rank}(A|B^{\mathsf{T}}).$$

Then $r(x)$ is equivalent to a polynomial of the form

$$s(x) = x^\mathsf{T} D x + C',$$

where D is a diagonal matrix with only ± 1 and zeros on the diagonal, with $r^+(D) \geq r^-(D)$, and C' is a constant.

Proof

By Corollary 8.4, to find a polynomial without linear part equivalent to

$$r(x) = x^\mathsf{T} A x + B x + C,$$

and only with quadratic terms of the form $\pm x_i^2$, is equivalent to finding an invertible real matrix a and a real column matrix b such that
(1) $a^\mathsf{T} A a = D$, D diagonal, with ± 1 and 0s on the diagonal,
(2) $(2b^\mathsf{T} A + B) a = 0$.

   The existence of a matrix a satisfying condition (1) is an immediate consequence of Proposition B.13, page 359.

   Since a is invertible, condition (2) is satisfied if and only if

$$2b^\mathsf{T} A + B = 0,$$

or, transposing, if and only if

$$2Ab = -B^\mathsf{T}.$$

   Hence, we can take as column matrix b satisfying the second condition any solution of the system

$$\boxed{2Ax = -B^\mathsf{T}} \tag{8.13}$$

This system has a solution, because we are assuming $\operatorname{rank}(A) = \operatorname{rank}(A|B^\mathsf{T})$.

   Notice that if $\det A \neq 0$, the condition on the ranks is automatically satisfied and, moreover, system (8.13) has a unique solution. This system is called the *equations of the center*, for reasons that will be revealed below; see Proposition 8.36, page 267.

   Summarizing, we have found an invertible real matrix a and a real column matrix b such that

$$A' = a^\mathsf{T} A a = D,$$

$$B' = (2b^\mathsf{T} A + B) a = 0,$$

$$C' = (b^\mathsf{T} A + B) b + C = \frac{1}{2} B b + C.$$

Hence, the given polynomial $r(x)$ is equivalent to

$$s(x) = x^\mathsf{T} D x + C',$$

with $C' = \frac{1}{2}Bb + C$. If $r^+(D) \geq r^-(D)$, we have finished. Otherwise, we need only change the signs, since $s(x)$ is equivalent to $-s(x)$, and this polynomial is given by

$$-s(x) = x^\mathsf{T}(-D)x - C',$$

and hence we have $r^+(-D) \geq r^-(-D)$. This completes the proof. $\qquad\square$

Observation 8.11

Observe that the condition on the rank that appears in the statement of Theorem 8.10 is invariant within the equivalence class. That is, if $r(x) = x^\mathsf{T} A x + B x + C$ satisfies $\mathrm{rank}(A) = \mathrm{rank}(A|B^\mathsf{T})$ and $s(x) = x^\mathsf{T} A' x + B' x + C'$ is equivalent to $r(x)$, then $\mathrm{rank}(A') = \mathrm{rank}(A'|B'^\mathsf{T})$.

In fact, by Theorem 8.9, we have $\mathrm{rank}(A) = \mathrm{rank}(A')$. On the other hand, equality (8.10) implies

$$\lambda(A'|B'^\mathsf{T}) = a^\mathsf{T}(A|B^\mathsf{T}) \begin{pmatrix} a & b \\ 0 & 1 \end{pmatrix}, \tag{8.14}$$

and, hence, by the invariance of the rank under multiplication by invertible matrices, $\mathrm{rank}(A'|B'^\mathsf{T}) = \mathrm{rank}(A|B^\mathsf{T})$. Hence, $\mathrm{rank}(A) = \mathrm{rank}(A|B^\mathsf{T})$ implies $\mathrm{rank}(A') = \mathrm{rank}(A'|B'^\mathsf{T})$.

8.8.1 Canonical Expression

We have seen that every polynomial $r(x) = x^\mathsf{T} A x + B x + C$, with

$$\mathrm{rank}(A) = \mathrm{rank}(A|B^\mathsf{T}),$$

is equivalent to a polynomial of the form

$$s(x) = x^\mathsf{T} D x + C',$$

where D is diagonal, with ± 1 and zeros on the diagonal, with at least as many 1s as -1s.

If we permute the elements of the diagonal of D we obtain a polynomial equivalent to $s(x)$, see Observation 8.6; hence, we shall assume that $s(x)$ is written as

$$s(x) = (x_1^2 - x_2^2) + \cdots + (x_{2i-1}^2 - x_{2i}^2) + x_{2i+1}^2 + \cdots + x_\rho^2 + C' \tag{8.15}$$

with $i = \text{index}(A) = \text{index}(D)$, $\rho = \text{rank}(A) = \text{rank}(D)$, and $C' \in \mathbb{R}$.
This expression includes the case

$$s(x) = (x_1^2 - x_2^2) + \cdots + (x_{2i-1}^2 - x_{2i}^2) + C',$$

when $\rho = 2i$.

If we reach this point with $C' = 0$, we can say that the given polynomial $r(x) = x^\mathsf{T} A x + B x + C$ is equivalent to

$$\boxed{s(x) = (x_1^2 - x_2^2) + \cdots + (x_{2i-1}^2 - x_{2i}^2) + x_{2i+1}^2 + \cdots + x_\rho^2} \tag{8.16}$$

If $C' > 0$, we make the change of variables

$$x_k' = \frac{x_k}{\sqrt{C'}},$$

and, by Definition 8.2, polynomial (8.15) is equivalent to

$$\boxed{t(x) = (x_1^2 - x_2^2) + \cdots + (x_{2i-1}^2 - x_{2i}^2) + x_{2i+1}^2 + \cdots + x_\rho^2 + 1} \tag{8.17}$$

If $C' < 0$, we make the change of variables

$$x_{2k-1}' = \frac{x_{2k}}{\sqrt{-C'}}, \quad k = 1, \ldots, i,$$

$$x_{2k}' = \frac{x_{2k-1}}{\sqrt{-C'}}, \quad k = 1, \ldots, i,$$

$$x_j' = \frac{x_j}{\sqrt{-C'}}, \quad j = 2i + 1, \ldots, \rho,$$

and, by Definition 8.2, polynomial (8.15) is equivalent to

$$\boxed{t(x) = (x_1^2 - x_2^2) + \cdots + (x_{2i-1}^2 - x_{2i}^2) - x_{2i+1}^2 - \cdots - x_\rho^2 + 1} \tag{8.18}$$

We observe that when $\rho = 2i$, the expressions (8.17) and (8.18) coincide. Both can be written as

$$t(x) = (x_1^2 - x_2^2) + \cdots + (x_{2i-1}^2 - x_{2i}^2) + 1.$$

To avoid repetitions, we shall consider this expression as a particular case of (8.17) with $\rho = 2i$, and when we write the expression (8.18), we shall assume $2i < \rho$.

8.9 Canonical Representatives with Linear Part

Theorem 8.12

Let

$$r(x) = x^{\mathsf{T}} A x + B x + C$$

be a quadratic polynomial with

$$\mathrm{rank}(A) \neq \mathrm{rank}(A|B^{\mathsf{T}}).$$

Then $r(x)$ is equivalent to a polynomial of the form

$$s(x) = x^{\mathsf{T}} D x + B' x,$$

where D is a diagonal matrix with ± 1 in the first ρ terms of the diagonal ($\rho = \mathrm{rank}(A)$) and zeros elsewhere, and $B' = (0, \ldots, 0, 1, 0, \ldots, 0)$ (the 1 in the $\rho + 1$-th position).

Proof

We know, by Proposition B.13, page 359, that there is an invertible real matrix a such that the matrix

$$D = a^{\mathsf{T}} A a$$

is diagonal with ± 1 in the first ρ terms of the diagonal and zeros elsewhere. Hence, the given polynomial $r(x)$ is equivalent to the polynomial

$$s(x) = x^{\mathsf{T}} A' x + B' x + C'$$

given by

$$A' = D,$$

$$B' = Ba,$$

$$C' = C,$$

(Corollary 8.4, with $\lambda = 1$, and $b = 0$), and it is also equivalent to the polynomial

$$s(x) = x^{\mathsf{T}} A' x + B' x + C'$$

given by

$$A' = -D,$$

$$B' = -Ba,$$

$$C' = -C,$$

(Corollary 8.4, with $\lambda = -1$, and $b = 0$). Hence, $r(x)$ is equivalent to a polynomial $s(x) = x^{\mathsf{T}} Dx + B'x + C'$ such that the diagonal matrix D of the quadratic part has at least as many 1s as -1s.

Permuting, if necessary, the elements of the diagonal of D, see Observation 8.6, we may assume that $r(x)$ is equivalent to the polynomial

$$s(x) = (x_1^2 - x_2^2) + \cdots + (x_{2i-1}^2 - x_{2i}^2) + x_{2i+1}^2 + \cdots + x_\rho^2 + b_1' x_1 + \cdots + b_n' x_n + C',$$

with $i = \text{index}(A) = \text{index}(D)$ and $\rho = \text{rank}(A) = \text{rank}(D)$.

Then we can make the change of variables

$$
\begin{aligned}
x_{2k-1}' &= x_{2k-1} + \frac{b_{2k-1}'}{2}, & k &= 1, \ldots, i, \\
x_{2k}' &= x_{2k} - \frac{b_{2k}'}{2}, & k &= 1, \ldots, i, \\
x_j' &= x_j + \frac{b_j'}{2}, & j &= 2i+1, \ldots, \rho, \\
x_j' &= x_j, & j &= \rho+1, \ldots, n,
\end{aligned}
\tag{8.19}
$$

suggested by the method of completing the squares, and we see that $s(x)$ is equivalent to the polynomial

$$t(x) = (x_1^2 - x_2^2) + \cdots + (x_{2i-1}^2 - x_{2i}^2) + x_{2i+1}^2 + \cdots + x_\rho^2$$
$$+ b_{\rho+1}' x_{\rho+1} + \cdots + b_n' x_n + C''$$

where C'' is a constant that can easily be computed. The polynomials $s(x)$ and $t(x)$ are equivalent because one can pass from one to the other by an affine change of variables.

Thus, we have successfully separated the quadratic variables from the linear ones.

The hypothesis on the rank, and the invariance of this hypothesis within the equivalence class (Observation 8.11), tell us that some b_j', $j = \rho+1, \ldots, n$ is different from zero.

Let us assume $b_{\rho+1}' \neq 0$. We make the affine change of variable

$$
\begin{aligned}
x_i' &= x_i, \quad i \neq \rho+1 \\
x_{\rho+1}' &= b_{\rho+1}' x_{\rho+1} + \cdots + b_n' x_n + C'',
\end{aligned}
\tag{8.20}
$$

and we obtain that $t(x)$ is equivalent to the polynomial

$$w(x) = (x_1^2 - x_2^2) + \cdots + (x_{2i-1}^2 - x_{2i}^2)$$

$$+ x_{2i+1}^2 + \cdots + x_\rho^2 + x_{\rho+1}. \tag{8.21}$$

This completes the proof.   □

8.9.1 Observations

Observation 1. The change of variables (8.19) can be written as $x' = ax + b$
with $a = \mathrm{id}$ and

$$b^\mathsf{T} = \left(\frac{b_1'}{2}, -\frac{b_2'}{2}, \ldots, \frac{b_{2i-1}'}{2}, -\frac{b_{2i}'}{2}, \frac{b_{2i+1}'}{2}, \ldots, \frac{b_\rho'}{2}, 0, \ldots, 0 \right).$$

Thus, $t(ax + b) = s(x)$.

Observation 2. The change of variables (8.20) can be written as $x' = ax + b$
with

$$a = \begin{pmatrix} I_\rho & O \\ O & M \end{pmatrix}, \tag{8.22}$$

where

$$M = \begin{pmatrix} b_{\rho+1}' & b_{\rho+2}' & \cdots & b_n' \\ 0 & 1 & \cdots & 0 \\ \vdots & \vdots & \ddots & \vdots \\ 0 & 0 & \cdots & 1 \end{pmatrix}, \qquad b = \begin{pmatrix} 0 \\ \vdots \\ C'' \\ \vdots \\ 0 \end{pmatrix}$$

(C'' in the $\rho + 1$-th position and zeros elsewhere).

Observation 3. If instead of supposing $b_{\rho+1}' \neq 0$ we suppose $b_j' \neq 0$, with $j \in \{\rho+2, \ldots, n\}$, we get

$$w(x) = (x_1^2 - x_2^2) + \cdots + (x_{2i-1}^2 - x_{2i}^2) + x_{2i+1}^2 + \cdots + x_\rho^2 + x_j.$$

But it is clear, merely changing the names of the variables, that this polynomial is equivalent to polynomial (8.21).

8.9.2 Canonical Expression

We have seen that every polynomial $r(x) = x^\mathsf{T} A x + Bx + C$, with

$$\mathrm{rank}(A) \neq \mathrm{rank}(A|B^\mathsf{T}),$$

is equivalent to

$$\boxed{w(x) = (x_1^2 - x_2^2) + \cdots + (x_{2i-1}^2 - x_{2i}^2) + x_{2i+1}^2 + \cdots + x_\rho^2 + x_{\rho+1}} \qquad (8.23)$$

with $i = \text{index}(A)$ and $\rho = \text{rank}(A)$. This expression includes the case $\rho = 2i$, where we have

$$w(x) = (x_1^2 - x_2^2) + \cdots + (x_{2i-1}^2 - x_{2i}^2) + x_{\rho+1}.$$

8.9.3 Conclusion

For convenience, we collect the results of this and the preceding Section 8.8 in the next theorem.

Theorem 8.13 (Classification Theorem)

Every quadratic polynomial in n variables is equivalent to one and only one of the following polynomials:

(I) $\quad r_I(x) = (x_1^2 - x_2^2) + \cdots + (x_{2i-1}^2 - x_{2i}^2) + x_{2i+1}^2 + \cdots + x_\rho^2,$

(II) $\quad r_{II}(x) = (x_1^2 - x_2^2) + \cdots + (x_{2i-1}^2 - x_{2i}^2) + x_{2i+1}^2 + \cdots + x_\rho^2 + 1,$

(III) $\quad r_{III}(x) = (x_1^2 - x_2^2) + \cdots + (x_{2i-1}^2 - x_{2i}^2) - x_{2i+1}^2 - \cdots - x_\rho^2 + 1,$

(IV) $\quad r_{IV}(x) = (x_1^2 - x_2^2) + \cdots + (x_{2i-1}^2 - x_{2i}^2) + x_{2i+1}^2 + \cdots + x_\rho^2 + x_{\rho+1},$

where $\rho > 0$, $0 \le 2i \le \rho \le n$ in cases (I) and (II), $0 \le 2i < \rho \le n$ in case (III), and $0 \le 2i \le \rho < n$ in case (IV).

Proof

We have seen, in Sections 8.8 and 8.9, that every quadratic polynomial in n variables is equivalent to one of the polynomials in this list.

The uniqueness is a consequence of the invariance of the rank and the index, Theorem 8.9, since in this list there are no two polynomials with the same invariants $(\rho, i, \tilde{\rho}, \tilde{i})$. In fact, it is sufficient to observe that a polynomial is of type (I) if and only if $\tilde{\rho} = \rho$; of type (II) if and only if $\tilde{\rho} = \rho + 1$ and $\tilde{i} = i$; of type (III) if and only if $\tilde{\rho} = \rho + 1$ and $\tilde{i} = i + 1$; and of type (IV) if and only if $\tilde{\rho} = \rho + 2$. Moreover, within each type, the quadratic rank and the quadratic index, ρ and i, uniquely determine the polynomial. $\qquad \square$

The polynomials of this list are called *canonical expressions* or *canonical representatives*.

Observe that case (IV) occurs only if $\rho < n$, and that, to avoid repetitions, we shall always consider that in case (III) we have $2i < \rho$. That is, the case where $2i = \rho$ belongs to case (II). It is also easy to see that in case (I) we have $\tilde{i} = i$, and that in case (IV) we have $\tilde{i} = i + 1$.

Corollary 8.14

Two quadratic polynomials are equivalent if and only if they have the same invariants $(\rho, i, \tilde{\rho}, \tilde{i})$.

Proof

The four numbers $(\rho, i, \tilde{\rho}, \tilde{i})$ determine a unique polynomial in the list of canonical representatives.   □

This result allows us to count, with a certain facility, the number of equivalence classes of quadratic polynomials.

8.10 The Number of Equivalence Classes of Quadratic Polynomials

We have seen, in the proof of Theorem 8.13, that the only possible 4-tuples are the following:

$$
\begin{array}{lll}
\text{(I)} & (\rho, i, \rho, i), & 2i \le \rho \le n, \\
\text{(II)} & (\rho, i, \rho + 1, i), & 2i \le \rho \le n, \\
\text{(III)} & (\rho, i, \rho + 1, i + 1), & 2i < \rho \le n, \\
\text{(IV)} & (\rho, i, \rho + 2, i + 1), & 2i \le \rho < n.
\end{array}
$$

Thus, to count the number of equivalence classes of quadratic polynomials we must count the number of 4-tuples in each of the cases (I), (II), (III), (IV) and add the results. The argument is slightly different in each case depending on whether the number of variables n is even or odd.

Type (I). $n = 2k + 1$. There are as many classes as pairs (ρ, i) with $0 \le 2i \le \rho \le n$. Hence, if $i = 0$, ρ varies between 1 and n. If $i = 1$, ρ varies between 2 and n. If $i = 2$, ρ varies between 4 and n. We increase the value of i until we reach $i = k - 1$. In this case ρ varies between $2k - 2$ and $2k + 1$. Finally, if

$i = k$, ρ varies between $2k$ and $2k + 1$. Observe that at each step the number of possible values of ρ decreases by two.

Therefore, the number of possible cases is

$$N(I) = n + (2k + (2k - 2) + \cdots + 6 + 4 + 2) = \frac{n^2 + 4n - 1}{4}.$$

Type (I). $n = 2k$. There are as many classes as pairs (ρ, i) with $0 \leq 2i \leq \rho \leq n$. Hence, if $i = 0$, ρ varies between 1 and n. If $i = 1$, ρ varies between 2 and n. If $i = 2$, ρ varies between 4 and n. We increase the value of i until we reach $i = k - 1$. In this case ρ varies between $2k - 2$ and $2k$. Finally, if $i = k$, ρ must be equal to $2k$. Observe that at each step the number of possible values of ρ decreases by two.

Therefore, the number of possible cases is

$$N(I) = n + (2k - 1 + (2k - 3) + \cdots + 5 + 3 + 1) = \frac{n^2 + 4n}{4}.$$

Type (II). There are as many classes as in case (I), since we must count pairs (ρ, i) with $0 \leq 2i \leq \rho \leq n$.

$$N(II) = N(I).$$

Type (III). $n = 2k + 1$. There are as many classes as in case (I), minus those coming from $\rho = 2i$. There are k classes of the latter type, because $i \in \{1, \ldots, k\}$. (The case $\rho = 2i = 0$ is not considered, because we always assume $\rho > 0$.) Hence,

$$N(III) = N(I) - k = \frac{n^2 + 2n + 1}{4}.$$

Type (III). $n = 2k$. By the same argument,

$$N(III) = N(I) - k = \frac{n^2 + 2n}{4}.$$

Type (IV). $n = 2k + 1$. There are as many classes as pairs (ρ, i) with $0 \leq 2i \leq \rho < n$. Hence, there are as many classes as in case (I), minus those coming from $\rho = n$. There are $k + 1$ classes of the latter type, because i varies between 0 and k.

$$N(IV) = N(I) - (k + 1) = \frac{n^2 + 2n - 3}{4}.$$

Type (IV). $n = 2k$. Analogously,

$$N(IV) = N(I) - (k + 1) = \frac{n^2 + 2n - 4}{4}.$$

Incidentally, the expression of the number $N = N(I) + N(II) + N(III) + N(IV)$ does not depend on whether n is even or odd. Carrying out this sum, we obtain the number N of equivalence classes of quadratic polynomials with n variables:

$$\boxed{N = n^2 + 3n - 1}$$

8.11 Regular Zeros

Polynomials $r(x)$ can be considered as maps

$$r : \mathbb{R}^n \longrightarrow \mathbb{R}.$$

Therefore, we can talk about their gradient, $\operatorname{grad} r(x)$, as the vector formed by the partial derivatives:

$$\operatorname{grad} r(x) = \left(\frac{\partial r}{\partial x_1}, \ldots, \frac{\partial r}{\partial x_n} \right).$$

Definition 8.15

A point $p \in \mathbb{R}^n$ is called a *regular point* of the polynomial $r(x)$ if $\operatorname{grad} r(p) \neq 0$.

Note that if $r(x) = x^{\mathsf{T}} A x + B x + C$, then

$$\boxed{\operatorname{grad} r(x) = 2x^{\mathsf{T}} A + B}$$

In fact, the i-th component of the gradient vector is

$$\frac{\partial}{\partial x_i}(x^{\mathsf{T}} A x + B x + C) = 2(0, \ldots, 1, \ldots, 0) A x + B_i = 2(x^{\mathsf{T}} A)_i + B_i.$$

Hence, $p \in \mathbb{R}^n$ is a regular point of $r(x)$ if and only if $2p^{\mathsf{T}} A + B \neq 0$, that is, if and only if p is not a solution of the system

$$2Ax = -B^{\mathsf{T}}.$$

This system is exactly the same as the system appearing in Section 8.8, (8.13), which we called the *equations of the center*. We shall study this system in Proposition 8.36, page 267.

Definition 8.16

If $p \in \mathbb{R}^n$ is a regular point of the quadratic polynomial $r(x)$ and $r(p) = 0$, we say that p is a *regular zero* of this polynomial.

If the polynomials $r(x)$ and $s(x)$ are equivalent, with $r(ax + b) = \lambda s(x)$, and x is a regular zero of $s(x)$, then $x' = ax + b$ is a regular zero of $r(x)$. In fact, applying the chain rule to the equality $r(ax + b) = \lambda s(x)$, we obtain

$$\mathrm{grad}(r(ax + b)) \cdot a = \lambda \, \mathrm{grad}(s(x)), \tag{8.24}$$

and hence, $\mathrm{grad}(s(x)) \neq 0$ if and only if $\mathrm{grad}(r(ax + b)) \neq 0$.

   Hence, if $r(x)$ and $s(x)$ are equivalent polynomials, we see that $r(x)$ *has regular zeros if and only if $s(x)$ has regular zeros.*

   Finally, we observe that the gradients of the four types of polynomials (I), (II), (III) and (IV), given in the Classification Theorem 8.13, are

$$\mathrm{grad}\, r_I(x) = (2x_1, -2x_2, \ldots, 2x_\rho, 0, \ldots, 0),$$

$$\mathrm{grad}\, r_{II}(x) = (2x_1, -2x_2, \ldots, 2x_\rho, 0, \ldots, 0),$$

$$\mathrm{grad}\, r_{III}(x) = (2x_1, -2x_2, \ldots, -2x_\rho, 0, \ldots, 0),$$

$$\mathrm{grad}\, r_{IV}(x) = (2x_1, -2x_2, \ldots, 2x_\rho, 1, \ldots, 0).$$

Thus, if a polynomial is equivalent to a polynomial of type (II), (III) or (IV), all its zeros are regular; if it is equivalent to a polynomial of type (I) with $i \neq 0$, then it has regular and non-regular zeros, and if $i = 0$, it does not have any regular zeros.

8.12 Affine Classification of Quadrics

We come back to affine spaces. Recall that the definition of equivalent polynomials was motivated by equality (8.9), page 231, which says that if f is an affinity of equation $x' = ax + b$, then

$$f(\mathcal{Q}(r(ax + b), \mathcal{R})) = \mathcal{Q}(r(x), \mathcal{R}).$$

   Therefore, the following definition is natural.

Definition 8.17

We say that two non-empty quadrics $\mathcal{Q}_1$ and $\mathcal{Q}_2$ of a real affine space $\mathbb{A}$ are *equivalent* if there exists a bijective affinity f such that $f(\mathcal{Q}_1) = \mathcal{Q}_2$.

Let us see what relationship there is between *equivalent polynomials* and *equivalent quadrics*.

Proposition 8.18

Let $\mathcal{R}$ be an affine frame of an affine space $\mathbb{A}$ of dimension n, and let $r(x)$ and $s(x)$ be equivalent polynomials in n variables. Then there is an affine frame $\mathcal{R}'$ such that

$$\mathcal{Q}(r(x), \mathcal{R}) = \mathcal{Q}(s(x), \mathcal{R}').$$

Proof

Since the polynomials are equivalent, there is, by Definition 8.2, an invertible $n \times n$ real matrix a, an $n \times 1$ real column matrix b, and a real number $\lambda \neq 0$, such that $r(ax + b) = \lambda s(x)$.

Equality (8.7), page 231, tells us that

$$\mathcal{Q}(r(x), \mathcal{R}) = \mathcal{Q}(\lambda s(x), \mathcal{R}') = \mathcal{Q}(s(x), \mathcal{R}'),$$

with

$$M(\mathcal{R}', \mathcal{R}) = \begin{pmatrix} a & b \\ 0 & 1 \end{pmatrix}.$$

This completes the proof. □

Theorem 8.19

Let $\mathbb{A}$ be an affine space of dimension n, $\mathcal{R}$ an affine frame of $\mathbb{A}$, and $r(x)$ and $s(x)$ equivalent polynomials in n variables.

Then, the quadrics $\mathcal{Q}(r(x), \mathcal{R})$ and $\mathcal{Q}(s(x), \mathcal{R})$ are equivalent (or empty).

Proof

Since $r(x)$ and $s(x)$ are equivalent, there is, by Definition 8.2, an invertible $n \times n$ real matrix a, an $n \times 1$ real column matrix b, and a real number $\lambda \in \mathbb{R}$, $\lambda \neq 0$, such that $r(ax + b) = \lambda s(x)$.

The equality (8.9), page 231, tells us directly that the affinity f of equation $x' = ax + b$ satisfies $f(\mathcal{Q}(r(x), \mathcal{R})) = \mathcal{Q}(s(x), \mathcal{R})$. Hence, these quadrics are equivalent (or empty). □

Corollary 8.20

Let $\mathcal{R}$ and $\mathcal{R}'$ be two affine frames in an affine space of dimension n, and let $r(x)$ be a polynomial in n variables. Then the quadrics $\mathcal{Q}(r(x), \mathcal{R})$ and $\mathcal{Q}(r(x), \mathcal{R}')$ are equivalent.

Proof

We know, by equality (8.7), page 231, that $\mathcal{Q}(r(x), \mathcal{R}) = \mathcal{Q}(r(ax + b), \mathcal{R}')$, with

$$M(\mathcal{R}', \mathcal{R}) = \begin{pmatrix} a & b \\ 0 & 1 \end{pmatrix}.$$

But, by the above theorem, this second quadric is equivalent to $\mathcal{Q}(r(x), \mathcal{R}')$. Hence, $\mathcal{Q}(r(x), \mathcal{R})$ and $\mathcal{Q}(r(x), \mathcal{R}')$ are equivalent. $\qquad\square$

The converse of Theorem 8.19 is also true, but we must recall that the definition of equivalent quadrics, Definition 8.17, has been given only for non-empty quadrics. Explicitly, we shall prove that *equivalent (and, therefore, non-empty) quadrics give rise to equivalent polynomials.*

But there may be empty quadrics (equal, therefore, as subsets of the affine space), defined by non-equivalent polynomials. For instance, in the affine space $\mathbb{R}^3$, the quadrics $x_1^2 + x_2^2 + 1 = 0$ and $x_1^2 + 1 = 0$ are empty, but the polynomials $x_1^2 + x_2^2 + 1$ and $x_1^2 + 1$ are not equivalent.

To study the converse of Theorem 8.19 we must introduce regular points.

Definition 8.21 (Regular points)

Let $\mathcal{Q} = \mathcal{Q}(r(x), \mathcal{R})$ be a quadric in an affine space $\mathbb{A}$ and let $P \in \mathcal{Q}$. We say that P is a *regular point* of $\mathcal{Q}$ if the point $(p_1, \ldots, p_n) \in \mathbb{R}^n$, where $(p_1, \ldots, p_n)$ are the coordinates of P in $\mathcal{R}$, is a *regular zero* of $r(x)$.

Thus, in order to determine if a point is regular we need to know, in principle, the affine frame and the polynomial defining the quadric.

This leads to the following natural question: If we have $\mathcal{Q} = \mathcal{Q}(r(x), \mathcal{R}) = \mathcal{Q}(s(x), \mathcal{R}')$, what is meant by the expression "P is a regular point of $\mathcal{Q}$"? Does it mean that the point $(p_1, \ldots, p_n) \in \mathbb{R}^n$, where $(p_1, \ldots, p_n)$ are the coordinates of P in $\mathcal{R}$, is a regular zero of $r(x)$, or that the point $(p'_1, \ldots, p'_n) \in \mathbb{R}^n$, where $(p'_1, \ldots, p'_n)$ are the coordinates of P in $\mathcal{R}'$, is a regular zero of $s(x)$?

The next proposition provides a partial answer to this question. See Observation 8.25.

Theorem 8.22 (Nullstellensatz)

Let $\mathbb{A}$ be an affine space of dimension n, and let $r(x)$ and $s(x)$ be two quadratic polynomials in n variables. Let $\mathcal{R}$ be an affine frame of $\mathbb{A}$. Let us assume that the quadric $\mathcal{Q}(r(x), \mathcal{R})$ has a regular point (with respect to $r(x)$) and also that

$$\mathcal{Q}(r(x), \mathcal{R}) = \mathcal{Q}(s(x), \mathcal{R}).$$

Then there is a $\lambda \in \mathbb{R}$, $\lambda \neq 0$, such that $r(x) = \lambda s(x)$.

Proof

This is an immediate consequence of the real Nullstellensatz, Theorem D.4, page 399.   □

Thus, once we fix the affine frame, the polynomial defining the quadric (with a regular point) is determined up to scalars, and hence the definition of *regular point* does not depend on which of these multiples is considered.

Theorem 8.23 (Converse of Theorem 8.19)

Let $\mathbb{A}$ be an affine space of dimension n, $\mathcal{R}$ an affine frame of $\mathbb{A}$ and $r(x)$ and $s(x)$ polynomials in n variables.

Let us assume that the quadrics $\mathcal{Q}(r(x), \mathcal{R})$ and $\mathcal{Q}(s(x), \mathcal{R})$ are equivalent (and, therefore, non-empty). Then $r(x)$ and $s(x)$ are equivalent.

Proof

We know that there is a bijective affinity $f : \mathbb{A} \longrightarrow \mathbb{A}$ such that $f(\mathcal{Q}(s(x), \mathcal{R})) = \mathcal{Q}(r(x), \mathcal{R})$. But we also know, by equality (8.9), page 231, that if f has equation $x' = ax + b$, then

$$f(\mathcal{Q}(s(x), \mathcal{R})) = \mathcal{Q}(t(x), \mathcal{R}), \quad \text{with } t(x) = s(a^{-1}(x - b)).$$

Hence, $\mathcal{Q}(r(x), \mathcal{R}) = \mathcal{Q}(t(x), \mathcal{R})$. Observe that the polynomials $s(x)$ and $t(x)$ are equivalent.

If some of the points of $\mathcal{Q}(r(x), \mathcal{R})$ are regular (with respect to $r(x)$) then, by Theorem 8.22, there is a $\lambda \neq 0$ such that $r(x) = \lambda t(x)$, and hence $r(x)$ and $s(x)$ are equivalent and we have finished.

If none of the points of $\mathcal{Q}(r(x), \mathcal{R})$ are regular (with respect to $r(x)$), then neither the points of $\mathcal{Q}(t(x), \mathcal{R})$ (Theorem 8.22) nor the points of $\mathcal{Q}(s(x), \mathcal{R})$ (with respect to $s(x)$) are regular, because $s(x)$ and $t(x)$ are equivalent.

Then, by the Classification Theorem and the expressions of the gradients given on page 249, both $r(x)$ and $s(x)$ are equivalent to a polynomial of type (I) with $i = 0$.

This means that $r(x)$ is equivalent to $r_I(x) = x_1^2 + \cdots + x_\rho^2$ and that $s(x)$ is equivalent to $r_I'(x) = x_1^2 + \cdots + x_{\rho'}^2$. By Theorem 8.19, there is a bijective affinity between $\mathcal{Q}(r(x), \mathcal{R})$ and $\mathcal{Q}(r_I(x), \mathcal{R})$ and a bijective affinity between $\mathcal{Q}(s(x), \mathcal{R})$ and $\mathcal{Q}(r_I'(x), \mathcal{R})$. Composing these affinities with f we obtain a bijective affinity between $\mathcal{Q}(r_I(x), \mathcal{R})$ and $\mathcal{Q}(r_I'(x), \mathcal{R})$.

But, since $r_I(x) = 0$ represents a hyperplane of dimension $n - \rho$ ($x_1 = x_2 = \cdots = x_\rho = 0$), and $r_I'(x) = 0$ represents a hyperplane of dimension $n - \rho'$ ($x_1 = x_2 = \cdots = x_{\rho'} = 0$), and bijective affinities take hyperplanes to hyperplanes of the same dimension, we must have $\rho = \rho'$ and $r(x)$ is equivalent to $s(x)$. This completes the proof.   □

Corollary 8.24 (Converse of Proposition 8.18)

Let $\mathbb{A}$ be an affine space of dimension n, and let $r(x)$ and $s(x)$ be polynomials in n variables. Let $\mathcal{R}$ and $\mathcal{R}'$ be affine frames such that

$$\mathcal{Q}(r(x), \mathcal{R}') = \mathcal{Q}(s(x), \mathcal{R}).$$

Let us assume that this quadric is non-empty. Then $r(x)$ and $s(x)$ are equivalent.

Proof

Since, by Corollary 8.20, $\mathcal{Q}(r(x), \mathcal{R}')$ is equivalent to $\mathcal{Q}(r(x), \mathcal{R})$, we have, by transitivity, that $\mathcal{Q}(r(x), \mathcal{R})$ is equivalent to $\mathcal{Q}(s(x), \mathcal{R})$. Hence, by the above theorem, $r(x)$ and $s(x)$ are equivalent.   □

Observation 8.25

In particular, it makes perfect sense to say that *a point $P \in \mathcal{Q}$ is regular*, because independently of the affine frame $\mathcal{R}$ and of the polynomial $r(x)$ used in the definition of $\mathcal{Q} = \mathcal{Q}(r(x), \mathcal{R})$, the point $(p_1, \ldots, p_n) \in \mathbb{R}^n$, where $(p_1, \ldots, p_n)$ are the coordinates of P in $\mathcal{R}$, is a regular zero of $r(x)$.

In fact, if we have $\mathcal{Q}(r(x), \mathcal{R}) = \mathcal{Q}(s(x), \mathcal{R}')$ and the relationship between the coordinates x in $\mathcal{R}$ and the coordinates x' in $\mathcal{R}'$ is given by $x = ax' + b$, then the expression of the gradients (8.24) tells us that x' is a regular zero of $s(x)$ if and only if x is a regular zero of $r(x)$.

Corollary 8.24, together with Proposition 8.9, allows us to define the rank, quadratic rank, index and quadratic index of a non-empty quadric as the rank, quadratic rank, index and quadratic index, respectively, of any polynomial defining the quadric.

If we do not impose the hypothesis that the quadrics are non-empty, there may be different polynomials, for instance $r(x) = x_1^2 + 1$ and $s(x) = x_1^2 + x_2^2 + 1$, defining the same quadric (the empty set), and with different quadratic rank (1 and 2, respectively).

Theorem 8.26 (Equivalence between polynomials and quadrics)

Let $\mathcal{Q}(r(x), \mathcal{R})$ and $\mathcal{Q}(s(x), \mathcal{R})$ be two non-empty quadrics in a real affine space $\mathbb{A}$. Then they are equivalent if and only if the polynomials defining them, $r(x)$ and $s(x)$, are equivalent.

Proof

This is a consequence of Theorems 8.19 and 8.23.   □

By Corollary 8.20, this result is also true even in the case where the two given quadrics do not refer to the same affine frame. Concretely, the non-empty quadrics $\mathcal{Q}(r(x), \mathcal{R})$ and $\mathcal{Q}(s(x), \mathcal{R}')$ of a real affine space $\mathbb{A}$ are equivalent if and only if the polynomials defining them, $r(x)$ and $s(x)$, are equivalent.

Corollary 8.27

Two non-empty quadrics are equivalent if and only if the quadratic polynomials defining them have the same invariants $(\rho, i, \tilde{\rho}, \tilde{i})$

Proof

This is a consequence of Theorems 8.26 and 8.14.   □

Theorem 8.28 (Classification Theorem)

Let $\mathcal{Q}(r(x), \mathcal{R})$ be a non-empty quadric of a real affine space $\mathbb{A}$ of dimension n. Then $\mathcal{Q}(r(x), \mathcal{R})$ is equivalent to one and only one of the quadrics given, in $\mathcal{R}$, by the following equations:

(I) $(x_1^2 - x_2^2) + \cdots + (x_{2i-1}^2 - x_{2i}^2) + x_{2i+1}^2 + \cdots + x_\rho^2 = 0$,

(II) $(x_1^2 - x_2^2) + \cdots + (x_{2i-1}^2 - x_{2i}^2) + x_{2i+1}^2 + \cdots + x_\rho^2 + 1 = 0$,

(III) $(x_1^2 - x_2^2) + \cdots + (x_{2i-1}^2 - x_{2i}^2) - x_{2i+1}^2 - \cdots - x_\rho^2 + 1 = 0$,

(IV) $(x_1^2 - x_2^2) + \cdots + (x_{2i-1}^2 - x_{2i}^2) + x_{2i+1}^2 + \cdots + x_\rho^2 + x_{\rho+1} = 0$,

where $\rho > 0$, $0 \le 2i \le \rho \le n$ in case (I), $0 < 2i \le \rho \le n$ in case (II), $0 \le 2i < \rho \le n$ in case (III), and $0 \le 2i \le \rho < n$ in case (IV).

Proof

This is an immediate consequence of Theorems 8.13 and 8.19. □

All quadrics of type (II) with $i = 0$, $x_1^2 + \cdots + x_\rho^2 + 1 = 0$, correspond to empty quadrics, and for this reason they have not been considered in the statement of the previous theorem.

This theorem can also be formulated in terms of a change of affine frame. Concretely, we have the following.

Theorem 8.29

Let $\mathcal{Q}(r(x), \mathcal{R})$ be a non-empty quadric of a real affine space $\mathbb{A}$ of dimension n. Then there is an affine frame $\mathcal{R}'$ in which the equation of $\mathcal{Q}(r(x), \mathcal{R})$ is one and only one of the equations given in the previous Theorem 8.28.

Proof

We know that $r(x)$ is equivalent to one and only one of the polynomials $r_0(x)$, of type (I), (II), (III) or (IV), given in Theorem 8.13. By Proposition 8.18, there is an affine frame $\mathcal{R}'$ such that $\mathcal{Q}(r(x), \mathcal{R}) = \mathcal{Q}(r_0(x), \mathcal{R}')$. □

When we have $\mathcal{Q}(r(x), \mathcal{R}) = \mathcal{Q}(r_0(x), \mathcal{R}')$, with $r_0(x)$ one of the *canonical representatives* given in Theorem 8.13, we say that $\mathcal{R}'$ is the *affine frame adapted to the quadric*, or simply the *adapted affine frame*.

8.12.1 Construction of the Adapted Affine Frame

In order to classify a quadric $\mathcal{Q}(r(x), \mathcal{R})$, we shall classify first the polynomial $r(x)$, finding explicitly the affine change $x' = ax + b$ that transforms $r(x)$ into one of the canonical representatives $r_0(x)$. If we have $\lambda r_0(x) = r(ax + b)$, then the adapted affine frame $\mathcal{R}'$ is given (see the proof of Proposition 8.18) by

$$M(\mathcal{R}',\mathcal{R}) = \begin{pmatrix} a & b \\ 0 & 1 \end{pmatrix}.$$

That is, the origin of $\mathcal{R}'$ has coordinates b in $\mathcal{R}$, and the vectors of the corresponding basis of $\mathcal{R}'$ have components given respectively by the columns of the matrix a.

Recall that we can find a directly by imposing that the matrix $a^{\mathsf{T}} A a$ is diagonal (where A is the matrix of the quadratic part of $r(x)$); and that, when $r_0(x)$ is of type (I), (II) or (III), b is given directly by the coordinates of a center of the quadric. If $r_0(x)$ is of type (IV) and, hence, it has no center, the only way to compute b is to follow the steps given in Section 8.9.

8.13 Affine Classification of Conics

Quadrics in affine spaces of dimension 2 are called *conics*. To classify them it is sufficient to apply Theorem 8.28 with $n = 2$, taking into account the restrictions on ρ and i.

Observe that we must have

$$1 \leq \rho \leq 2,$$

$$0 \leq i \leq 1,$$

$$1 \leq \tilde{\rho} \leq 3,$$

$$0 \leq \tilde{i} \leq 1,$$

and, as we pointed out on page 246, if $\tilde{\rho} = \rho$, then $\tilde{i} = i$; if $\tilde{\rho} = \rho + 1$, then $i \leq \tilde{i} \leq i + 1$, and if $\tilde{\rho} = \rho + 2$, then $\tilde{i} = i + 1$. This results in Table 8.1.

Observe that if $r(x) = x^{\mathsf{T}} A x + B x + C$, and we let

$$d = \det A,$$

$$D = \det \begin{pmatrix} A & B^{\mathsf{T}}/2 \\ B/2 & C \end{pmatrix} = \det \tilde{A},$$

then we have

$$\rho = 2 \quad \text{if and only if} \quad d \neq 0, \qquad \rho = 1 \quad \text{if and only if} \quad d = 0,$$
$$\tilde{\rho} = 3 \quad \text{if and only if} \quad D \neq 0, \qquad i = 1 \quad \text{if and only if} \quad d < 0.$$

Note that the sign of d is an invariant, since the formula $\lambda A' = a^{\mathsf{T}} A a$ implies (we are in dimension two) $\lambda^2 \det A' = (\det a)^2 \det A$, and hence $\det A$ and $\det A'$ have the same sign.

	ρ	i	$\tilde{\rho}$	$\tilde{i}$	Equation	Name
(I)	2	0	2	0	$x_1^2 + x_2^2 = 0$	Point
(I)	2	1	2	1	$x_1^2 - x_2^2 = 0$	Two str. lines
(I)	1	0	1	0	$x_1^2 = 0$	Double str. line
(II)	2	1	3	1	$x_1^2 - x_2^2 + 1 = 0$	Hyperbola
(II)	2	0	3	0	$x_1^2 + x_2^2 + 1 = 0$	Empty
(II)	1	0	2	0	$x_1^2 + 1 = 0$	Empty
(III)	2	0	3	1	$-x_1^2 - x_2^2 + 1 = 0$	Ellipse
(III)	1	0	2	1	$-x_1^2 + 1 = 0$	Two parallel str. lines
(IV)	1	0	3	1	$x_1^2 + x_2 = 0$	Parabola

Table 8.1. Conics

Analogously, formula (8.10) tells us that whether D vanishes or not is an invariant, but its sign is not invariant because, repeating the same argument, we obtain a λ^3, which has the same sign as λ. Concretely we have

$$\lambda^3 \det \tilde{A}' = \det \tilde{A}(\det C)^2,$$

because $\tilde{A}$ is a 3×3 matrix, and we cannot ensure that $\det \tilde{A}'$ and $\det \tilde{A}$ have the same sign.

This comment justifies the classification of conics in terms of the sign of d and whether or not D is zero.

However it is clear that in higher dimensions we cannot reduce the four invariants $(\rho, i, \tilde{\rho}, \tilde{i})$ to the study of only two determinants.

The relationship between the four invariants and the two determinants d and D is explicitly set out in Table 8.2.

	ρ	i	$\tilde{\rho}$	$\tilde{i}$	d	D	Name
(I)	2	0	2	0	$d > 0$	$D = 0$	Point
(I)	2	1	2	1	$d < 0$	$D = 0$	Two str. lines
(I)	1	0	1	0	$d = 0$	$D = 0$	Double str. line
(II)	2	1	3	1	$d < 0$	$D \neq 0$	Hyperbola
(II)	2	0	3	0	$d > 0$	$D \neq 0$	Empty (imaginary ellipse)
(II)	1	0	2	0	$d = 0$	$D = 0$	Empty (imaginary str. lines)
(III)	2	0	3	1	$d > 0$	$D \neq 0$	Ellipse
(III)	1	0	2	1	$d = 0$	$D = 0$	Two parallel str. lines
(IV)	1	0	3	1	$d = 0$	$D \neq 0$	Parabola

Table 8.2. Conics

Thus, *omitting empty conics*, the classification can be quickly obtained by computing only two determinants.

$$
D \neq 0 \quad
\begin{cases}
d > 0 & \text{Ellipse} \\
d < 0 & \text{Hyperbola} \\
d = 0 & \text{Parabola}
\end{cases}
$$

$$
D = 0 \quad
\begin{cases}
d > 0 & \text{Point} \\
d < 0 & \text{Two intersecting str. lines} \\
d = 0 & \text{Two parallel or equal str. lines}
\end{cases}
$$

Hence, with the exception of parallel or equal straight lines, non-empty conics are completely determined by these two determinants. See Figures 8.1 and 8.2 for the case $D \neq 0$.

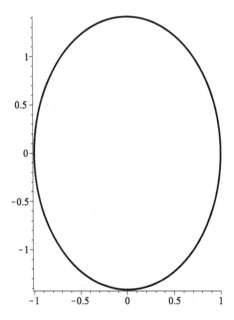

Figure 8.1. Ellipse

A detailed study of conics in the Euclidean affine plane $\mathbb{R}^2$ from an elementary geometrical point of view can be found in [8]. For instance, one can find an elementary proof that conics with $D \neq 0$ can be defined as sets of points of the plane such that the quotient of the distances to a given point (the focus) and to a given straight line (the directrix) is constant. This constant is lower than 1 for ellipses; greater than 1 for hyperbolas; and equal to 1 for parabolas. There are also equivalent definitions of conics, methods of constructing tangents, the relationship with cross-sections of a cone of $\mathbb{R}^3$, etc.

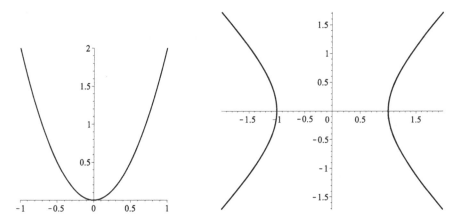

Figure 8.2. Parabola and hyperbola

Conics also appear in a natural way when studying the orbits of planets. We recommend [27] for a simple demonstration, using only Newton's laws, that the orbits of the planets are conics.

8.14 Affine Classification of Quadrics in Dimension Three

We apply Theorem 8.28 with $n = 3$, taking into account the following restrictions on ρ and i:

$$1 \leq \rho \leq 3, \qquad 0 \leq i \leq 1,$$
$$1 \leq \tilde{\rho} \leq 4, \qquad 0 \leq \tilde{i} \leq 2.$$

As we remarked on page 246, if $\tilde{\rho} = \rho$, then $\tilde{i} = i$; if $\tilde{\rho} = \rho + 1$, then $i \leq \tilde{i} \leq i + 1$, and if $\tilde{\rho} = \rho + 2$, then $\tilde{i} = i + 1$. This results in Table 8.3 (see also Figure 8.3).

Example 8.30

Classify the quadric of the affine space $\mathbb{R}^3$ given in the canonical affine frame by $r(x) = x_1^2 - 2x_2^2 + 2x_1x_2 + 4x_1x_3 + 8x_3 + 1 = 0$ and find the adapted affine frame.

	ρ	i	$\tilde{\rho}$	$\tilde{i}$	Equation	Name
(I)	3	0	3	0	$x_1^2 + x_2^2 + x_3^2 = 0$	Point
(I)	3	1	3	1	$x_1^2 + x_2^2 - x_3^2 = 0$	Cone
(I)	2	0	2	0	$x_1^2 + x_2^2 = 0$	Straight line
(I)	2	1	2	1	$x_1^2 - x_2^2 = 0$	Two planes
(I)	1	0	1	0	$x_1^2 = 0$	Double plane
(II)	3	0	4	0	$x_1^2 + x_2^2 + x_3^2 + 1 = 0$	Empty
(II)	3	1	4	1	$x_1^2 + x_2^2 - x_3^2 + 1 = 0$	Hyperboloid of two sheets
(II)	2	0	3	0	$x_1^2 + x_2^2 + 1 = 0$	Empty
(II)	2	1	3	1	$x_1^2 - x_2^2 + 1 = 0$	Hyperbolic cylinder
(II)	1	0	2	0	$x_1^2 + 1 = 0$	Empty
(III)	3	1	4	2	$x_1^2 - x_2^2 - x_3^2 + 1 = 0$	Hyperboloid of one sheet
(III)	3	0	4	1	$-x_1^2 - x_2^2 - x_3^2 + 1 = 0$	Ellipsoid
(III)	2	0	3	1	$-x_1^2 - x_2^2 + 1 = 0$	Elliptic cylinder
(III)	1	0	2	1	$-x_1^2 + 1 = 0$	Parallel planes
(IV)	2	0	4	1	$x_1^2 + x_2^2 + z = 0$	Elliptic paraboloid
(IV)	2	1	4	2	$x_1^2 - x_2^2 + z = 0$	Hyperbolic paraboloid
(IV)	1	0	3	1	$x_1^2 + z = 0$	Parabolic cylinder

Table 8.3. Quadrics

Solution

The matrix of the quadratic part is

$$A = \begin{pmatrix} 1 & 1 & 2 \\ 1 & -2 & 0 \\ 2 & 0 & 0 \end{pmatrix},$$

which has rank $\rho = 3$. The characteristic polynomial of A is $-x^3 - x^2 + 7x + 8$, which has one change of sign between consecutive non-zero coefficients, and hence $i = 1$.

The enlarged matrix is

$$\tilde{A} = \begin{pmatrix} 1 & 1 & 2 & 0 \\ 1 & -2 & 0 & 0 \\ 2 & 0 & 0 & 4 \\ 0 & 0 & 4 & 1 \end{pmatrix},$$

which has rank $\tilde{\rho} = 4$. The characteristic polynomial of $\tilde{A}$ is $-x^4 + 24x^2 + 17x - 72$, which has two changes of sign between consecutive non-zero coeffi-

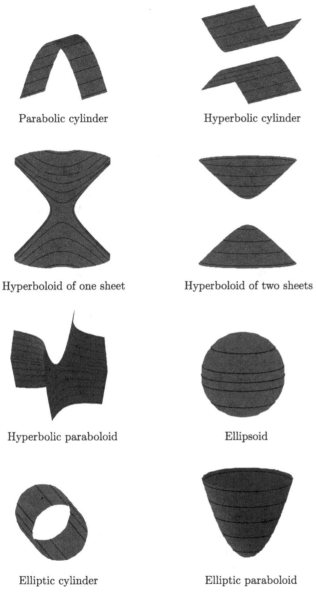

Parabolic cylinder

Hyperbolic cylinder

Hyperboloid of one sheet

Hyperboloid of two sheets

Hyperbolic paraboloid

Ellipsoid

Elliptic cylinder

Elliptic paraboloid

Figure 8.3. Quadrics in $\mathbb{R}^3$

cients, and hence $\tilde{i} = 2$. Looking at the table we see that the given quadric, which has invariants $(\rho, i, \tilde{\rho}, \tilde{i}) = (3, 1, 4, 2)$, is a hyperboloid of one sheet.

Let us find the affine frame in which this quadric has equation $x_1^2 - x_2^2 - x_3^2 + 1 = 0$.

Since $\operatorname{rank}(A) = \operatorname{rank}(A|B^{\mathsf{T}}) = 3$ (observe that $B = (0,0,1)$), this quadric has a unique center, given by

$$2 \begin{pmatrix} 1 & 1 & 2 \\ 1 & -2 & 0 \\ 2 & 0 & 0 \end{pmatrix} \begin{pmatrix} x \\ y \\ z \end{pmatrix} = - \begin{pmatrix} 0 \\ 0 \\ 8 \end{pmatrix}.$$

Solving this system one obtains that the center is the point

$$C = (-2, -1, 3/2).$$

This point will be the origin of the affine frame that we are looking for.

To find the basis, we diagonalize the matrix A. We have

$$\left(\begin{array}{ccc} 1 & 1 & 2 \\ 1 & -2 & 0 \\ 2 & 0 & 0 \\ \hline 1 & 0 & 0 \\ 0 & 1 & 0 \\ 0 & 0 & 1 \end{array} \right) \rightarrow \left(\begin{array}{ccc} 1 & 1 & 2 \\ 0 & -3 & -2 \\ 0 & -2 & -4 \\ \hline 1 & 0 & 0 \\ 0 & 1 & 0 \\ 0 & 0 & 1 \end{array} \right) \rightarrow \left(\begin{array}{ccc} 1 & 0 & 0 \\ 0 & -3 & -2 \\ 0 & -2 & -4 \\ \hline 1 & -1 & -2 \\ 0 & 1 & 0 \\ 0 & 0 & 1 \end{array} \right),$$

where the first transformation is $F_2 \rightarrow F_2 - F_1$, $F_3 \rightarrow F_3 - 2F_1$, and the second $C_2 \rightarrow C_2 - C_1$, $C_3 \rightarrow C_3 - 2C_1$. Next we have

$$\left(\begin{array}{ccc} 1 & 0 & 0 \\ 0 & -3 & -2 \\ 0 & -2 & -4 \\ \hline 1 & -1 & -2 \\ 0 & 1 & 0 \\ 0 & 0 & 1 \end{array} \right) \rightarrow \left(\begin{array}{ccc} 1 & 0 & 0 \\ 0 & -3 & -2 \\ 0 & 0 & -8/3 \\ \hline 1 & -1 & -2 \\ 0 & 1 & 0 \\ 0 & 0 & 1 \end{array} \right) \rightarrow \left(\begin{array}{ccc} 1 & 0 & 0 \\ 0 & -3 & 0 \\ 0 & 0 & -8/3 \\ \hline 1 & -1 & -4/3 \\ 0 & 1 & -2/3 \\ 0 & 0 & 1 \end{array} \right),$$

where the first transformation is $F_3 \rightarrow F_3 - \frac{2}{3} F_2$, and the second $C_3 \rightarrow C_3 - \frac{2}{3} C_2$.

The basis in which A diagonalizes is given by the columns of the lower matrix that has been transformed from the identity matrix. That is, $u = (1,0,0), v = (-1,1,0), w = (-4/3, -2/3, 1)$. Consider the affine frame $\mathcal{R} = \{C; \mathcal{B}\}$ with $\mathcal{B} = (u, v, w)$.

The relationship between the initial coordinates (x_1, x_2, x_3) and the coordinates (x_1', x_2', x_3') in this new affine frame is

$$\begin{pmatrix} x_1 + 2 \\ x_2 + 1 \\ x_3 - 3/2 \end{pmatrix} = \begin{pmatrix} 1 & -1 & -4/3 \\ 0 & 1 & -2/3 \\ 0 & 0 & 1 \end{pmatrix} \begin{pmatrix} x_1' \\ x_2' \\ x_3' \end{pmatrix}.$$

Substituting these values into the initial equation of the quadric, we get

$$x_1'^2 - 3x_2'^2 - 8/3x_3'^2 + 7 = 0,$$

which can be written as

$$\left(\frac{x_1'}{\sqrt{7}}\right)^2 - \left(\frac{\sqrt{3}}{\sqrt{7}}x_2'\right)^2 - \left(\frac{\sqrt{8}}{\sqrt{21}}x_3'\right)^2 + 1 = 0,$$

or as

$$\bar{x}_1{}^2 - \bar{x}_2{}^2 - \bar{x}_3{}^2 + 1 = 0,$$

with

$$\bar{x}_1 = \frac{x_1'}{\sqrt{7}}, \qquad \bar{x}_2 = \frac{\sqrt{3}}{\sqrt{7}}x_2', \qquad \bar{x}_3 = \frac{\sqrt{8}}{\sqrt{21}}x_3'.$$

Hence, the adapted affine frame is $\mathcal{R}' = \{C; \mathcal{B}'\}$, where C is the center of the quadric, and $\mathcal{B}' = (\bar{u}, \bar{v}, \bar{w})$ with

$$\bar{u} = \sqrt{7}u, \qquad \bar{v} = \frac{\sqrt{7}}{\sqrt{3}}v, \qquad \bar{w} = \frac{\sqrt{21}}{\sqrt{8}}w.$$

□

Example 8.31

Classify the quadric of an affine space $\mathbb{A}$ of dimension 3 given, in an affine frame $\mathcal{R}$, by $r(x) = 3x_1^2 - x_3^2 - 2x_1x_3 + 4x_1 + x_2 + x_3 + 1 = 0$, and find the adapted affine frame.

Solution

Modifying slightly the method used in the above problem, we shall use the method of completing the squares.

$$3x_1^2 - x_3^2 - 2x_1x_3 + 4x_1 + x_2 + x_3 + 1$$

$$= 3\left(x_1 - \frac{z}{3}\right)^2 - \frac{x_3^2}{3} - x_3^2 + 4x_1 + x_2 + x_3 + 1$$

$$= 3x_1'^2 - \frac{4}{3}x_3^2 + 4x_1' + \frac{7}{3}x_3 + x_2 + 1 \qquad \left(x_1' = x_1 - \frac{x_3}{3}\right)$$

$$= 3\left(x_1' + \frac{2}{3}\right)^2 - \frac{4}{3} - \frac{4}{3}\left(x_3 - \frac{7}{8}\right)^2 + \frac{49}{48} + x_2 + 1$$

$$= x_1''^2 - x_3'^2 + x_2',$$

with

$$x_1'' = \sqrt{3}\left(x_1 - \frac{x_3}{3} + \frac{2}{3}\right),$$

$$x_2' = x_2 + \frac{11}{16},$$

$$x_3' = \frac{2}{\sqrt{3}}\left(x_3 - \frac{7}{8}\right).$$

Hence, the relationship between the polynomials $r(x)$ and $r_0(x) = x_1^2 - x_3^2 + x_2$, is $r(x) = r_0(ax + b)$ with

$$a = \begin{pmatrix} \sqrt{3} & 0 & -\sqrt{3}/3 \\ 0 & 1 & 0 \\ 0 & 0 & 2/\sqrt{3} \end{pmatrix}, \qquad b = \begin{pmatrix} 2\sqrt{3}/3 \\ 11/16 \\ -14/8\sqrt{3} \end{pmatrix}.$$

Equivalently,

$$r(a^{-1}(x - b)) = r_0(x).$$

Hence, as we explained in the Section "Construction of the adapted affine frame" on page 255, the adapted affine frame is determined by

$$M(\mathcal{R}', \mathcal{R}) = \begin{pmatrix} a^{-1} & -a^{-1}b \\ 0 & 1 \end{pmatrix},$$

where $\mathcal{R}$ is the initial affine frame.

The calculation gives

$$a^{-1} = \begin{pmatrix} \sqrt{3}/3 & 0 & \sqrt{3}/6 \\ 0 & 1 & 0 \\ 0 & 0 & \sqrt{3}/2 \end{pmatrix}, \qquad -a^{-1}b = \begin{pmatrix} -3/8 \\ -11/16 \\ 7/8 \end{pmatrix}.$$

The origin of $\mathcal{R}'$ is the point C given, in $\mathcal{R}$, by $C = (-3/8, -11/16, 7/8)$ and the vectors of the basis are

$$u = (\sqrt{3}/3, 0, 0), \qquad v = (0, 1, 0), \qquad w = (\sqrt{3}/6, 0, \sqrt{3}/2),$$

and in this affine frame the quadric has equation $x_1^2 - x_3^2 + x_2 = 0$. It is a hyperbolic paraboloid.   □

8.15 Quadrics Without Regular Points

As we said in the classification of polynomials, page 249, if a quadric is equivalent to a quadric of type (II), (III) or (IV), all its points are regular. If it is equivalent to a quadric of type (I) with $i \neq 0$, it has regular and non-regular points.

If a quadric is equivalent to a quadric of type (I), with $i = 0$, it does not have any regular points. These quadrics are the only ones that are linear varieties:

$$x_1^2 + \cdots + x_\rho^2 = 0$$

is the linear variety with equations $x_1 = \cdots = x_\rho = 0$. Since bijective affinities take linear varieties to linear varieties of the same dimension, any quadric equivalent to a quadric of type (I), with $i = 0$, is a linear variety of dimension $n - \rho$.

The Classification Theorem tells us that quadrics of type (I) with $i = 0$ are the only ones contained in a linear variety. We give a direct proof of this in the following proposition.

Proposition 8.32

Let us assume that a quadric $\mathcal{Q} = \mathcal{Q}(r(x), \mathcal{R})$ of an affine space $\mathbb{A}$ is contained in a linear variety $L \neq \mathbb{A}$. Then $\mathcal{Q}$ does not have any regular points.

Proof

Every linear variety L is contained in a hyperplane Π, that is, in a linear variety of codimension 1. Let $s(x) = a_1 x_1 + \cdots + a_n x_n + c = 0$ be the equation of Π. By hypothesis, every zero of $r(x)$ is a zero of $s^2(x)$. Suppose $r(x)$ has a regular zero, then, by Theorem D.4, page 399, $r(x) = \lambda s^2(x)$. In particular, $\mathcal{Q} = L = \Pi$. But $s^2(x)$ has no regular zeros, because $\operatorname{grad} s^2(x) = 2s(x)(a_1, \ldots, a_n)$, which vanishes when $s(x)$ vanishes. Contradiction.

Therefore, $\mathcal{Q}$ does not have any regular points.   □

Hence, quadrics contained in linear varieties are linear varieties. In this case, the equations of this linear variety can be found from the quadratic polynomial defining the quadric. ·

Proposition 8.33

Let us assume that a quadric $\mathcal{Q} = \mathcal{Q}(r(x), \mathcal{R})$ of an affine space $\mathbb{A}$, with $r(x) = x^\mathsf{T} A x + B x + C$, is contained in a linear variety $L \neq \mathbb{A}$. Then $\mathcal{Q}$ is the linear variety with equations

$$\begin{cases} Bx + 2C = 0, \\ 2Ax + B^{\mathsf{T}} = 0. \end{cases}$$

Proof

By Proposition 8.32, $\operatorname{grad} r(x) = 2Ax + B^{\mathsf{T}} = 0$ on the points of $\mathcal{Q}$. Substituting this equality into the expression for $r(x)$ we obtain

$$x^{\mathsf{T}}Ax + Bx + C = -\frac{1}{2}x^{\mathsf{T}}B^{\mathsf{T}} + Bx + C = \frac{1}{2}Bx + C.$$

Hence, $r(x) = 0$ if and only if $2Ax + B^{\mathsf{T}} = 0$ and $Bx + 2C = 0$.   □

For instance, in $\mathbb{R}^3$, the quadric $x_1^2 + x_2^2 = 0$ is, as a set of points, the straight line $x_1 = x_2 = 0$, because the first equation says nothing, since $B = 0$ and $C = 0$, and the second equation is

$$2Ax + B^{\mathsf{T}} = 2\begin{pmatrix} 1 & 0 \\ 0 & 1 \end{pmatrix}\begin{pmatrix} x_1 \\ x_2 \end{pmatrix} = \begin{pmatrix} 0 \\ 0 \end{pmatrix}.$$

However, a circle is a quadric of $\mathbb{R}^2$, but it is not a quadric of $\mathbb{R}^3$. See Exercises 8.8 and 8.9, page 281.

8.16 Quadrics with Center

Definition 8.34

We say that a point $G \in \mathbb{A}$ is a *center* of a quadric $\mathcal{Q}$ if

$$S_G(\mathcal{Q}) = \mathcal{Q},$$

where S_G is the central symmetry with respect to G.

Let us recall that the central symmetry with respect to a point G, S_G, is the affinity taking every point $X \in \mathbb{A}$ to the point $X' = S_G(X)$ such that G is the midpoint of the line segment XX'. In coordinates, S_G is given by the equations

$$x' = 2g - x,$$

where g is the column matrix formed by the coordinates of G. See the definition of central symmetry on page 71.

Notice that the center may or may not belong to the quadric.

Proposition 8.35

Let $\mathcal{Q} = \mathcal{Q}(r(x), \mathcal{R})$ be a non-empty quadric with $r(x) = x^{\mathsf{T}}Ax + Bx + C$. Let us assume that $\operatorname{rank}(A) = \operatorname{rank}(A|B^{\mathsf{T}})$. Then $\mathcal{Q}$ has a linear variety of centers of dimension $n - \rho$, where $\rho = \operatorname{rank}(A)$.

Proof

By the hypothesis on the rank we know that, in some affine frame, $\mathcal{Q}$ is given by an equation of type (I), (II) or (III). The central symmetry S_G, where G is any point of the type

$$G = (0, \dots, 0, a_{\rho+1}, \dots, a_n),$$

leaves $\mathcal{Q}$ invariant. In fact, this symmetry is given by

$$(x_1, \dots, x_n) \mapsto (-x_1, \dots, -x_\rho, 2a_{\rho+1} - x_{\rho+1}, \dots, 2a_n - x_n),$$

and it is clear that this transformation leaves invariant equations of type (I), (II) or (III). $\qquad\square$

However, if a quadric is given by an equation of type (IV), it is not invariant under any central symmetry.

Observe that the quadric $\mathcal{Q}(r(x), \mathcal{R})$ has a center if the polynomial $r(x)$ is equivalent to a polynomial without linear part, and that if a quadric has more than one center, it has infinitely many centers.

Proposition 8.36 (Equations of the center)

Let $\mathcal{Q} = \mathcal{Q}(r(x), \mathcal{R})$ be a non-empty quadric with $r(x) = x^{\mathsf{T}}Ax + Bx + C$. Let $G \in \mathbb{A}$, and let $(g_1, \dots, g_n)$ be the coordinates of G in $\mathcal{R}$. Then these coordinates are a solution of the system

$$2Ax = -B^{\mathsf{T}}$$

if and only if G is a center of $\mathcal{Q}$.

Proof

Let us first assume that $2Ag = -B^{\mathsf{T}}$, where g denotes the column matrix formed by the coordinates of G in $\mathcal{R}$. Let X be an arbitrary point of $\mathcal{Q}$. We want to show that the point X', symmetric to X with respect to G, also belongs to the quadric.

The coordinates of X' satisfy the matricial equality

$$x' = 2g - x;$$

hence,

$$
\begin{aligned}
x'^\mathsf{T} A x' &+ B x' + C \\
&= (2g - x)^\mathsf{T} A(2g - x) + B(2g - x) + C \\
&= x^\mathsf{T} A x - 4g^\mathsf{T} A x + 4g^\mathsf{T} A g + 2Bg - Bx + C \\
&= -4g^\mathsf{T} A x + 4g^\mathsf{T} A g + 2Bg - 2Bx \\
&= 2(2g^\mathsf{T} A + B)(g - x) \\
&= 0,
\end{aligned}
$$

since the condition $2Ag = -B^\mathsf{T}$ is equivalent, transposing, to $2g^\mathsf{T} A = -B$.

Conversely, let us assume that $G \in \mathbb{A}$ is a point such that $S_G(\mathcal{Q}) = \mathcal{Q}$. Since the equation of S_G is $x' = 2g - x$, the above equality implies that if x satisfies

$$x^\mathsf{T} A x + B x + C = 0,$$

then

$$x'^\mathsf{T} A x' + B x' + C = 0,$$

with $x' = 2g - x$.

By the above calculation, we have

$$(2g^\mathsf{T} A + B)(g - x) = 0 \tag{8.25}$$

for each point x such that $r(x) = 0$.

But this implies $2Ag + B^\mathsf{T} = 0$. Indeed, if $2Ag + B^\mathsf{T} \neq 0$, (8.25) is the equation of a hyperplane, and hence $\mathcal{Q}$ would be contained in this hyperplane. By Proposition 8.33, $\mathcal{Q}$ is contained in the linear variety with equations $2Ax + B^\mathsf{T} = 0$. It is clear that the center of symmetry of a quadric contained in a linear variety must be in this linear variety. Hence, we must have $2Ag + B^\mathsf{T} = 0$, a contradiction. $\qquad\square$

Observe that a quadric has a center if and only if $\mathrm{rank}(A) = \mathrm{rank}(A|B^\mathsf{T})$. That is, *a quadric has a center if and only if it is of type* (I), (II) *or* (III).

Definition 8.37

Let $Q = Q(r(x), \mathcal{R})$ be a non-empty quadric with $r(x) = x^\mathsf{T} A x + B x + C$. We say that Q is *non-degenerate* if $\det A \neq 0$.

Observe that, by Propositions 8.9 and 8.24, this definition neither depends on the polynomial nor on the affine frame used to define Q. In fact, the condition $\det A \neq 0$ is equivalent to the condition that Q has maximum quadratic rank; see page 254.

Corollary 8.38

Let $Q = Q(r(x), \mathcal{R})$ be a non-empty quadric with $r(x) = x^\mathsf{T} A x + B x + C$. Then Q has a unique center if and only if it is non-degenerate.

Proof

The system

$$2Ax = -B^\mathsf{T}$$

has a unique solution if and only if $\det A \neq 0$. □

Corollary 8.39

Let $Q = Q(r(x), \mathcal{R})$ be a non-empty quadric with $r(x) = x^\mathsf{T} A x + B x + C$. Then a point $G \in \mathbb{A}$ is a center of Q if and only if

$$\operatorname{grad} r(g) = 0,$$

where g denotes the coordinates of G in $\mathcal{R}$.

Proof

We know that $\operatorname{grad} r(x) = B + 2x^\mathsf{T} A$, see page 248. □

In practice, to calculate the center, we just solve the system

$$\boxed{\text{Equations of the center:} \quad \frac{\partial r(x)}{\partial x_i} = 0, \quad i = 1, \dots, n.}$$

This is very natural. Consider, for example, the family of ellipses in the Euclidean affine plane $\mathbb{R}^2$,

$$r_t(x) = \frac{x_1^2}{a^2} + \frac{x_2^2}{b^2} = t, \quad 0 < t \le 1.$$

The vector

$$\operatorname{grad} r_t(x) = \left(\frac{\partial r_t}{\partial x_1}, \frac{\partial r_t}{\partial x_2} \right)$$

is the gradient vector of this family, and so it is perpendicular to each ellipse in the family. These ellipses, as t tends to zero, contract onto the center. Hence, on the center, this gradient vector must vanish.

8.17 Tangent Cone

In this section we only consider quadrics possessing a regular point.

Definition 8.40 (Tangent straight line)

We say that a straight line $l \colon P + \langle v \rangle$ is tangent to a quadric $\mathcal{Q} = \mathcal{Q}(r(x), \mathcal{R})$, with $r(x) = x^{\mathsf{T}} A x + B x + C$, when $l \cap \mathcal{Q}$ is a point, and $v^{\mathsf{T}} A v \ne 0$.

As usual, when we write $v^{\mathsf{T}} A v$ we are identifying the vector v of the vector space associated to the affine space $\mathbb{A}$ with the column matrix formed by its components in the basis $\mathcal{B}$ associated to $\mathcal{R}$.

The condition $v^{\mathsf{T}} A v \ne 0$ neither depends on the polynomial nor on the affine frame used to define $\mathcal{Q}$. Indeed, let us assume that $\mathcal{Q}(r(x), \mathcal{R}) = \mathcal{Q}(s(x), \mathcal{R}')$, with $s(x) = x^{\mathsf{T}} A' x + B' x + C'$. Since $\mathcal{Q}$ has a regular point, there exists a $\lambda \in \mathbb{R}$, $\lambda \ne 0$ such that $\lambda A' = a^{\mathsf{T}} A a$ (see Theorem 8.22 and (8.7)).

But the relationship between the components of v in the bases $\mathcal{B}$ and $\mathcal{B}'$, associated respectively to $\mathcal{R}$ and $\mathcal{R}'$, is $v = a v'$, with $a = M(\mathcal{B}', \mathcal{B})$. Hence,

$$v^{\mathsf{T}} A v = (a v')^{\mathsf{T}} A a v = v'^{\mathsf{T}} A' v',$$

and the condition $v^{\mathsf{T}} A v \ne 0$ is independent of the polynomial and of the affine frame.

We can also say that a straight line l is tangent to the quadric $\mathcal{Q}$ when $l \cap \mathcal{Q}$ reduces to a *double* point. In fact, to calculate $l \cap \mathcal{Q}$ we must find a $\lambda \in \mathbb{R}$ such that the point $P + \lambda v$ belongs to the quadric. That is, denoting by p the column matrix of the components of P, we must solve the equation

$$(p + \lambda v)^{\mathsf{T}} A(p + \lambda v) + B(p + \lambda v) + C$$

$$= \lambda^2 v^{\mathsf{T}} A v + \lambda \operatorname{grad} r(p)(v) + r(p)$$

$$= 0, \tag{8.26}$$

and this equation is of second degree if and only if $v^{\mathsf{T}} A v \neq 0$.

Hence, the definition of tangent straight line we have just given is equivalent to the following.

Definition 8.41 (Tangent straight line)

We say that a straight line l is tangent to the quadric $\mathcal{Q}$ when $l \cap \mathcal{Q}$ reduces to a double point (see Figure 8.4).

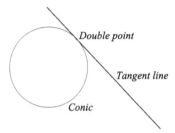

Figure 8.4.   Tangent line

Observe that if A is positive definite, the condition $v^{\mathsf{T}} A v \neq 0$ is automatically satisfied. When $v^{\mathsf{T}} A v = 0$ we say that v is an *isotropic* vector with respect to A (or that l is an isotropic straight line with respect to A); see the definition of an isotropic vector of a bilinear symmetric map on page 378. Thus we can say that *l is tangent to $\mathcal{Q}$ when $l \cap \mathcal{Q}$ reduces to a point and l is non-isotropic with respect to A.*

Since (8.26) has only one solution, we must have

$$(\operatorname{grad}(r(p))(v))^2 = 4(r(p))(v^{\mathsf{T}} A v) \quad (v^{\mathsf{T}} A v \neq 0). \tag{8.27}$$

In this case, the point $l \cap \mathcal{Q}$ is the point $P + \lambda v$ with

$$\lambda = \frac{-\operatorname{grad}(r(p))(v)}{2(v^{\mathsf{T}} A v)}.$$

In particular, it is easy to see that this λ satisfies

$$\operatorname{grad}(r(p + \lambda v))(v) = 0.$$

If $\mathbb{A}$ is the Euclidean affine space $\mathbb{R}^n$, the above equality tells us that the product of a row matrix by a column matrix is equal to zero, which in other words tells us that the ordinary scalar product of the vectors $\mathrm{grad}(r(p+\lambda v))$ and v is zero, and hence, that *the direction vector of the tangent straight line is orthogonal to the gradient vector at the contact point*. This justifies the name *tangent straight line*, since the tangent line is contained in the tangent plane (the plane orthogonal to the gradient) at the contact point.

Observation 8.42

The given definition of tangent straight line makes sense even if the gradient at the contact point vanishes. For instance, in the Euclidean plane $\mathbb{R}^2$, consider the conic $\mathcal{C}\colon x_1^2 - x_2^2 = 0$ and a point $P = (a, b)$ with $a^2 - b^2 \neq 0$. The straight lines $ay - bx = 0$ and $x - y + b - a = 0$ pass through P and cut $\mathcal{C}$, respectively, in a single point, see Figure 8.5.

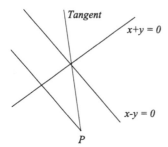

Figure 8.5.   Tangent line in a singular point

The first line is tangent, because the condition $v^{\mathsf{T}} A v \neq 0$ is

$$\begin{pmatrix} a & b \end{pmatrix} \begin{pmatrix} 1 & 0 \\ 0 & -1 \end{pmatrix} \begin{pmatrix} a \\ b \end{pmatrix} = a^2 - b^2 \neq 0,$$

and at the contact point $(0, 0)$ the gradient vanishes; however, the second line is not tangent, since the condition $v^{\mathsf{T}} A v \neq 0$ is not satisfied, because the direction vector of the straight line is $v = (1, 1)$ and, hence,

$$v^{\mathsf{T}} A v = \begin{pmatrix} 1 & 1 \end{pmatrix} \begin{pmatrix} 1 & 0 \\ 0 & -1 \end{pmatrix} \begin{pmatrix} 1 \\ 1 \end{pmatrix} = 1 - 1 = 0.$$

Note that if we remove the condition $v^{\mathsf{T}} A v \neq 0$ in (8.27), there are straight lines which are not tangent, but which nevertheless satisfy the equation. These are, explicitly, all those straight lines $l\colon P + \langle v \rangle$ with

$$\mathrm{grad}\, r(p)(v) = 0, \quad v^{\mathsf{T}} A v = 0.$$

If, in addition to these two conditions, $r(p) = 0$, the straight line l is contained in the quadric, since (8.26) has infinitely many solutions; if $r(p) \neq 0$, the straight line l does not cut the quadric, since (8.26) has no solution. In this case (i.e., $\operatorname{grad} r(p)(v) = 0$, $v^{\mathsf{T}} A v = 0$, $r(p) \neq 0$) we say that l is an *asymptote* of the quadric.

Example 8.43 $(r(p) = 0,\ \operatorname{grad} r(p)(v) = 0,\ v^{\mathsf{T}} A v = 0)$

In the affine plane $\mathbb{R}^2$, the straight line l: $P + \langle v \rangle$ with $P = (1, 1)$ and $v = (1, 1)$, is contained in the conic $x_1^2 - x_2^2 = 0$. It is not tangent, according to our definition, but it satisfies (8.27), without the condition $v^{\mathsf{T}} A v \neq 0$, since we have

$$\operatorname{grad} r(p)(v) = \begin{pmatrix} 2 & -2 \end{pmatrix} \begin{pmatrix} 1 \\ 1 \end{pmatrix} = 0,$$

$$v^{\mathsf{T}} A v = \begin{pmatrix} 1 & 1 \end{pmatrix} \begin{pmatrix} 1 & 0 \\ 0 & -1 \end{pmatrix} \begin{pmatrix} 1 \\ 1 \end{pmatrix} = 0.$$

Example 8.44 $(\operatorname{grad} r(p)(v) = 0,\ v^{\mathsf{T}} A v = 0)$

In the affine plane $\mathbb{R}^2$, the straight line l: $P + \langle v \rangle$ with $P = (0, 0)$ and $v = (1, 1)$ does not cut the hyperbola $x_1^2 - x_2^2 - 1 = 0$ but it satisfies (8.27), without the condition $v^{\mathsf{T}} A v \neq 0$, since we have

$$\operatorname{grad} r(p)(v) = \begin{pmatrix} 0 & 0 \end{pmatrix} \begin{pmatrix} 1 \\ 1 \end{pmatrix} = 0,$$

$$v^{\mathsf{T}} A v = \begin{pmatrix} 1 & 1 \end{pmatrix} \begin{pmatrix} 1 & 0 \\ 0 & 1 \end{pmatrix} \begin{pmatrix} 1 \\ 1 \end{pmatrix} = 0.$$

Notice that l is the asymptote of the hyperbola.

Example 8.45 $(\operatorname{grad} r(p)(v) = 0,\ v^{\mathsf{T}} A v = 0)$

In the affine space $\mathbb{R}^3$, the straight line l: $P + \langle v \rangle$, with $P = (0, 0, 0)$ and $v = (0, 0, 1)$, does not cut the cylinder $x_1^2 + x_2^2 - 1 = 0$, but it satisfies (8.27), without the condition $v^{\mathsf{T}} A v \neq 0$, since we have

$$\operatorname{grad} r(p)(v) = \begin{pmatrix} 0 & 0 & 0 \end{pmatrix} \begin{pmatrix} 0 \\ 0 \\ 1 \end{pmatrix} = 0,$$

$$v^{\mathsf{T}}Av = \begin{pmatrix} 0 & 0 & 1 \end{pmatrix} \begin{pmatrix} 1 & 0 & 0 \\ 0 & 1 & 0 \\ 0 & 0 & 0 \end{pmatrix} \begin{pmatrix} 0 \\ 0 \\ 1 \end{pmatrix} = 0.$$

Observe that l is the axis of the cylinder.

Definition 8.46 (Tangent cone)

The *tangent cone* to a quadric Q from a point P is the set T_P formed by P and by the points of all straight lines through P tangent, asymptotic or contained in the quadric.

Next we shall see that the tangent cone to a quadric is also a quadric and we shall study its equation.

The direction vectors of all these straight lines through P must satisfy (8.27). Substituting v by $(x-p)/\lambda$ in (8.27), we see that the coordinates x of the points of the straight lines through P tangent, asymptotic or contained in the quadric, satisfy

$$\boxed{\text{Tangent cone:} \quad (\operatorname{grad} r(p)(x-p))^2 = 4r(p)((x-p)^{\mathsf{T}}A(x-p))} \qquad (8.28)$$

This equation is of second degree, and therefore it represents a quadric, if we exclude the points such that $r(p) = \operatorname{grad} r(p) = 0$, since in this case the equation reduces to $0 = 0$. That is, when we consider the tangent cone to a quadric from a point, we are assuming either that this point does not belong to the quadric $(r(p) \neq 0)$, or that it is a regular point of the quadric $(r(p) = 0, \operatorname{grad} r(p) \neq 0)$. See Exercise 8.16, page 282.

Conversely, any point $x \neq p$ satisfying (8.28) is a point of the tangent cone. However $x = p$ is also a solution of (8.28) (this case was not considered above, because we supposed $v \neq \vec{0}$), which is in agreement with the fact that $P \in T_P$.

For instance, the tangent cone to the circle $x_1^2 + x_2^2 - 1 = 0$ from the point $P = (0,0)$ is $x_1^2 + x_2^2 = 0$, that is, $T_P = P$, although there is no straight line through P tangent, asymptotic or contained in the circle.

It can happen that the tangent cone does not contain any tangent straight lines. For instance, the tangent cone to the hyperboloid $x_1^2 + x_2^2 - x_3^2 + 1 = 0$ from the point $P = (0,0,0)$ is the cone $x_1^2 + x_2^2 - x_3^2 = 0$, which does not contain any straight line tangent to the hyperboloid (they are tangent at infinity, see Figure 8.6). See Exercise 8.15, page 281.

Since $P \in T_P$ is a center of T_P (the symmetry with respect to P leaves T_P invariant), T_P is a quadric of type (I), the only types of quadrics containing their center. For instance, in dimension 3, the tangent cone can be a cone, but

Figure 8.6.   Tangent cone

can also be a point, a straight line, a plane or two planes. See Exercise 8.15, page 281.

Observation 8.47 (Polar hyperplane)

The above calculations can be slightly simplified by recalling, as mentioned in Section 8.2.1, that the equation $r(x) = 0$ can be written as

$$\bar{x}^\mathsf{T} M \bar{x} = 0, \tag{8.29}$$

where $\bar{x} = (x_1, \ldots, x_n, 1)$, and M is the matrix associated to the polynomial, that is,

$$M = \begin{pmatrix} A & B^\mathsf{T}/2 \\ B/2 & C \end{pmatrix}. \qquad\bullet \tag{8.30}$$

Let

$$\bar{x} = (x_1, \ldots, x_n, 1),$$
$$\bar{p} = (p_1, \ldots, p_n, 1),$$
$$\bar{v} = (v_1, \ldots, v_n, 0).$$

Then the equation of the straight line $P + \langle v \rangle$ can be written as

$$\bar{x} = \bar{p} + \lambda \bar{v}.$$

The intersection of the straight line with the quadric can be obtained by substituting $\bar{x} = \bar{p} + \lambda \bar{v}$ into (8.29). We get

$$\bar{x}^{\mathsf{T}} M \bar{x} = (\bar{p} + \lambda \bar{v})^{\mathsf{T}} M (\bar{p} + \lambda \bar{v}) = 0.$$

This equation, quadratic in λ, must have a unique solution, and therefore its discriminant must vanish. That is,

$$(\bar{p}^{\mathsf{T}} M \bar{v})^2 - (\bar{p}^{\mathsf{T}} M \bar{p})(\bar{v}^{\mathsf{T}} M \bar{v}) = 0. \tag{8.31}$$

Replacing $\bar{v}$ by $(\bar{x} - \bar{p})/\lambda$ we find that the coordinates x of the points of the straight lines through P tangent to the quadric $\bar{x}^{\mathsf{T}} M \bar{x} = 0$ satisfy the equation

$$\boxed{\text{Tangent cone:} \quad (\bar{p}^{\mathsf{T}} M \bar{x})^2 - (\bar{p}^{\mathsf{T}} M \bar{p})(\bar{x}^{\mathsf{T}} M \bar{x}) = 0} \tag{8.32}$$

If we now substitute M by its expression (8.30), we obtain

$$\left(p^{\mathsf{T}} A x + \frac{1}{2} B(p + x) + C \right)^2 = r(p) r(x),$$

which is precisely (8.28), written in a slightly different form.

When P is a point of the quadric $(\bar{p}^{\mathsf{T}} M \bar{p} = 0)$, the equation of the tangent cone from P reduces to

$$\boxed{\bar{p}^{\mathsf{T}} M \bar{x} = 0}$$

which is a linear variety called a *tangent hyperplane*.

If P does not belong to the quadric, the hyperplane $\bar{p}^{\mathsf{T}} M \bar{x} = 0$ is said to be the *polar hyperplane* of P with respect to the quadric. The intersection of the polar hyperplane with the quadric is formed by points of the tangent cone, see Figure 8.7. This is obvious, because these points satisfy $\bar{p}^{\mathsf{T}} M \bar{x} = 0$ and $\bar{x}^{\mathsf{T}} M \bar{x} = 0$. See Exercise 8.17, page 282.

The reader well versed in Projective Geometry will have recognized in this section the projective treatment of quadrics.

Example 8.48

Find, in an affine space $\mathbb{A}$ of dimension 3, the tangent cone to the quadric given in some affine frame $\mathcal{R}$ by $x_1^2 + x_2^2 + x_3^2 - 1 = 0$, from the point $P = (2, 0, 0)$.

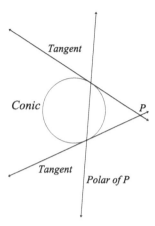

Figure 8.7. Polar line

Solution

In this case

$$M = \begin{pmatrix} 1 & 0 & 0 & 0 \\ 0 & 1 & 0 & 0 \\ 0 & 0 & 1 & 0 \\ 0 & 0 & 0 & -1 \end{pmatrix}.$$

Since $\bar{p} = (2, 0, 0, 1)$, we have $\bar{p}^T M \bar{p} = 3$, and hence the equation is

$$\left(\begin{pmatrix} 2 & 0 & 0 & 1 \end{pmatrix} \begin{pmatrix} 1 & 0 & 0 & 0 \\ 0 & 1 & 0 & 0 \\ 0 & 0 & 1 & 0 \\ 0 & 0 & 0 & -1 \end{pmatrix} \begin{pmatrix} x \\ y \\ z \\ 1 \end{pmatrix} \right)^2 = 3(x_1^2 + x_2^2 + x_3^2 - 1).$$

This gives $x_1^2 - 4x_1 + 4 = 3(x_2^2 + x_3^2)$. See Figure 8.8. □

Example 8.49

Find, in an affine space $\mathbb{A}$ of dimension 2, the tangent cone to the quadric given in some affine frame $\mathcal{R}$ by $x_1^2 + 2x_2^2 - 1 = 0$, from the point $P = (2, 4)$.

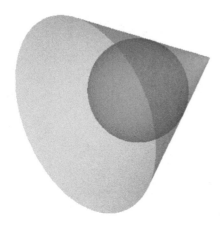

Figure 8.8. Tangent cone

Solution

In this case

$$M = \begin{pmatrix} 1 & 0 & 0 \\ 0 & 2 & 0 \\ 0 & 0 & -1 \end{pmatrix}.$$

Since $\bar{p} = (2, 4, 1)$, we have $\bar{p}^{\mathsf{T}} M \bar{p} = 35$, and hence the equation is

$$\left((2\ \ 4\ \ 1) \begin{pmatrix} 1 & 0 & 0 \\ 0 & 2 & 0 \\ 0 & 0 & -1 \end{pmatrix} \begin{pmatrix} x_1 \\ x_2 \\ 1 \end{pmatrix} \right)^2 = 35(x_1^2 + 2x_2^2 - 1).$$

This gives $-31x_1^2 - 6x_2^2 + 32x_1x_2 - 4x_1 - 16x_2 + 36 = 0$. But this quadratic polynomial is the product of two linear polynomials:

$$-31x_1^2 - 6x_2^2 + 32x_1x_2 - 4x_1 - 16x_2 + 36$$

$$= \left((16 - \sqrt{70})x_1 - 6x_2 - (8 - 2\sqrt{70}) \right)$$

$$\cdot \left((16 + \sqrt{70})x_1 - 6x_2 - (8 + 2\sqrt{70}) \right),$$

so that the tangent cone is reduced to the two tangent straight lines

$$(16 - \sqrt{70})x_1 - 6x_2 - (8 - 2\sqrt{70}) = 0,$$

$$(16 + \sqrt{70})x_1 - 6x_2 - (8 + 2\sqrt{70}) = 0.$$

The polar straight line is $2x_1 + 8x_2 - 1 = 0$. See Figure 8.9.   □

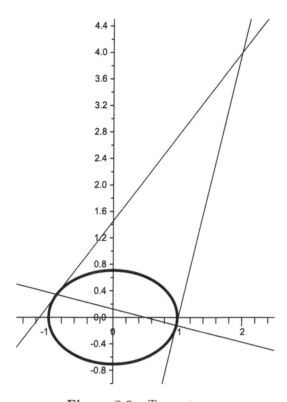

Figure 8.9. Tangent cone

EXERCISES

8.1. Let us consider, in an affine space $\mathbb{A}$ of dimension 2, the conic given in some affine frame $\mathcal{R}$ by

$$2x_1^2 + 8x_2^2 - 4x_1x_2 + \sqrt{2}x_2 = 0.$$

(a) Is the conic an ellipse, a hyperbola or a parabola?
(b) Find a central symmetry leaving the conic invariant.
(c) Find an axial symmetry leaving the conic invariant.

8.2. Let us consider, in an affine space $\mathbb{A}$ of dimension 2, the conic given in some affine frame $\mathcal{R}$ by

$$3x_1^2 + 7x_2^2 + 6x_1x_2 + x_1 + x_2 = 0.$$

(a) Find a 2×2 symmetric matrix A and a 1×2 matrix B such

that

$$3x_1^2 + 7x_2^2 + 6x_1x_2 + x_1 + x_2 = \begin{pmatrix} x_1 & x_2 \end{pmatrix} A \begin{pmatrix} x_1 \\ x_2 \end{pmatrix} + B \begin{pmatrix} x_1 \\ x_2 \end{pmatrix}.$$

(b) Find the center of the conic, that is, find a point $C = (c_1, c_2)$ such that

$$2A \begin{pmatrix} c_1 \\ c_2 \end{pmatrix} = -B^\mathsf{T}.$$

(c) Find a 2×2 invertible matrix a such that

$$a^\mathsf{T} A a = \begin{pmatrix} 1 & 0 \\ 0 & 1 \end{pmatrix}.$$

(d) Prove that the affinity

$$\begin{pmatrix} x' \\ y' \end{pmatrix} = a^{-1} \begin{pmatrix} x \\ y \end{pmatrix} - a^{-1} \begin{pmatrix} c_1 \\ c_2 \end{pmatrix}$$

transforms the initial conic into the conic given by $x_1^2 + x_2^2 = 1$.

(e) Find, with respect to $\mathcal{R}$, a central symmetry that leaves the initial conic invariant.

(f) Find, with respect to $\mathcal{R}$, an axial symmetry that leaves the initial conic invariant.

8.3. Classify the conics of an affine plane $\mathbb{A}$ given in an affine frame $\mathcal{R}$ by

$$2x_1^2 - x_2^2 - x_1 - x_2 - 3 = 0,$$

$$-3x_1^2 + 9x_2^2 + 6x_1x_2 - 4x_1 - 1 = 0,$$

$$x_1x_2 + 1 = 0.$$

Find, if it exists, the center of each of these conics.

8.4. Classify the quadrics of an affine space $\mathbb{A}$ of dimension three given in an affine frame $\mathcal{R}$ by

$$2x_1^2 - x_2^2 - x_1 - x_2 - 3 = 0,$$

$$-3x_1^2 + 9x_2^2 + 6x_1x_2 - 4x_1 - 1 = 0,$$

$$x_1x_2 + 1 = 0.$$

Find, if it exists, the center of each of these quadrics.

8.5. Classify the quadrics of an affine space $\mathbb{A}$ of dimension three given in an affine frame $\mathcal{R}$ by

$$x_1^2 - x_2^2 - 4x_1x_3 - 8x_2x_3 - 3x_2 + x_3 - 1 = 0,$$

$$-x_1^2 + x_3^2 + 6x_1x_2 - 4x_1 - x_2 = 0,$$

$$x_3^2 + x_1x_2 + 1 = 0.$$

Find, if it exists, the center of each of these quadrics.

8.6. Given the polynomials $r(x) = x_1^2 + x_1x_2 - x_1 + 2$ and $s(x) = 2x_1^2 + 4x_2^2 - 6x_1x_2 + 2$, find two affine frames $\mathcal{R}$ and $\mathcal{R}'$ of the affine plane $\mathbb{R}^2$ such that

$$\mathcal{Q}(r(x), \mathcal{R}) = \mathcal{Q}(s(x), \mathcal{R}').$$

8.7. Characterize the points of the affine plane $\mathbb{R}^2$ such that the conic through them and through the points $(0,0), (1,0), (0,1), (2,2)$ is an ellipse. (This is adapted from an exercise in [4].)

8.8. Is the sphere S^2 of $\mathbb{R}^3$ a quadric of $\mathbb{R}^4$?

8.9. Study the compatibility of the system of equations considered in Proposition 8.33.

8.10. Find the affine classification of real quadrics in dimension four.

8.11. Using Proposition B.12, page 358, study the classification of quadrics on a complex affine space. Show that every quadric of a complex affine space of dimension n is equivalent to one and only one of the quadrics given, in some affine frame, by the following equations:

(I) $x_1^2 + \cdots + x_\rho^2 = 0$,

(II/III) $x_1^2 + \cdots + x_\rho^2 + 1 = 0$,

(IV) $x_1^2 + \cdots + x_\rho^2 + x_{\rho+1} = 0$,

where $0 < \rho \le n$ in cases (I) and (II/III), and $0 < \rho < n$ in case (IV).

8.12. Given two conics in different planes of the affine space $\mathbb{R}^3$, with two common points, and given a point not contained in these planes, prove that there is a unique quadric containing the conics and the point.

Find the quadric containing the point $(1,1,1)$ and the circles $x_1^2 + x_3^2 = 1$, $x_2 = 0$ and $x_2^2 + x_3^2 = 1$, $x_1 = 0$. Find a quadric containing the circle $(x_1 - 1)^2 + x_3^2 = 1$, $x_2 = 0$, the parabola $x_2 = x_3^2$, $x_1 = 0$ and the point $(1,1,1)$, see Figure 8.10.

8.13. Given a hyperboloid of one sheet, find planes cutting it in two straight lines, in an ellipse, in a hyperbola or in a parabola. Given a hyperboloid of two sheets, find planes cutting it in an ellipse, a hyperbola or in a parabola. Given an elliptic paraboloid, find planes cutting it in an ellipse or in a parabola. Given a hyperbolic paraboloid, find planes cutting it in a hyperbola or in a parabola.

8.14. In the affine space $\mathbb{R}^3$, find the tangent cone to the quadric $x_1^2 + x_2^2 - x_3^2 + 1 = 0$ from the point $(5,5,5)$. See Figure 8.11.

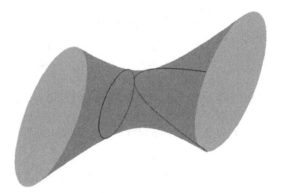

Figure 8.10. Quadric containing two conics

Figure 8.11. Tangent cone

8.15. Find the tangent cone to the parabolic cylinder of $\mathbb{R}^3$ given by the equation $x_2 = x_1^2$ from the point $(1,0,0)$. Find the tangent cone to a sphere of $\mathbb{R}^3$ from its center. Find the tangent cone to the quadric of $\mathbb{R}^3$ given by $x_1^2 - 1 = 0$ from the point $(3,0,0)$. Find the tangent cone to a hyperboloid of one sheet from a point on its axis. Find the tangent cones to $x_1^2 - x_2^2 = 0$ from the points $(0,0)$ and $(1,1)$.

8.16. Prove that (8.28), defining the tangent cone to a quadric from a point, is of second degree, except if the point is a non-regular point of the quadric, or when the quadric is a double hyperplane (this case was not considered earlier, because the tangent cone has been defined only for quadrics with a regular point).

8.17. In the notation of Observation 8.47, page 275, prove that the equation $\bar{p}^{\mathsf{T}} M \bar{x} = 0$ represents a hyperplane (the *polar hyperplane* of the point P with respect to the quadric $\bar{x}^{\mathsf{T}} M \bar{x} = 0$) if and only if $\operatorname{grad} r(p) \neq 0$. Observe that the condition $\operatorname{grad} r(p) \neq 0$ is equivalent to the condition that the point $p \in \mathbb{R}^n$ does not belong to the intersection of n hyperplanes. Find the polar hyperplane in all cases (quadric and point) considered in Exercise 8.15.

.

Orthogonal Classification of Quadrics

9.1 Introduction

In the previous chapter we regarded as "equal" a quadric and its transformation under an affinity. In this chapter we similarly identify a quadric and its transformation under a Euclidean motion. We will essentially use a well known result from algebra, the *Spectral Theorem* (also known as the *Simultaneous Diagonalization Theorem*, because it gives a basis with respect to which two given symmetric bilinear maps simultaneously diagonalize), see Theorem C.8, page 387.

9.2 Orthogonally Equivalent Quadratic Polynomials

In this and in the following sections, until Section 9.7, we will not mention affine spaces, focusing instead on polynomials. We define an equivalence relation between polynomials very similar to the relation we used in the previous chapter, Definition 8.2, but with the condition that the invertible matrix a appearing there is now an orthogonal matrix. This will allow us to say that two quadrics are *orthogonally equivalent* if there is a Euclidean motion (not an arbitrary affinity) mapping one of them onto the other.

A. Reventós Tarrida, *Affine Maps, Euclidean Motions and Quadrics*,
Springer Undergraduate Mathematics Series,
DOI 10.1007/978-0-85729-710-5_9, © Springer-Verlag London Limited 2011

Definition 9.1

The polynomials in n variables

$$r(x) = x^\mathsf{T} Ax + Bx + C,$$

$$s(x) = x^\mathsf{T} A'x + B'x + C',$$

are *orthogonally equivalent* if there is an orthogonal $a \in \mathcal{M}_{n \times n}(\mathbb{R})$, a column matrix $b \in \mathcal{M}_{n \times 1}(\mathbb{R})$, and a real number $\lambda \neq 0$, such that

$$r(ax + b) = \lambda s(x).$$

That is, two polynomials are orthogonally equivalent if we can formally transform one of them into a multiple of the other by an orthogonal change of variables.

Proposition 9.2

The polynomials in n variables

$$r(x) = x^\mathsf{T} Ax + Bx + C,$$

$$s(x) = x^\mathsf{T} A'x + B'x + C',$$

are orthogonally equivalent if and only if there is an orthogonal $a \in \mathcal{M}_{n \times n}(\mathbb{R})$, a column matrix $b \in \mathcal{M}_{n \times 1}(\mathbb{R})$, and a real number $\lambda \neq 0$, such that

$$\begin{pmatrix} a^\mathsf{T} & 0 \\ b^\mathsf{T} & 1 \end{pmatrix} \begin{pmatrix} A & B^\mathsf{T}/2 \\ B/2 & C \end{pmatrix} \begin{pmatrix} a & b \\ 0 & 1 \end{pmatrix} = \lambda \begin{pmatrix} A' & B'^\mathsf{T}/2 \\ B'/2 & C' \end{pmatrix}. \tag{9.1}$$

Proof

The proof is as in Proposition 8.3, page 232.   □

Therefore, two polynomials are orthogonally equivalent if and only if the symmetric matrices associated to them are, up to a scalar factor, conjugated by a matrix that can be interpreted as the matrix of a Euclidean motion in an orthonormal affine frame.

This relation is clearly an equivalence relation.

Corollary 9.3

The polynomials in n variables

$$r(x) = x^{\mathsf{T}} A x + B x + C,$$

$$s(x) = x^{\mathsf{T}} A' x + B' x + C'$$

are orthogonally equivalent if and only if there is an orthogonal $a \in \mathcal{M}_{n \times n}(\mathbb{R})$, a column matrix $b \in \mathcal{M}_{n \times 1}(\mathbb{R})$, and a real number $\lambda \neq 0$, such that

$$\lambda A' = a^{\mathsf{T}} A a,$$

$$\lambda B' = 2 b^{\mathsf{T}} A a + B a,$$

$$\lambda C' = b^{\mathsf{T}} A b + B b + C.$$

Proof

Simply multiply the matrices of (9.1). □

Notice that $r(x)$ is orthogonally equivalent to $\mu r(x)$, for all $\mu \in \mathbb{R}$, $\mu \neq 0$, as can be seen by taking $a = I_n$, $b = 0$ and $\lambda = \frac{1}{\mu}$.

Corollary 9.4

The quadratic polynomials in n variables and without linear part

$$r(x) = x^{\mathsf{T}} A x + C,$$

$$s(x) = x^{\mathsf{T}} A' x + C'$$

are orthogonally equivalent if and only if there is an orthogonal $a \in \mathcal{M}_{n \times n}(\mathbb{R})$ and a real number $\lambda \neq 0$ such that $\lambda C' = C$ and $\lambda A' = a^{\mathsf{T}} A a$.

Proof

The proof is as in Corollary 8.5, page 234. □

In particular, polynomials of the form $r(x) = x^{\mathsf{T}} A x$ cannot be equivalent to polynomials of the form $r(x) = x^{\mathsf{T}} A' x + C$, with $C \neq 0$.

Proposition 9.5

If the quadratic polynomials in n variables

$$r(x) = x^\mathsf{T} A x + B x + C,$$

$$s(x) = x^\mathsf{T} A' x + B' x + C',$$

are orthogonally equivalent, then the eigenvalues of A and A' are proportional.

Proof

The condition $\lambda A' = a^\mathsf{T} A a$ can be written as $\lambda A' = a^{-1} A a$, since a is orthogonal; and it is well known that A and $a^{-1} A a$ have the same characteristic polynomial (see, for instance [8], page 332).   □

Example 9.6

The polynomials

$$r(x) = (x_1 - 1)^2 + (x_2 - 1)^2 - 1,$$

$$s(x) = 2x_1^2 + 2x_2^2 - 2$$

are orthogonally equivalent, since if we make the change of variables

$$x_1 = x_1' + 1,$$

$$x_2 = x_2' + 1,$$

we have

$$r(x) = r(x_1' + 1, x_2' + 1) = x_1'^2 + x_2'^2 - 1 = \frac{1}{2} s(x').$$

This can also be seen directly from Proposition 9.2. It is sufficient to observe that the above change of variables can be written as

$$\begin{pmatrix} x_1 \\ x_2 \\ 1 \end{pmatrix} = \begin{pmatrix} 1 & 0 & 1 \\ 0 & 1 & 1 \\ 0 & 0 & 1 \end{pmatrix} \begin{pmatrix} x_1' \\ x_2' \\ 1 \end{pmatrix}. \tag{9.2}$$

But, since

$$r(x) = \begin{pmatrix} x_1 & x_2 & 1 \end{pmatrix} \begin{pmatrix} 1 & 0 & -1 \\ 0 & 1 & -1 \\ -1 & -1 & 1 \end{pmatrix} \begin{pmatrix} x_1 \\ x_2 \\ 1 \end{pmatrix},$$

and

$$s(x) = \begin{pmatrix} x_1 & x_2 & 1 \end{pmatrix} \begin{pmatrix} 2 & 0 & 0 \\ 0 & 2 & 0 \\ 0 & 0 & -2 \end{pmatrix} \begin{pmatrix} x_1 \\ x_2 \\ 1 \end{pmatrix},$$

substituting (9.2) into the expression of $r(x)$, one has

$$\begin{pmatrix} 1 & 0 & 0 \\ 0 & 1 & 0 \\ 1 & 1 & 1 \end{pmatrix} \begin{pmatrix} 1 & 0 & -1 \\ 0 & 1 & -1 \\ -1 & -1 & 1 \end{pmatrix} \begin{pmatrix} 1 & 0 & 1 \\ 0 & 1 & 1 \\ 0 & 0 & 1 \end{pmatrix} = \frac{1}{2} \begin{pmatrix} 2 & 0 & 0 \\ 0 & 2 & 0 \\ 0 & 0 & -2 \end{pmatrix},$$

which is exactly condition (9.1) with $\lambda = \frac{1}{2}$.

9.2.1 Invariants

Since orthogonally equivalent polynomials are equivalent in the sense given in Definition 8.3, it is clear that orthogonally equivalent polynomials have the same invariants $(\rho, i, \tilde{\rho}, \tilde{i})$. However, there are polynomials with the same invariants $(\rho, i, \tilde{\rho}, \tilde{i})$ that are not orthogonally equivalent, for instance, $x_1^2 + x_2^2 - 1$ and $x_1^2 + x_2^2 - 4$.

9.3 Criterion for Ordering

This section is purely technical, but has been introduced to facilitate the reading of the next section.

Definition 9.7 (Criterion for ordering)

We say that the non-zero real numbers $d_1 \geq \cdots \geq d_\rho$ are *well ordered* if one of the following four conditions is satisfied:

1. There are more positive d_is than negative d_is.
2. The number of positive d_is is equal to the number of negative d_is and $d_1 > |d_\rho|$.
3. The number of positive d_is is equal to the number of negative d_is and, if $d_j = |d_{\rho+1-j}|$, for $j = 1, \ldots, t-1$, with $t \leq \rho/2$, then $d_t > |d_{\rho+1-t}|$.
4. The number of positive d_is is equal to the number of negative d_is and $d_j = |d_{\rho+1-j}|$, for $j = 1, \ldots, \rho/2$.

Obviously, this definition is of interest only when the number of positive d_is is equal to the number of negative d_is, and this can only happen when ρ is even.

The four cases considered in the above definition can be condensed to only two by making use of the lexicographic order used in dictionaries.

Definition 9.8

We say that the non-zero real numbers $d_1 \geq \cdots \geq d_\rho$ are *well ordered* if one of the following two conditions is satisfied:

1. There are more positive d_is than negative d_is.
2. The number of positive d_is is equal to the number of negative d_is and $(d_1, \ldots, d_r) \geq (-d_\rho, \ldots, -d_{r+1})$, with $r = \rho/2$, in the lexicographic order.

In particular, if $d_1 \geq \cdots \geq d_\rho$ are well ordered, then $d_1 > 0$.

For instance, $8 \geq 4 \geq -1$ are well ordered, but the numbers that we obtain by multiplying by -1, $1 \geq -4 \geq -8$, are not.

$4 \geq 3 \geq 2 \geq -1 \geq -3 \geq -4$ are well ordered, since $4 = |-4|$, $3 = |-3|$, $2 > |-1|$. However, the numbers that we obtain by multiplying by -1, $4 \geq 3 \geq 1 \geq -2 \geq -3 \geq -4$ are not.

The numbers $8 \geq 4 \geq -4 \geq -8$ are well ordered, and they coincide with the numbers that we obtain multiplying by -1. The numbers $4 \geq 4 \geq -4 \geq -4$ are well ordered.

These examples suggest the following lemma.

Lemma 9.9

Given real numbers $d_1 \geq \cdots \geq d_\rho$, then $d_1 \geq \cdots \geq d_\rho$ or $-d_\rho \geq \cdots \geq -d_1$ are well ordered. If they are both well ordered, then they are equal.

Proof

It is evident that if the number of $d_j > 0$ is different from the number of $d_j < 0$ one of the two orderings has more positive terms than negative. If there are as many positive d_js as negative, then either $d_j = |d_{\rho+1-j}|$, for $j = 1, \ldots, \rho/2$, in which case both orderings coincide, or there is a first term satisfying $d_t \neq |d_{\rho+1-t}|$, in which case one and only one of the two orderings will satisfy $d_t \geq |d_{\rho+1-t}|$. $\qquad\square$

9.4 Canonical Representatives Without Linear Part

In this and in the following section we shall study the equivalence classes of polynomials corresponding to the equivalence relation of polynomials introduced in Definition 9.1. We shall see that within each class there is an especially simple polynomial, which we shall call the canonical representative of the class.

Theorem 9.10

Let $r(x) = x^\mathsf{T} Ax + Bx + C$ be a quadratic polynomial with

$$\operatorname{rank}(A) = \operatorname{rank}(A|B^\mathsf{T}).$$

Then $r(x)$ is orthogonally equivalent to a polynomial of the form

$$s(x) = x^\mathsf{T} Dx + C',$$

where D is a diagonal matrix of the same rank as A, and C' is a constant.

Proof

In order to find a polynomial without linear part equivalent to

$$r(x) = x^\mathsf{T} Ax + Bx + C,$$

it is sufficient to find, by Corollary 9.3 with $\lambda = 1$, a real orthogonal matrix a and a real column matrix b such that

$$D = a^\mathsf{T} Aa, \quad \text{with } D \text{ diagonal,}$$

$$0 = (2b^\mathsf{T} A + B)a,$$

$$C' = b^\mathsf{T} Ab + Bb + C = C + \frac{1}{2} Bb.$$

The third condition is irrelevant, because no condition is imposed on C'. Since a is invertible, the second condition holds if and only if

$$2b^\mathsf{T} A + B = 0,$$

or, transposing, if and only if

$$2Ab = -B^\mathsf{T}.$$

Hence, for the column matrix b we can take any solution of the system

$$\boxed{2Ax = -B^\mathsf{T}} \tag{9.3}$$

This system has a solution, because we are assuming $\operatorname{rank}(A) = \operatorname{rank}(A|B^\mathsf{T})$.

Observe that if $\det A \neq 0$, the condition on the ranks is automatically satisfied, and the system (9.3) has a unique solution. We have already mentioned that this system is called the *system of equations of the center*.

The existence of a real orthogonal matrix a satisfying the first condition ($a^\mathsf{T} Aa$ diagonal) is a direct consequence of the matricial version of the Spectral

Theorem, see Theorem C.11, page 389. This theorem says precisely that given a real symmetric matrix A, there is a real orthogonal matrix a such that the matrix

$$D = a^{\mathsf{T}} A a$$

is diagonal. This completes the proof. $\qquad\qquad\qquad\qquad\qquad\qquad\qquad\qquad\square$

Corollary 9.11

Let $r(x) = x^{\mathsf{T}} A x + B x + C$ be a quadratic polynomial with

$$\operatorname{rank}(A) = \operatorname{rank}(A|B^{\mathsf{T}}).$$

Then $r(x)$ is either orthogonally equivalent to a polynomial of the form

$$s(x) = x_1^2 + d_2 x_2^2 + \cdots + d_\rho x_\rho^2,$$

with $1 \geq d_2 \geq \cdots \geq d_\rho$ well ordered, or to a polynomial of the form

$$t(x) = d_1 x_1^2 + d_2 x_2^2 + \cdots + d_\rho x_\rho^2 + 1,$$

with $d_1 \geq \cdots \geq d_\rho$.

Proof

Permuting if necessary the variables and taking into account that a polynomial is orthogonally equivalent to its negative, we may assume, by the above theorem and Lemma 9.9, that $r(x)$ is equivalent to

$$s'(x) = d_1' x_1^2 + \cdots + d_\rho' x_\rho^2 + C', \tag{9.4}$$

with $d_1' \geq \cdots \geq d_\rho'$ well ordered, and $C' \in \mathbb{R}$ ($C' = \pm(C + \frac{1}{2} B b)$). Notice that polynomials differing only by a permutation of the variables are orthogonally equivalent, since the matrix a performing this permutation is obtained by permuting the corresponding rows in the identity matrix, see Observation 8.6, and such a matrix is orthogonal.

Now we have two cases, depending on the vanishing of C'.

First case, $C' = 0$. Dividing (9.4) by d_1', we obtain that the initial polynomial is orthogonally equivalent to

$$\boxed{s(x) = x_1^2 + d_2 x_2^2 + \cdots + d_\rho x_\rho^2} \tag{9.5}$$

with $1 \geq d_2 \geq \cdots \geq d_\rho$ well ordered.

Second case, $C' > 0$. Dividing (9.4) by C', we obtain that the initial polynomial is orthogonally equivalent to

$$\boxed{t(x) = d_1 x_1^2 + \cdots + d_\rho x_\rho^2 + 1} \tag{9.6}$$

with $d_1 \geq \cdots \geq d_\rho$ well ordered.

Third case, $C' < 0$. Dividing (9.4) by C', we obtain that the initial polynomial is orthogonally equivalent to

$$t(x) = \bar{d}_1 x_1^2 + \cdots + \bar{d}_\rho x_\rho^2 + 1$$

with $\bar{d}_1 \leq \cdots \leq \bar{d}_\rho$ (not, in general, well ordered). Replacing the variables x_i by $x_{\rho+1-i}$ and $\bar{d}_i$ by $d_{\rho+1-i}$, $i = 1, \ldots, \rho$, we have

$$\boxed{t(x) = d_1 x_1^2 + \cdots + d_\rho x_\rho^2 + 1} \tag{9.7}$$

with $d_1 \geq d_2 \geq \cdots \geq d_\rho$, and this completes the proof. $\qquad\square$

Finally observe that, in the notation introduced in Theorem 8.13, polynomials orthogonally equivalent to polynomials of the form (9.5) are of type (I). Polynomials orthogonally equivalent to polynomials of the form (9.7) with $\rho = 2i$, or with $\rho \neq 2i$ but with more d_js positive than negative, are of type (II), since in this case $\tilde{i} = i$; and if there are fewer d_js positive than negative, they are of type (III), since in this case $\tilde{i} = i + 1$.

9.5 Canonical Representatives with Linear Part

We begin with a technical lemma that we will need later. It refers to a special class of polynomials, those in which the variables appearing in the quadratic part do not appear in the linear part.

Lemma 9.12 (Separated variables)

Let us consider the polynomial

$$r(x) = d_1 x_1^2 + \cdots + d_\rho x_\rho^2 + b_{\rho+1} x_{\rho+1} + \cdots + b_n x_n + C.$$

Assume that some b_i, for instance $b_{\rho+1}$, is different from zero. Then, $r(x)$ is orthogonally equivalent to

$$s(x) = d_1 x_1^2 + \cdots + d_\rho x_\rho^2 + M x_{\rho+1},$$

where

$$M = \sqrt{b_{\rho+1}^2 + \cdots + b_n^2}.$$

Proof

The idea is to make the change of variables

$$
\begin{aligned}
x_i' &= x_i, \quad i \le \rho, \\
x_{\rho+1}' &= M^{-1}(b_{\rho+1}x_{\rho+1} + \cdots + b_n x_n + C), \\
x_j' &= \sum_{k=\rho+1}^{n} p_{kj} x_k, \quad j = \rho+2, \ldots, n,
\end{aligned}
\tag{9.8}
$$

in such a way that the matrix

$$
P = \begin{pmatrix}
b_{\rho+1}/M & p_{\rho+1,\rho+2} & \cdots & p_{\rho+1,n} \\
\vdots & \vdots & & \vdots \\
b_n/M & p_{n,\rho+2} & \cdots & p_{n,n}
\end{pmatrix}
\tag{9.9}
$$

is orthogonal.

For this, we assume that we are in $\mathbb{R}^{n-\rho}$, with the ordinary scalar product, and coordinates $x_{\rho+1}, \ldots, x_n$, and we find an orthonormal basis of the hyperplane $H : b_{\rho+1}x_{\rho+1} + \cdots + b_n x_n = 0$. Let this basis be

$$u_j = (p_{\rho+1,j}, \ldots, p_{n,j}), \quad j = \rho+2, \ldots, n.$$

Now we complete this orthonormal basis of H to an orthonormal basis of $\mathbb{R}^{n-\rho}$ adding as first vector the unit direction vector of the hyperplane H,

$$u_{\rho+1} = M^{-1}(b_{\rho+1}, \ldots, b_n).$$

Thus, the matrix P defined in (9.9) is the matrix of the change of basis between orthonormal bases, and therefore it is an orthogonal matrix. That is, $PP^{\mathsf{T}} = I_{n-\rho}$.

The change of variables (9.8) can be written as

$$x' = \begin{pmatrix} I_\rho & O \\ O & P^{\mathsf{T}} \end{pmatrix} x + \tilde{C}$$

where $\tilde{C}$ is the column matrix

$$\tilde{C} = \begin{pmatrix} 0 \\ \vdots \\ 0 \\ C \\ 0 \\ \vdots \\ 0 \end{pmatrix},$$

with C in the position $\rho + 1$, and zeros elsewhere.

It is therefore a change of variables of the form $x' = ax + b$ with a orthogonal, and it is evident that if we replace x_i by x'_i in the initial polynomial $r(x)$ we obtain

$$s(x') = d_1 (x'_1)^2 + \cdots + d_\rho (x'_r)^2 + M x'_{\rho+1},$$

and, hence, $r(x)$ is equivalent to $s(x)$, completing the proof. □

Theorem 9.13

Let

$$r(x) = x^\mathsf{T} A x + B x + C$$

be a quadratic polynomial with

$$\mathrm{rank}(A) \neq \mathrm{rank}(A|B^\mathsf{T}).$$

Then $r(x)$ is orthogonally equivalent to a polynomial of the form

$$s(x) = d_1 x_1^2 + \cdots + d_\rho x_\rho^2 + x_{\rho+1}, \tag{9.10}$$

with $d_1 \geq d_2 \geq \cdots \geq d_\rho$ well ordered.

Proof

Observe that we must have $\rho < n$ and $\mathrm{rank}(A|B^\mathsf{T}) = \rho + 1$. We know, by the Spectral Theorem C.11, page 389, that there is a real orthogonal matrix a such that $D = a^\mathsf{T} A a$ is diagonal, and of the form

$$
D = \begin{pmatrix}
d_1' & & & & & \\
 & \ddots & & & & \\
 & & d_\rho' & & & \\
 & & & 0 & & \\
 & & & & \ddots & \\
 & & & & & 0
\end{pmatrix}, \tag{9.11}
$$

with $d_i' \neq 0$ and $\rho = \mathrm{rank}(A)$, where $d_1' \geq \cdots \geq d_\rho'$.

The given polynomial $r(x)$ is equivalent (with $\lambda = 1$ and $b = 0$) to

$$
r(x) = x^\mathsf{T} A' x + B' x + C',
$$

with

$$
A' = a^\mathsf{T} A a = D,
$$

$$
B' = (2b^\mathsf{T} A + B)a = Ba,
$$

$$
C' = (b^\mathsf{T} A + B)b + C = C.
$$

That is, $r(x)$ is equivalent to

$$
r(x) = d_1' x_1^2 + \cdots + d_\rho' x_\rho^2 + b_1' x_1 + \cdots + b_n' x_n + C.
$$

Since a polynomial is orthogonally equivalent to its negative, we may assume, by Lemma 9.9, that $d_1 \geq \cdots \geq d_\rho$ are well ordered.

The coefficients b_i' are the components of B' and they are given by the equation $B' = Ba$ (or $B' = -Ba$ if we have changed the sign of $r(x)$), and therefore they depend on the matrix a used to diagonalize A (this matrix is not unique).

Then we can make the orthogonal change of variables

$$
x_i' = x_i + \frac{b_i'}{2d_i'}, \quad i = 1, \ldots, \rho,
$$

$$
x_k' = x_k, \quad k = \rho + 1, \ldots, n, \tag{9.12}
$$

suggested by the method of completing the squares, and we see that $r(x)$ is equivalent to the polynomial

$$
s(x) = d_1' x_1^2 + \cdots + d_\rho' x_\rho^2 + b_{\rho+1}' x_{\rho+1} + \cdots + b_n' x_n + C_2, \tag{9.13}
$$

for some constant C_2 that can easily be computed. Hence, we have separated the quadratic variables from the linear ones.

The hypothesis on the rank, and the invariance of this hypothesis within each equivalence class (Observation 8.11), tells us that some of the coefficients b'_j, $j = \rho + 1, \ldots, n$, are different from zero.

We cannot continue with the same argument as in Section 8.9, because the change (8.20), page 243, is not orthogonal. However, we can apply Lemma 9.12.

Applying this lemma, we see that the polynomial $s(x)$ given in (9.13) is equivalent to the polynomial

$$t(x) = d'_1 x_1^2 + \cdots + d'_\rho x_\rho^2 + M x_{\rho+1},$$

with

$$M = \sqrt{(b'_{\rho+1})^2 + \cdots + (b'_n)^2}.$$

Dividing by $M > 0$, $t(x)$ is equivalent to

$$\boxed{s(x) = d_1 x_1^2 + \cdots + d_\rho x_\rho^2 + x_{\rho+1}}$$

with $d'_i = M d_i$, $i = 1, \ldots, \rho$, and $d_1 \geq d_2 \geq \cdots \geq d_\rho$ well ordered. This completes the proof. $\Box$

Polynomials orthogonally equivalent to polynomials with linear part satisfy $\tilde{\rho} = \rho + 2$ and they are, therefore, of type (IV).

9.5.1 Observations

Observation 1. The change of variables (9.12) can be written as $x' = ax + b$ with $a = I_n$ and

$$b^{\mathsf{T}} = \left(\frac{b'_1}{2d_1}, \ldots, \frac{b'_\rho}{2d_\rho}, 0, \ldots, 0 \right).$$

Observation 2. If, instead of supposing $b'_{\rho+1} \neq 0$, we suppose $b'_j \neq 0$, with $j \in \{\rho + 2, \ldots, n\}$, we obtain

$$r(x) = d'_1 x_1^2 + \cdots + d'_\rho x_\rho^2 + x_j.$$

But it is clear, by a simple change of variables, that this polynomial is equivalent to (9.10). We have already noted in Observation 8.6 that the matrix a performing the permutation of $x_{\rho+1}$ with x_j is the matrix obtained by permuting the rows $\rho + 1$ and j in the identity matrix, and this matrix is orthogonal.

Observation 3. If the polynomials

$$r(x) = d_1 x_1^2 + \cdots + d_\rho x_\rho^2 + M x_{\rho+1},$$

$$s(x) = d_1 x_1^2 + \cdots + d_\rho x_\rho^2 + M' x_{\rho+1},$$

with M and M' positive, are orthogonally equivalent, then $M = M'$.

If $\rho + 1 = n$, this is an immediate consequence of equality (9.1), page 286, because under these conditions we must have $\lambda = 1$, and hence, taking determinants on both sides of (9.1), we have

$$\det \begin{pmatrix} d_1 & & & 0 & 0 \\ & \ddots & & \vdots & \vdots \\ & & d_\rho & & \\ & & & 0 & M/2 \\ 0 & \cdots & & M/2 & 0 \end{pmatrix}$$

$$= \lambda^{n+1} \det \begin{pmatrix} d_1 & & & 0 & 0 \\ & \ddots & & \vdots & \vdots \\ & & d_\rho & & \\ & & & 0 & M'/2 \\ 0 & \cdots & & M'/2 & 0 \end{pmatrix}$$

and from this we deduce $M = M'$.

If $\rho + 1 < n$ these determinants are zero, and we cannot use this argument. However, we can make the following observation (which is also valid if $\rho + 1 = n$). If $r(x)$ and $s(x)$ are orthogonally equivalent, there is a real orthogonal matrix a and a real column matrix b such that $\lambda a D = Da$, where D is the matrix (9.11), and $(2b^{\mathsf{T}} D + B)a = \lambda B'$, where $B = (0, \ldots, M, \ldots, 0)$ and $B' = (0, \ldots, M', \ldots, 0)$ (all terms are zero, except the term in the $\rho + 1$-th position).

But, $\lambda a D = Da$ implies $\lambda = 1$ and also (using the block product of matrices) that a is of the form

$$a = \begin{pmatrix} P & O \\ O & Q \end{pmatrix},$$

with $P = (p_{ij})$ a $\rho \times \rho$ orthogonal matrix, and $Q = (q_{ij})$ a $(n - \rho) \times (n - \rho)$ orthogonal matrix.

Thus, the equality $(2b^{\mathsf{T}} D + B)a = \lambda B'$, which is equivalent to

$$(2b_1 d_1, \ldots, 2b_\rho d_\rho, M, 0, \ldots, 0)a = \lambda(0, \ldots, 0, M', 0, \ldots, 0),$$

implies (by equating the components from the $\rho + 1$-th position onward)

$$Mq_{\rho+1,\rho+1} = \lambda M',$$

$$Mq_{\rho+1,j} = 0, \quad j = \rho+2,\ldots,n.$$

Since Q is orthogonal, we have

$$q_{\rho+1,\rho+1}^2 + \cdots + q_{\rho+1,n}^2 = 1,$$

and, therefore, $q_{\rho+1,\rho+1}^2 = 1$. Since M and M' are positive, we must have $M = M'$, and this completes the proof.

9.5.2 Conclusion

For convenience, we collect the results of this and the preceding Section 9.4 in the following theorem.

Theorem 9.14 (Classification Theorem)

Every quadratic polynomial in n variables is orthogonally equivalent to one and only one of the following polynomials:

  (I) $r(x) = x_1^2 + d_2 x_2^2 + \cdots + d_\rho x_\rho^2$ with $1 \geq d_2 \geq \cdots \geq d_\rho$, well ordered,

(II/III) $r(x) = d_1 x_1^2 + \cdots + d_\rho x_\rho^2 + 1$ with $d_1 \geq d_2 \geq \cdots \geq d_\rho$,

  (IV) $r(x) = d_1 x_1^2 + \cdots + d_\rho x_\rho^2 + x_{\rho+1}$ with $d_1 \geq d_2 \geq \cdots \geq d_\rho$, well ordered,

where $0 < \rho \leq n$ in cases (I) and (II/III), and $0 < \rho < n$ in case (IV).

Proof

We have seen, in Sections 9.4 and 9.5, that every quadratic polynomial in n variables is orthogonally equivalent to one of the polynomials in this list.

It remains to show that two different polynomials of this list are not orthogonally equivalent. If they do not have the same invariants $(\rho, i, \tilde{\rho}, \tilde{i})$ we are done. Let us assume, therefore, that $r(x)$ and $s(x)$ are polynomials in this list with the same invariants $(\rho, i, \tilde{\rho}, \tilde{i})$. In particular, $r(x)$ and $s(x)$ must be of the same type (I), (II/III) or (IV). We study each of the three cases.

$r(x)$ and $s(x)$ are both of type (I). We have $r(x) = x_1^2 + d_2 x_2^2 + \cdots + d_\rho x_\rho^2$, with $1 = d_1 \geq d_2 \geq \cdots \geq d_\rho$ well ordered; and $s(x) = x_1^2 + d_2' x_2^2 + \cdots + d_\rho' x_\rho^2$, with $1 = d_1' \geq d_2' \geq \cdots \geq d_\rho'$ well ordered.

If $r(x)$ and $s(x)$ were orthogonally equivalent, the matrices

$$\begin{pmatrix} 1 & 0 & 0 & 0 \\ 0 & d_2 & 0 & 0 \\ 0 & 0 & \ddots & 0 \\ 0 & 0 & 0 & d_\rho \end{pmatrix}, \qquad \begin{pmatrix} 1 & 0 & 0 & 0 \\ 0 & d_2' & 0 & 0 \\ 0 & 0 & \ddots & 0 \\ 0 & 0 & 0 & d_\rho' \end{pmatrix}$$

would have proportional eigenvalues; see Proposition 9.5.

By Lemma 9.9, the scalar transforming a well ordered set of real numbers into another well ordered set of real numbers must be positive; but as the largest number in these orderings is in both cases equal to 1, this scalar must be 1, and hence, $d_i' = d_i$, for $i = 2, \ldots, \rho$. Thus, $r(x) = s(x)$.

$r(x)$ *and* $s(x)$ *are both of type* (II/III). We have $r(x) = d_1 x_1^2 + \cdots + d_\rho x_\rho^2 + 1$ with $d_1 \geq \cdots \geq d_\rho$, and $s(x) = d_1' x_1^2 + \cdots + d_\rho' x_\rho^2 + 1$ with $d_1' \geq \cdots \geq d_\rho'$.

If $r(x)$ and $s(x)$ were orthogonally equivalent, by Corollary 9.4, the matrices

$$\begin{pmatrix} d_1 & 0 & 0 & 0 \\ 0 & d_2 & 0 & 0 \\ 0 & 0 & \ddots & 0 \\ 0 & 0 & 0 & d_\rho \end{pmatrix}, \qquad \begin{pmatrix} d_1' & 0 & 0 & 0 \\ 0 & d_2' & 0 & 0 \\ 0 & 0 & \ddots & 0 \\ 0 & 0 & 0 & d_\rho' \end{pmatrix},$$

would have the same eigenvalues. Thus, $d_i' = d_i$, for $i = 1, \ldots, \rho$, and hence $r(x) = s(x)$.

$r(x)$ *and* $s(x)$ *are both of type* (IV). We have $r(x) = d_1 x_1^2 + d_2 x_2^2 + \cdots + d_\rho x_\rho^2 + x_{\rho+1}$, with $d_1 \geq d_2 \geq \cdots \geq d_\rho$ well ordered; and $s(x) = d_1' x_1^2 + d_2' x_2^2 + \cdots + d_\rho' x_\rho^2 + x_{\rho+1}$, with $d_1' \geq d_2' \geq \cdots \geq d_\rho'$ well ordered.

If $r(x)$ and $s(x)$ were orthogonally equivalent, the matrices

$$\begin{pmatrix} d_1 & 0 & 0 & 0 \\ 0 & d_2 & 0 & 0 \\ 0 & 0 & \ddots & 0 \\ 0 & 0 & 0 & d_\rho \end{pmatrix}, \qquad \begin{pmatrix} d_1' & 0 & 0 & 0 \\ 0 & d_2' & 0 & 0 \\ 0 & 0 & \ddots & 0 \\ 0 & 0 & 0 & d_\rho' \end{pmatrix},$$

would have proportional eigenvalues; see Proposition 9.5.

By Lemma 9.9, the scalar transforming a well ordered set of real numbers into another well ordered set of real numbers must be positive, and hence $d_j' = \lambda d_j$, for $j = 1, \ldots, \rho$, $\lambda > 0$. Since $r(x)$ is orthogonally equivalent to $\lambda r(x)$, we have that the polynomials

$$\lambda r(x) = \lambda d_1 x_1^2 + \cdots + \lambda d_\rho x_\rho^2 + \lambda x_{\rho+1},$$

$$s(x) = \lambda d_1 x_1^2 + \cdots + \lambda d_\rho x_\rho^2 + x_{\rho+1}$$

are equivalent. From Observation 3 on page 297 applied to these two polynomials, we see that $\lambda = 1$, that is, $d'_j = d_j$, for $j = 1, \ldots, \rho$, and hence $r(x) = s(x)$. This completes the proof. $\qquad\square$

Notice that polynomials of type (II/III) are of affine type (II) when there are at least as many positive d_js as negative d_js, and they are of affine type (III) when there are fewer positive d_js than negative d_js. The type (I) and (IV) polynomials are of affine type (I) and (IV), respectively.

The polynomials of this list are called *canonical expressions* or *canonical representatives*. The coefficients d_i appearing in the canonical representative of a given polynomial $r(x) = x^{\mathsf{T}} A x + B x + C$ are multiples of the eigenvalues of A. That is, we can order the eigenvalues a_j of A in such a way that there is a $\lambda \in \mathbb{R}$, $\lambda > 0$, with $\lambda a_j = d_j$, $j = 1, \ldots, \rho$.

9.6 Invariants

The invariants $(\rho, i, \tilde{\rho}, \tilde{i})$ associated to each polynomial were sufficient for the affine classification, but they are not sufficient for the orthogonal classification. To each polynomial $r(x) = x^{\mathsf{T}} A x + B x + C$ we associate the element of $\mathbb{R}^{4+\rho}$,

$$(\rho, i, \tilde{\rho}, \tilde{i}, d_1, \ldots, d_\rho),$$

where the d_j are the coefficients of the canonical representative of $r(x)$, ordered as in the Classification Theorem 9.14.

Proposition 9.15

Two quadratic polynomials are orthogonally equivalent if and only if they have the same invariants $(\rho, i, \tilde{\rho}, \tilde{i}, d_1, \ldots, d_\rho)$.

Proof

If they are orthogonally equivalent they have the same canonical representative and, therefore, the same invariants. Conversely, if they have the same invariants, that means that they have the same canonical representative and, therefore, they are orthogonally equivalent. $\qquad\square$

Example 9.16

In Table 9.1 we give the invariants of some polynomials.

Type	Polynomial	Invariants
(I)	$r_1(x) = 4x_1^2 - x_2^2$	$(2, 1, 2, 1, 1, -1/4)$
(I)	$r_2(x) = 4x_1^2 - 8x_2^2$	$(2, 1, 2, 1, 1, -1/2)$
(I)	$r_3(x) = -4x_1^2 + x_2^2$	$(2, 1, 2, 1, 1, -1/4)$
(II)	$r_4(x) = 4x_1^2 - x_2^2 + 1$	$(2, 1, 3, 1, 4, -1)$
(II)	$r_5(x) = 4x_1^2 - 8x_2^2 + 1$	$(2, 1, 3, 1, 4, -8)$
(II)	$r_6(x) = -4x_1^2 + 8x_2^2 + 1$	$(2, 1, 3, 1, 8, -4)$
(II)	$r_7(x) = x_1^2 - x_2^2 + 1$	$(2, 1, 3, 1, 1, -1)$
(II)	$r_8(x) = -x_1^2 + x_2^2 + 1$	$(2, 1, 3, 1, 1, -1)$
(II)	$r_9(x) = -4x_1^2 + x_2^2 + x_3^2 + 1$	$(3, 1, 4, 1, 1, 1, -4)$
(III)	$r_{10}(x) = 4x_1^2 - x_2^2 - x_3^2 + 1$	$(3, 1, 4, 2, 4, -1, -1)$
(IV)	$r_{11}(x) = -4x_1^2 + x_2^2 + x_3^2 + x_4$	$(3, 1, 5, 2, 1, 1, -4)$
(IV)	$r_{12}(x) = 4x_1^2 - x_2^2 - x_3^2 + x_4$	$(3, 1, 5, 2, 1, 1, -4)$

Table 9.1. Invariants

Notice that $r_1(x)$ and $r_3(x)$ are orthogonally equivalent, as are $r_{11}(x)$ and $r_{12}(x)$, and $r_7(x)$ and $r_8(x)$, respectively. However, $r_5(x)$ and $r_6(x)$, although they only differ in the sign of the quadratic part, do not have the same invariants; and, although they are of the same affine type, they are not orthogonally equivalent. Likewise for $r_9(x)$ and $r_{10}(x)$.

Example 9.17

In Table 9.2 we present the canonical representatives of some polynomials.

Type	Polynomial	Canonical	Invariants
(I)	$x_1^2 + 5x_2^2 - 2x_1 + 20x_2 + 21$	$x_1^2 + (1/5)x_2^2$	$(2, 0, 2, 0, 1, 1/5)$
(II)	$4x_1^2 - x_2^2 + x_1$	$16x_1^2 - 64x_2^2 + 1$	$(2, 1, 3, 1, 16, -64)$
(III)	$x_1^2 + x_2^2 - 1/5$	$-5x_1^2 - 5x_2^2 + 1$	$(2, 0, 3, 1, -5, -5)$
(IV)	$x_1^2 + x_2^2 + 2x_1 + x_3 + 1$	$x_1^2 + x_2^2 + x_3$	$(2, 0, 4, 1, 1, 1)$

Table 9.2. Canonical representatives

9.7 Orthogonal Classification of Quadrics

Let us return to Euclidean affine spaces and study the relationship between quadrics associated to orthogonally equivalent polynomials. This section is an adaptation of Section 8.12 to the orthogonal case.

Definition 9.18

We say that two non-empty quadrics $\mathcal{Q}_1$ and $\mathcal{Q}_2$ of a Euclidean affine space $\mathbb{A}$ are orthogonally equivalent if and only if there is a Euclidean motion f such that $f(\mathcal{Q}_1) = \mathcal{Q}_2$.

Let us see what relationship there is between *orthogonally equivalent polynomials* and *orthogonally equivalent quadrics*.

Proposition 9.19

Let $\mathcal{R}$ be an orthonormal affine frame of a Euclidean affine space $\mathbb{A}$ of dimension n, and let $r(x)$ and $s(x)$ be orthogonally equivalent polynomials in n variables. Then there is an orthonormal affine frame $\mathcal{R}'$ such that

$$\mathcal{Q}(r(x), \mathcal{R}) = \mathcal{Q}(s(x), \mathcal{R}').$$

Proof

The proof is as in Proposition 8.18, taking into account that the matrix of the change of basis between orthonormal bases is an orthogonal matrix.   □

Theorem 9.20

Let $\mathbb{A}$ be a Euclidean affine space of dimension n, $\mathcal{R}$ an orthonormal affine frame of $\mathbb{A}$, and let $r(x)$ and $s(x)$ be orthogonally equivalent polynomials in n variables.

Then the quadrics $\mathcal{Q}(r(x), \mathcal{R})$ and $\mathcal{Q}(s(x), \mathcal{R})$ are orthogonally equivalent (or empty).

Proof

The proof is as in Theorem 8.19, taking into account that the affinity constructed there is now a Euclidean motion.   □

Corollary 9.21

Let $\mathcal{R}$ and $\mathcal{R}'$ be two orthonormal affine frames in a Euclidean affine space of dimension n, and let $r(x)$ be a polynomial in n variables. Then the quadrics $\mathcal{Q}(r(x), \mathcal{R})$ and $\mathcal{Q}(r(x), \mathcal{R}')$ are orthogonally equivalent.

Proof

Compare Corollary 8.20.   □

The converse of Theorem 9.20 is also true. Concretely we shall prove that *orthogonally equivalent (and, therefore, non-empty) quadrics give rise to orthogonally equivalent polynomials.*

Theorem 9.22 (Converse of Theorem 9.20)

Let $\mathbb{A}$ be a Euclidean affine space of dimension n, $\mathcal{R}$ an orthonormal affine frame in $\mathbb{A}$ and $r(x)$ and $s(x)$ polynomials in n variables.

Let us assume that the quadrics $\mathcal{Q}(r(x), \mathcal{R})$ and $\mathcal{Q}(s(x), \mathcal{R})$ are orthogonally equivalent (and, therefore, non-empty) and that at least one of the points of $\mathcal{Q}(r(x), \mathcal{R})$ is regular. Then $r(x)$ and $s(x)$ are orthogonally equivalent.

Proof

The proof is as in the first part of the proof of Theorem 8.23, where it is assumed that there is a regular point, taking into account that the affinity constructed there is now a Euclidean motion.   □

Note that the hypothesis on regular points does not appear in the statement of Theorem 8.23, but the proof is divided into two parts depending on whether there is a regular point or not. The argument used there for the case where there are no regular points cannot be adapted to the orthogonal case. For instance, in the Euclidean affine space $\mathbb{R}^2$, the polynomials $r(x) = x_1^2 + x_2^2$ and $s(x) = 7x_1^2 + 5x_2^2$ are not orthogonally equivalent, but the quadrics given by $x_1^2 + x_2^2 = 0$ and $7x_1^2 + 5x_2^2 = 0$ are orthogonally equivalent.

Corollary 9.23 (Converse of Proposition 9.19)

Let $\mathbb{A}$ be a Euclidean affine space of dimension n, and let $r(x)$ and $s(x)$ be polynomials in n variables. Let $\mathcal{R}$ and $\mathcal{R}'$ be orthonormal affine frames such

that

$$\mathcal{Q}(r(x), \mathcal{R}') = \mathcal{Q}(s(x), \mathcal{R}).$$

Let us assume that this quadric is non-empty, and that at least one of its points is regular. Then $r(x)$ and $s(x)$ are orthogonally equivalent.

Proof

Compare Corollary 8.24. □

Theorem 9.24 (Equivalence between polynomials and quadrics)

Let $\mathcal{Q}(r(x), \mathcal{R})$ and $\mathcal{Q}(s(x), \mathcal{R})$ be two non-empty quadrics of a Euclidean affine space $\mathbb{A}$, with $\mathcal{R}$ orthonormal, and let us assume that at least one of the points of $\mathcal{Q}(r(x), \mathcal{R})$ is regular. Then these two quadrics are orthogonally equivalent if and only if the polynomials defining them, $r(x)$ and $s(x)$, are orthogonally equivalent.

Proof

This is a consequence of Theorems 9.20 and 9.22. □

Corollary 9.25

Two non-empty quadrics, with at least one regular point, are orthogonally equivalent if and only if the polynomials defining them have the same invariants $(\rho, i, \tilde{\rho}, \tilde{i}, d_1, \ldots, d_\rho)$.

Proof

This is a consequence of Theorems 9.15 and 9.24. □

Theorem 9.26 (Classification Theorem)

Every non-empty quadric of a Euclidean affine space $\mathbb{A}$ of dimension n, with at least one regular point, is orthogonally equivalent to one and only one of the quadrics given, in a certain orthonormal affine frame, by the following equations:

(I) $x_1^2 + d_2 x_2^2 + \cdots + d_\rho x_\rho^2 = 0$, $1 \geq d_2 \geq \cdots \geq d_\rho$ well ordered,

(II/III) $d_1 x_1^2 + \cdots + d_\rho x_\rho^2 + 1 = 0$, $d_1 \geq d_2 \geq \cdots \geq d_\rho$,

(IV) $d_1 x_1^2 + \cdots + d_\rho x_\rho^2 + x_{\rho+1} = 0$, $d_1 \geq d_2 \geq \cdots \geq d_\rho$ well ordered,

where $0 < \rho \leq n$ in cases (I) and (II/III), and $0 < \rho < n$ in case (IV).

Proof

This is a consequence of Theorems 9.14 and 9.24.   □

We have already noted on page 249 that the only quadrics without regular points are those of type (I) and index $i = 0$. This means that in the list given in this theorem, the case (I) with all d_js positive is omitted. Case (II/III) with all d_js positive (in particular $i = 0$) has also been omitted, because it corresponds to empty quadrics.

If we wanted a complete list, including empty quadrics and quadrics without regular points, then we would have to add the two cases (I) and (II/III) with $i = 0$, but at the expense of a loss of faithfulness.

This theorem can also be stated in terms of a change of affine frame as follows.

Theorem 9.27

Let $\mathcal{Q}(r(x), \mathcal{R})$ be a non-empty quadric of a real affine space $\mathbb{A}$ of dimension n, with $\mathcal{R}$ an orthonormal affine frame, and assume that $\mathcal{Q}(r(x), \mathcal{R})$ contains at least one regular point. Then there is an orthonormal affine frame $\mathcal{R}'$ in which the equation of $\mathcal{Q}$ is one and only one of the equations given in the above Theorem 9.26.

Proof

We know that $r(x)$ is orthogonally equivalent to one and only one of the polynomials $r_0(x)$ of type (I), (II), (III) or (IV) given in Theorem 9.26. By Proposition 9.19, there is an orthonormal affine frame $\mathcal{R}'$ such that $\mathcal{Q}(r(x), \mathcal{R}) = \mathcal{Q}(r_0(x), \mathcal{R}')$.   □

We say that $\mathcal{R}'$ is the *orthonormal affine frame adapted to the quadric*, since the polynomial defining the quadric in this affine frame is the simplest possible; it is one of the *canonical representatives* given in Theorem 9.14.

9.7.1 Construction of the Adapted Orthonormal Affine Frame

In order to classify orthogonally a quadric $Q(r(x), \mathcal{R})$ we shall first classify the polynomial, finding explicitly the equations $x' = ax + b$, with a orthogonal, transforming $r(x)$ into one of the canonical polynomials $r_0(x)$. If we have $\lambda r_0(x) = r(ax + b)$, then the adapted affine frame $\mathcal{R}'$ is given (see the proof of Proposition 8.18) by

$$M(\mathcal{R}', \mathcal{R}) = \begin{pmatrix} a & b \\ 0 & 1 \end{pmatrix}.$$

The coordinates of the center, in $\mathcal{R}$, are the components of b; and the components in $\mathcal{B}$ (the basis of $\mathcal{R}$) of the vectors of the basis are given by the columns of the matrix a. These vectors are the normalized eigenvectors of the matrix A corresponding to the quadratic part of the polynomial.

If $r_0(x)$ is of type (I) or (II/III), then (see Section 8.8) the origin of $\mathcal{R}'$ is a center of the quadric. If $r_0(x)$ is of type (IV) and hence, it has no center, we must compute b following the steps of Section 9.4. See Example 9.29.

9.8 Orthogonal Classification of Conics

We apply Theorem 9.26 to the case $n = 2$. In cases (I) and (II/III) we can have $\rho = 1$ or $\rho = 2$. In case (IV) we must have $\rho = 1$.

Type (I) with $\rho = 1$: $x_1^2 = 0$.

Type (I) with $\rho = 2$: $x_1^2 + d_2 x_2^2 = 0$, with $1 \geq d_2$ well ordered. In particular, $|d_2| \leq 1$.

Type (II/III) with $\rho = 1$: $d_1 x_1^2 + 1 = 0$, with $d_1 > 0$.

Type (II/III) with $\rho = 2$: $d_1 x_1^2 + d_2 x_2^2 + 1 = 0$, with $d_1 \geq d_2$.

Type (IV) with $\rho = 1$: $d_1 x_1^2 + x_2 = 0$, with $d_1 > 0$.

The different values of the index are determined by the relative number of positive versus negative d_js. We collect all possible cases in Table 9.3.

Observe that in cases (I) and (II) with all d_js positive (in particular, with index $i = 0$), different values of d_j, and hence non-equivalent polynomials, give rise to the same quadric (a point, a straight line or the empty set). These quadrics do not contain regular points.

ρ	i	$\tilde{\rho}$	$\tilde{i}$	d_1	d_2	Equation	Name
(I)	2	0	2	0	1	$1 \geq d_2 > 0$	$x_1^2 + d_2 x_2^2 = 0$	Point
(I)	2	1	2	1	1	$0 > d_2 \geq -1$	$x_1^2 + d_2 x_2^2 = 0$	Two lines
(I)	1	0	1	0	1		$x_1^2 = 0$	Double line
(II)	2	1	3	1	> 0	< 0	$d_1 x_1^2 + d_2 x_2^2 + 1 = 0$	Hyperbola
(II)	2	0	3	0	$\geq d_2$	> 0	$d_1 x_1^2 + d_2 x_2^2 + 1 = 0$	Empty
(II)	1	0	2	0	> 0		$d_1 x_1^2 + 1 = 0$	Empty
(III)	2	0	3	1	< 0	$\leq d_1$	$d_1 x_1^2 + d_2 x_2^2 + 1 = 0$	Ellipse
(III)	1	0	2	1	< 0		$d_1 x_1^2 + 1 = 0$	Parallel lines
(IV)	1	0	3	1	> 0		$d_1 x_1^2 + x_2 = 0$	Parabola

Table 9.3. Conics

Example 9.28

Classify orthogonally the conic of a Euclidean affine plane $\mathbb{A}$ given in an orthonormal affine frame $\mathcal{R}$ by $r(x) = 5x_1^2 - x_2^2 - 6\sqrt{3}x_1x_2 + 4 = 0$.

Solution

The eigenvalues of the matrix of the quadratic part

$$A = \begin{pmatrix} 5 & -3\sqrt{3} \\ -3\sqrt{3} & -1 \end{pmatrix},$$

are $\lambda_1 = 8$ and $\lambda_2 = -4$. The corresponding normalized eigenvectors are $u_1 = \frac{1}{2}(-\sqrt{3}, 1)$ and $u_2 = \frac{1}{2}(1, \sqrt{3})$. The equations of the center

$$\frac{\partial r(x)}{\partial x_1} = 0, \qquad \frac{\partial r(x)}{\partial x_2} = 0,$$

give $x_1 = x_2 = 0$. Hence, the adapted affine frame is

$$\mathcal{R}' = \{(0,0); (u_1, u_2)\}.$$

In this affine frame, the equation of the conic is $r(ax + b) = 0$, where $b = 0$ (coordinates of the center), and a is the matrix with columns the components of u_1 and u_2. We obtain $2x_1^2 - x_2^2 + 1 = 0$; a hyperbola. $\square$

9.9 Orthogonal Classification of Quadrics in Dimension Three

We apply Theorem 9.26 to the case $n = 3$. In cases (I) and (II/III) we can have $\rho = 1$, $\rho = 2$ or $\rho = 3$. In case (IV) we must have $\rho = 1$ or $\rho = 2$.

Type (I) with $\rho = 1$: $x_1^2 = 0$.

Type (I) with $\rho = 2$: $x_1^2 + d_2 x_2^2 = 0$, with $1 \geq d_2$ well ordered. In particular $|d_2| \leq 1$.

Type (I) with $\rho = 3$: $x_1^2 + d_2 x_2^2 + d_3 x_3^2 = 0$, with $1 \geq d_2 \geq d_3$ well ordered. In particular d_2 is positive.

Type (II/III) with $\rho = 1$: $d_1 x_1^2 + 1 = 0$, with $d_1 > 0$.

Type (II/III) with $\rho = 2$: $d_1 x_1^2 + d_2 x_2^2 + 1 = 0$, with $d_1 \geq d_2$.

Type (II/III) with $\rho = 3$: $d_1 x_1^2 + d_2 x_2^2 + d_3 x_3^2 + 1 = 0$, with $d_1 \geq d_2 \geq d_3$.

Type (IV) with $\rho = 1$: $d_1 x_1^2 + x_2 = 0$, with $d_1 > 0$.

Type (IV) with $\rho = 2$: $d_1 x_1^2 + d_2 x_2^2 + x_3 = 0$, with $d_1 \geq d_2$ well ordered. In particular $|d_2| \leq d_1$.

The different values of the index are determined by the relative number of positive versus negative d_js. We collect all possible cases in Table 9.4, where the equation is always $d_1 x_1^2 + d_2 x_2^2 + d_3 x_3^2 + 1 = 0$ with the corresponding restrictions on d_i.

Observe that in cases (I) and (II/III) with all d_js positive (in particular, index $i = 0$), different values of d_j, in fact non-equivalent polynomials, give rise to the same quadric (a point, a straight line, a plane or an empty set). These quadrics do not contain regular points.

The abbreviated names in the final column of the table are introduced only for typographical reasons. They have the following meanings: 2P = two intersecting planes; P^2 = double plane; H2 = hyperboloid of two sheets; HC = hyperbolic cylinder; H1 = hyperboloid of one sheet; E = ellipsoid; EC = elliptic cylinder; PP = parallel planes; EP = elliptic paraboloid; HP = hyperbolic paraboloid; PC = parabolic cylinder.

Example 9.29

Let $\mathcal{R} = \{P; \mathcal{B}\}$ be an orthonormal affine frame of a Euclidean affine space $\mathbb{A}$ of dimension three. Classify orthogonally the quadric given in $\mathcal{R}$ by $r(x) = 3x_1^2 - x_3^2 - 2x_1x_3 + 4x_1 + x_2 + x_3 + 1 = 0$ and find the adapted orthonormal affine frame.

	ρ	i	$\tilde{\rho}$	$\tilde{i}$	d_1	d_2	d_3	Name
(I)	3	0	3	0	1	$1 \geq d_2 \geq d_3$	> 0	Point
(I)	3	1	3	1	1	$1 \geq d_2 > 0$	< 0	Cone
(I)	2	0	2	0	1	$1 \geq d_2 > 0$	0	Str. line
(I)	2	1	2	1	1	$0 > d_2 \geq -1$	0	2P
(I)	1	0	1	0	1	0	0	P^2
(II)	3	0	4	0	$\geq d_2$	$\geq d_3$	> 0	Empty
(II)	3	1	4	1	$\geq d_2$	> 0	< 0	H2
(II)	2	0	3	0	$\geq d_2$	> 0	0	Empty
(II)	2	1	3	1	> 0	< 0	0	HC
(II)	1	0	2	0	> 0	0	0	Empty
(III)	3	1	4	2	> 0	< 0	$\leq d_2$	H1
(III)	3	0	4	1	< 0	$\leq d_1$	$\leq d_2$	E
(III)	2	0	3	1	< 0	$\leq d_1$	0	EC
(III)	1	0	2	1	< 0	0	0	PP
(IV)	2	0	4	1	$\geq d_2$	> 0	1	EP
(IV)	2	1	4	2	> 0	$0 > d_2 \geq -d_1$	1	HP
(IV)	1	0	3	1	> 0	0	1	PC

Table 9.4. Quadrics

Solution

The matrix of the quadratic part is

$$A = \begin{pmatrix} 3 & 0 & -1 \\ 0 & 0 & 0 \\ -1 & 0 & -1 \end{pmatrix}.$$

The eigenvalues are $\lambda_1 = 1 + \sqrt{5}$, $\lambda_2 = 1 - \sqrt{5}$ and $\lambda_3 = 0$. The normalized eigenvectors are, respectively,

$$u_1 = a(1,0,2 - \sqrt{5}), \qquad u_2 = b(1,0,2 + \sqrt{5}), \qquad u_3 = a(0,1,0)$$

with

$$a = \frac{1}{\sqrt{10 - 4\sqrt{5}}}, \qquad b = \frac{1}{\sqrt{10 + 4\sqrt{5}}}.$$

Then (u_1, u_2, u_3) is the basis of the adapted orthonormal affine frame that we are looking for. To find the origin of this affine frame we observe that this quadric has no center, because rank $A \neq$ rank$(A|B^{\mathsf{T}})$, where $B = (4,1,1)$. Hence we are in the situation described in Section 9.5. Following this section we make the change of variables

$$\begin{pmatrix} x_1 \\ x_2 \\ x_3 \end{pmatrix} = \begin{pmatrix} a & b & 0 \\ 0 & 0 & 1 \\ -ac^{-1} & bc & 0 \end{pmatrix} \begin{pmatrix} x_1' \\ x_2' \\ x_3' \end{pmatrix}, \quad \text{with } c = 2 + \sqrt{5}. \tag{9.14}$$

Substituting these values into $r(x)$ we obtain

$$\begin{aligned} r(x) &= 3(ax_1' + bx_2')^2 - (-ac^{-1}x_1' + bcx_2')^2 \\ &\quad - 2(ax_1' + bx_2')(-ac^{-1}x_1' + bcx_2') \\ &\quad + 4(ax_1' + bx_2') + x_3' + (-ac^{-1}x_1' + bcx_2') + 1 \\ &= \lambda_1 x_1'^2 + \lambda_2 x_2'^2 + dx_1' + ex_2' + x_3' + 1, \end{aligned}$$

with

$$d = (6 - \sqrt{5})a = \frac{6 - \sqrt{5}}{\sqrt{10 - 4\sqrt{5}}}, \qquad e = (6 + \sqrt{5})b = \frac{6 + \sqrt{5}}{\sqrt{10 + 4\sqrt{5}}}.$$

Now we complete squares

$$\begin{aligned} &\lambda_1 x_1'^2 + \lambda_2 x_2'^2 + dx_1' + ex_2' + x_3' + 1 \\ &= \lambda_1 \left(x_1' + \frac{d}{2\lambda_1} \right)^2 - \frac{d^2}{4\lambda_1} + \lambda_2 \left(x_2' + \frac{e}{2\lambda_2} \right)^2 - \frac{e^2}{4\lambda_2} + x_3' + 1 \\ &= \lambda_1 x_1''^2 + \lambda_2 x_2''^2 + x_3'', \end{aligned}$$

with

$$x_1'' = x_1' + \frac{d}{2\lambda_1},$$

$$x_2'' = x_2' + \frac{e}{2\lambda_2},$$

$$x_3'' = x_3' - \frac{d^2}{4\lambda_1} - \frac{e^2}{4\lambda_2} + 1.$$

Solving for x_1', x_2', x_3' and substituting them into (9.14) we obtain the global change of coordinates

$$\begin{pmatrix} x_1 \\ x_2 \\ x_3 \end{pmatrix} = \bar{a} \begin{pmatrix} x_1'' \\ x_2'' \\ x_3'' \end{pmatrix} + \bar{a} \begin{pmatrix} -\dfrac{d}{2\lambda_1} \\ -\dfrac{e}{2\lambda_2} \\ \dfrac{d^2}{4\lambda_1} + \dfrac{e^2}{4\lambda_2} - 1 \end{pmatrix},$$

with

$$\bar{a} = \begin{pmatrix} a & b & 0 \\ 0 & 0 & 1 \\ -ac^{-1} & bc & 0 \end{pmatrix}.$$

We write this change as $x = \bar{a}x'' + \bar{b}$, and so $r(\bar{a}x'' + \bar{b}) = r_0(x'')$, where $r_0(x) = \lambda_1 x_1^2 + \lambda_2 x_2^2 + x_3$ is the canonical representative. By the Section "Construction of the adapted orthonormal affine frame" on page 307, the adapted affine frame $\mathcal{R}'$ is given by

$$M(\mathcal{R}', \mathcal{R}) = \begin{pmatrix} \bar{a} & \bar{b} \\ 0 & 1 \end{pmatrix},$$

where $\mathcal{R}$ is the affine frame in which the quadric was given.

Hence, the basis is formed by the vectors whose components in $\mathcal{B}$ are given by the columns of $\bar{a}$; and the origin C is the point with coordinates $\bar{b}$, that is,

$$C = \left(\frac{ad}{2\lambda_1} - \frac{be}{2\lambda_2}, \frac{d^2}{4\lambda_1} + \frac{e^2}{4\lambda_2} - 1, -\frac{ad}{2c\lambda_1} - \frac{bec}{2\lambda_2} \right) = \left(-\frac{3}{8}, -\frac{11}{16}, \frac{7}{8} \right).$$

Observe that the straight line $C + \langle u_3 \rangle$ is an axis of symmetry. This orthogonal symmetry is given in $\mathcal{R}$ by

$$S(x_1, x_2, x_3) = \left(-\frac{3}{4} - x_1, x_2, \frac{7}{4} - x_3 \right).$$

Compare this example with Example 8.31, page 263.   □

9.10 Symmetries of a Quadric

Looking at the classification, one can see that any map with equations given in the adapted affine frame by

$$S(x_1, \ldots, x_n) = (\pm x_1, \ldots, \pm x_n)$$

leaves any quadric of type (I) or (II/III) invariant.

Once we fix the signs "$\pm$", we have $S \circ S = \mathrm{id}$, that is, S is a symmetry. Since the affine frame is orthonormal, S is an orthogonal symmetry.

For instance,

$$S(x_1, \ldots, x_n) = (x_1, -x_2, \ldots, -x_n)$$

is an axial symmetry with respect to the straight line through the origin of the adapted affine frame (a center of the quadric) with direction vector the first vector of the adapted basis (an eigenvector of the symmetric matrix associated to the quadratic part of $r(x)$). A straight line through the center of the quadric with direction vector an eigenvector of the quadratic part is a *symmetry axis of the quadric*.

If

$$S(x_1, \ldots, x_n) = (x_1, x_2, -x_3, \ldots, -x_n),$$

we have a mirror symmetry with respect to the plane through the origin of the adapted affine frame directed by two eigenvectors of the quadratic part. A plane through the center directed by the space generated by two eigenvectors of the quadratic part is a *plane of symmetry of the quadric*.

If

$$S(x_1, \ldots, x_n) = (-x_1, \ldots, -x_n),$$

we have a central symmetry with respect to the origin.

Finally, observe that any map with equations given, in the adapted affine frame, by

$$S(x_1, \ldots, x_n) = (\pm x_1, \ldots, \pm x_{\rho-1}, x_\rho, \pm x_{\rho+1}, \ldots, \pm x_n),$$

where ρ is the quadratic rank, leaves invariant any quadric of type (IV). In particular, the straight line through the origin of the adapted affine frame (these quadrics have no center) and direction vector the ρ-th eigenvector, is a symmetry axis of the quadric.

For instance, the axial symmetry $S(x_1, x_2, x_3) = (-x_1, -x_2, x_3)$, leaves invariant the elliptic paraboloid $x_1^2 + x_2^2 + x_3 = 0$.

Example 9.30

Find the symmetry axes of the quadric of the Euclidean affine space $\mathbb{R}^3$ given in the canonical affine frame by $r(x) = 4x_1^2 + 4x_2^2 + 4x_3^2 + 4x_1 x_3 + x_3 - 1 = 0$. Find the equations of the symmetry with respect to one of these axes.

Solution

This polynomial can be written as $x^{\mathsf{T}} A x + B x + C$, with

$$A = \begin{pmatrix} 4 & 0 & 2 \\ 0 & 4 & 0 \\ 2 & 0 & 4 \end{pmatrix}, \qquad B = (0, 0, 1), \qquad C = -1.$$

Since $\det(A) \neq 0$, we have $\rho = 3$. Its characteristic polynomial is $p_A(x) = -x^3 + 12x_1^2 - 44x + 48$, which has three changes of sign, and hence $i = 0$.

The matrix

$$\tilde{A} = \begin{pmatrix} 4 & 0 & 2 & 0 \\ 0 & 4 & 0 & 0 \\ 2 & 0 & 4 & 1 \\ 0 & 0 & 1 & -1 \end{pmatrix}$$

has rank 4, and hence $\tilde{\rho} = 4$. Its characteristic polynomial is $p_{\tilde{A}}(x) = x^4 + x^3 + 31x^2 + 4x - 64$, which has one change of sign, and hence $\tilde{i} = 1$. Hence, the invariants are $(\rho, i, \tilde{\rho}, \tilde{i}) = (3, 0, 4, 1)$, and the quadric is, therefore, an ellipsoid.

We want to find an orthonormal affine frame in which the equation of this ellipsoid is

$$d_1 x_1^2 + d_2 x_2^2 + d_3 x_3^2 + 1 = 0,$$

with $d_3 \leq d_2 \leq d_1 < 0$.

The origin will be the center of the quadric. We calculate this center by solving the system $2Ax = -B^{\mathsf{T}}$, that is,

$$\begin{pmatrix} 8 & 0 & 4 \\ 0 & 8 & 0 \\ 4 & 0 & 8 \end{pmatrix} \begin{pmatrix} x_1 \\ x_2 \\ x_3 \end{pmatrix} = - \begin{pmatrix} 0 \\ 0 \\ 1 \end{pmatrix}.$$

We obtain $C = (1/12, 0, -1/6)$.

The orthonormal basis we are looking for is formed by the normalized eigenvectors of A.

The eigenvalues of A are the roots of the characteristic polynomial $p_A(x)$. These roots are $\lambda_1 = 2$, $\lambda_2 = 4$ and $\lambda_3 = 6$. The corresponding eigenvectors are $u_1 = (1, 0, -1)$, $u_2 = (0, 1, 0)$ and $u_3 = (1, 0, 1)$.

This answers the question posed in the problem: the symmetry axes of the quadric are the straight lines $C + \langle u_i \rangle$, $i = 1, 2, 3$. For completion, we continue the classification.

Consider the orthonormal affine frame $\mathcal{R} = \{C; \mathcal{B}\}$, where $\mathcal{B} = (v_1, v_2, v_3)$ with $v_1 = 1/\sqrt{2}\, u_1$, $v_2 = u_2$ and $v_3 = 1/\sqrt{2}\, u_3$.

The relationship between the initial coordinates (x_1, x_2, x_3) and the coordinates (x_1', x_2', x_3') in $\mathcal{R}$ is given by

$$\begin{pmatrix} x_1 - 1/12 \\ x_2 \\ x_3 + 1/6 \end{pmatrix} = \begin{pmatrix} \frac{1}{\sqrt{2}} & 0 & \frac{1}{\sqrt{2}} \\ 0 & 1 & 0 \\ -\frac{1}{\sqrt{2}} & 0 & \frac{1}{\sqrt{2}} \end{pmatrix} \begin{pmatrix} x_1' \\ x_2' \\ x_3' \end{pmatrix}.$$

Substituting these values into the equation $4x_1^2 + 4x_2^2 + 4x_3^2 + 4x_1 x_3 + x_3 -$

$1 = 0$, we obtain

$$2(x_1')^2 + 4(x_2')^2 + 6(x_3')^2 - 13/12 = 0,$$

that is,

$$-\frac{24}{13}(x_1')^2 - \frac{48}{13}(x_2')^2 - \frac{72}{13}(x_3')^2 + 1 = 0.$$

These coefficients, $d_1 = -\frac{24}{13}$, $d_2 = -\frac{48}{13}$, $d_3 = -\frac{72}{13}$, already satisfy the condition $d_3 \leq d_2 \leq d_1 < 0$. Otherwise, we need only permute the variables.

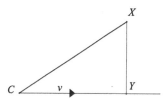

Figure 9.1.   Orthogonal symmetry

Now we find the orthogonal symmetry with respect to $l\colon C + \langle v_3 \rangle$. Since v_3 is a unit vector, the symmetry of a point X is given by (see Figure 9.1)

$$S(X) = X + 2\overrightarrow{XY}, \quad \text{with } Y = C + (\overrightarrow{CX} \cdot v_3)v_3.$$

Substituting, we obtain

$$Y = \left(\frac{x_1}{2} + \frac{x_3}{2} + \frac{1}{8}, 0, \frac{x_1}{2} + \frac{x_3}{2} - \frac{1}{8} \right),$$

and, hence,

$$S(x_1, x_2, x_3) = \left(x_3 + \frac{1}{4}, x_2, x_1 - \frac{1}{4} \right).$$

$\square$

EXERCISES

9.1. Find the canonical representatives of the following polynomials:

$$r(x) = x_1^2 - 3x_2^2 + 8x_1 x_2 + x_1 + x_2 + 1,$$

$$s(x) = 5x_1^2 + x_2^2 + 2x_1 x_2 - 3x_1 + x_2 + x_3 + 1,$$

$$t(x) = 5x_1^2 + x_2^2 + 2x_1 x_2 + 4x_1 + 2.$$

9.2. Find the invariants $(\rho, i, \tilde{\rho}, \tilde{i}, d_1, \ldots, d_\rho)$ of the following polynomials:

$$r(x) = -2x_1^2 + x_2^2 + 2x_1 x_2 + x_1 + 9,$$

$$s(x) = -2x_1^2 + x_2^2 + x_1 + x_2,$$

$$t(x) = x_1^2 + 2x_1 x_2 + x_2 - 8.$$

9.3. Classify the conic of a Euclidean affine plane given in an orthonormal affine frame by

$$2x_1^2 + 8x_2^2 - 4x_1 x_2 + \sqrt{2}x_2 = 0;$$

that is, find its canonical expression. Find an adapted affine frame.

9.4. Consider the conic of the Euclidean affine space $\mathbb{R}^2$ given by

$$3x_1^2 + 7x_2^2 + 6x_1 x_2 + x_1 + x_2 = 0.$$

(a) Find a symmetric 2×2 matrix A and a 1×2 matrix B such that

$$3x_1^2 + 7x_2^2 + 6x_1 x_2 + x_1 + x_2 = (x_1\ x_2)\, A \begin{pmatrix} cx_1 \\ x_2 \end{pmatrix} + B \begin{pmatrix} cx_1 \\ x_2 \end{pmatrix}.$$

(b) Find the center of the conic, that is, find a point $C = (c_1, c_2)$ such that

$$2A \begin{pmatrix} c_1 \\ c_2 \end{pmatrix} = -B^{\mathsf{T}}.$$

(c) Find an orthogonal 2×2 matrix a such that $D = a^{\mathsf{T}} A a$ is a diagonal matrix.

(d) Prove that the Euclidean motion of the Euclidean affine space $\mathbb{R}^2$

$$\begin{pmatrix} x' \\ y' \end{pmatrix} = a^{-1} \begin{pmatrix} x_1 \\ x_2 \end{pmatrix} - a^{-1} \begin{pmatrix} c_1 \\ c_2 \end{pmatrix}$$

transforms the given conic into an ellipse with center the origin. See Exercise 8.2 of Chapter 8, page 279.

9.5. Classify orthogonally the conics of a Euclidean affine plane $\mathbb{A}$ given in an orthonormal affine frame $\mathcal{R}$ by

$$2x_1^2 - x_2^2 - x_1 - x_2 - 3 = 0,$$

$$-3x_1^2 + 9x_2^2 + 6x_1 x_2 - 4x_1 - 1 = 0,$$

$$x_1 x_2 + 1 = 0.$$

Find, in each case, the center.

9.6. Classify orthogonally the quadrics of a Euclidean affine space $\mathbb{A}$ of dimension three given in an orthonormal affine frame $\mathcal{R}$ by

$$2x_1^2 - x_2^2 - x_1 - x_2 - 3 = 0,$$

$$-3x_1^2 + 9x_2^2 + 6x_1x_2 - 4x_1 - 1 = 0,$$

$$x_1x_2 + 1 = 0.$$

Find, in each case, the center.

9.7. Classify orthogonally the quadrics of a Euclidean affine space $\mathbb{A}$ of dimension three given in an orthonormal affine frame $\mathcal{R}$ by

$$x_1^2 - x_2^2 - 4x_1x_3 - 8x_2x_3 - 3x_2 + x_3 - 1 = 0,$$

$$-x_1^2 + x_3^2 + 6x_1x_2 - 4x_1 - x_2 = 0,$$

$$x_3^2 + x_1x_2 + 1 = 0.$$

Find, in each case, the center.

9.8. Given the polynomials

$$r(x) = x_1^2 - 4x_2^2 + x_1x_2 - x_2,$$

$$s(x) = -x_1^2 - 2x_2^2 - 5x_1x_2 - \frac{\sqrt{2}}{2}x_1 - \frac{\sqrt{2}}{2}x_2,$$

find two orthonormal affine frames $\mathcal{R}$ and $\mathcal{R}'$ of the Euclidean affine space $\mathbb{R}^2$ such that

$$Q(r(x), \mathcal{R}) = Q(s(x), \mathcal{R}').$$

9.9. (Ruled quadrics) Prove that through any point (a_1, a_2, a_3) on the hyperbolic paraboloid HP: $d_1x_1^2 + d_2x_2^2 + z = 0$ ($d_1 > 0$, $d_2 < 0$) pass two straight lines lying entirely on it. In fact, the direction vector is given by

$$(d_2(d_1a_1^2 \pm a_2a_1\sqrt{d_1}\sqrt{-d_2}), d_1a_1(a_2d_2 \mp a_1\sqrt{d_1}\sqrt{-d_2}), 2d_1d_2a_1a_3).$$

9.10. (Ruled quadrics) Prove that through every point (a_1, a_2, a_3) on the one sheet hyperboloid $H1$: $d_1x_1^2 + d_2x_2^2 + d_3x_3^2 + 1 = 0$ ($d_1 > 0$, $d_2 < 0$, $d_3 < 0$) pass two straight lines lying entirely on it. In fact, the direction vector (u_1, u_2, u_3) (determined up to a scalar factor) can be found by solving the system

$$d_1u_1^2 + d_2u_2^2 + d_3u_3^2 = 0,$$

$$d_1a_1u_1 + d_2a_2u_2 + d_3a_3u_3 = 0.$$

9.11. Let P be a point in the interior of an ellipsoid of the Euclidean affine space $\mathbb{R}^3$. Draw three orthogonal straight lines through P, and let A_1, A_2 be the intersection points of the first line with the ellipsoid; B_1, B_2 the intersection points of the second line with the ellipsoid; and C_1, C_2 the intersection points of the third line with the ellipsoid. Prove that the sum

$$\frac{1}{PA_1 \cdot PA_2} + \frac{1}{PB_1 \cdot PB_2} + \frac{1}{PA_1 \cdot PA_2}$$

does not depend on the three orthogonal straight lines chosen. (This exercise is taken from [4].)

A
Vector Spaces with Scalar Product

A.1 Introduction

The aim of this chapter is to "copy" the metric properties of the vector space $\mathbb{R}^n$ in arbitrary $\mathbb{R}$-vector spaces of finite dimension.

After recalling the notion of the "standard" scalar product on $\mathbb{R}^n$, we study bilinear maps as a natural generalization of this scalar product to arbitrary $\mathbb{R}$-vector spaces. We associate matrices to these bilinear maps in a way that is similar, but necessarily different, to the way that matrices are associated to linear maps.

Next we introduce the notion of a *scalar product on an $\mathbb{R}$-vector space*, as a particular case of a bilinear map, and the concepts of norm, cosine of the angle between two vectors and orthogonality.

We shall also study maps preserving a scalar product. These maps are linear and the matrix associated to each of them, in an orthonormal basis, has the curious property that its inverse and its transpose coincide ($A^{\mathsf{T}} = A^{-1}$). This kind of matrix is said to be orthogonal; see Section A.8, page 341.

A.1.1 The Standard Scalar Product on $\mathbb{R}^n$

Definition A.1

The *standard scalar product* of two vectors u and v of $\mathbb{R}^n$ is the number $u \cdot v$ given by

$$u \cdot v = u_1 v_1 + u_2 v_2 + \cdots + u_n v_n,$$

where $u = (u_1, u_2, \ldots, u_n)$ and $v = (v_1, v_2, \ldots, v_n)$.

The scalar product enables us to define the distance between points of $\mathbb{R}^n$, the angle between vectors, etc.

Concretely, the *modulus* or *norm* of a vector $u \in \mathbb{R}^n$ is the number $|u|$ given by

$$|u| = \sqrt{u \cdot u}.$$

The Cauchy-Schwarz inequality holds:

$$(u \cdot v)^2 \le |u|^2 |v|^2,$$

for all $u, v \in \mathbb{R}^n$. The *cosine* of the angle between two non-zero vectors $u, v \in \mathbb{R}^n$ is defined as

$$\cos(u, v) = \frac{u \cdot v}{|u| \cdot |v|}.$$

The *distance* between two points P and Q of $\mathbb{R}^n$ is the number

$$d(P, Q) = |\overrightarrow{PQ}|,$$

where $\overrightarrow{PQ} = Q - P$. The properties of distance given in Proposition 5.5, page 160, are satisfied.

A.2 Bilinear Maps

Note that the existence of the canonical basis of $\mathbb{R}^n$ plays a key role in the definition of the standard scalar product. Since an arbitrary $\mathbb{R}$-vector space has no canonical basis, we must find an alternative definition of scalar product in the general case. What we will do is observe that the standard scalar product on $\mathbb{R}^n$ can be considered as a map from $\mathbb{R}^n \times \mathbb{R}^n$ to $\mathbb{R}$ satisfying certain properties, and then we shall define a *scalar product* on an $\mathbb{R}$-vector space E to be any map from $E \times E$ to $\mathbb{R}$ satisfying these same properties.

Definition A.2

A *bilinear map* φ on an $\mathbb{R}$-vector space E is a map

$$\varphi : E \times E \longrightarrow \mathbb{R}$$

which satisfies the following, for all $u, v, w \in E$ and for all $\lambda \in \mathbb{R}$:

(1) $\varphi(u + v, w) = \varphi(u, w) + \varphi(v, w)$,
$\varphi(u, v + w) = \varphi(u, v) + \varphi(u, w)$.

(2) $\varphi(\lambda u, v) = \lambda \varphi(u, v)$,
$\varphi(u, \lambda v) = \lambda \varphi(u, v)$.

Observe that the standard scalar product is a bilinear map on $\mathbb{R}^n$.

A.2.1 The Matrix of a Bilinear Map

Let $\mathcal{B} = (e_1, \ldots, e_n)$ be a basis of the $\mathbb{R}$-vector space E. To each bilinear map φ on E we associate a matrix $A \in \mathcal{M}_{n \times n}(\mathbb{R})$ in the following way:

$$A = (a_{ij}), \quad \text{with } a_{ij} = \varphi(e_i, e_j).$$

That is,

$$A = \begin{pmatrix} a_{11} & a_{12} & \cdots & a_{1n} \\ a_{21} & a_{22} & \cdots & a_{2n} \\ \vdots & \vdots & & \vdots \\ a_{n1} & a_{n2} & \cdots & a_{nn} \end{pmatrix}.$$

φ determines, and is determined by, A. Indeed, let

$$u = \sum_{i=1}^{n} u_i e_i, \qquad v = \sum_{j=1}^{n} v_j e_j.$$

Then

$$\varphi(u, v) = \varphi \left(\sum_{i=1}^{n} u_i e_i, \sum_{j=1}^{n} v_j e_j \right)$$

$$= \sum_{i=1}^{n} \sum_{j=1}^{n} u_i v_j \varphi(e_i, e_j)$$

$$= \sum_{i=1}^{n} \sum_{j=1}^{n} u_i a_{ij} v_j$$

$$= \begin{pmatrix} u_1 & \cdots & u_n \end{pmatrix} \begin{pmatrix} a_{11} & a_{12} & \cdots & a_{1n} \\ a_{21} & a_{22} & \cdots & a_{2n} \\ \vdots & \vdots & & \vdots \\ a_{n1} & a_{n2} & \cdots & a_{nn} \end{pmatrix} \begin{pmatrix} v_1 \\ \vdots \\ v_n \end{pmatrix}. \tag{A.1}$$

The matrix A is called the *matrix of φ with respect to the basis $\mathcal{B}$*, and will be denoted by $M(\varphi, \mathcal{B})$.

Using this notation and the notation used for the components of vectors, see page 143, formula (A.1) is rewritten as

$$\varphi(u, v) = C(u, \mathcal{B})^{\mathsf{T}} M(\varphi, \mathcal{B}) C(v, \mathcal{B}). \tag{A.2}$$

A.2.2 Change of Basis

Let $\mathcal{B}_1 = (e_1, \ldots, e_n)$ and $\mathcal{B}_2 = (u_1, \ldots, u_n)$ be two bases of E, and let φ be a bilinear map on an $\mathbb{R}$-vector space E. What is the relationship between the matrices $M(\varphi, \mathcal{B}_1)$ and $M(\varphi, \mathcal{B}_2)$?

Let

$$C = M(\mathcal{B}_2, \mathcal{B}_1)$$

be the matrix of the change of basis, that is,

$$C = (c_{ir}) \quad \text{with } u_r = \sum_i c_{ir} e_i.$$

To simplify the notation, we set

$$A = (a_{ij}) = M(\varphi, \mathcal{B}_1), \qquad B = (b_{ij}) = M(\varphi, \mathcal{B}_2).$$

We also use the notation $[M]_{ij}$ to denote the (i, j)-th entry of the matrix M. So, if $M = (m_{ij})$, we have $[M]_{ij} = m_{ij}$.

Thus,

$$b_{rs} = \varphi(u_r, u_s) = \sum_{i,j} c_{ir} c_{js} \varphi(e_i, e_j) = \sum_{i,j} c_{ir} c_{js} a_{ij}$$

$$= \sum_{i,j} c_{ir} a_{ij} c_{js} = \sum_i c_{ir} [AC]_{is} = \sum_i [C^{\mathsf{T}}]_{ri} [AC]_{is} = [C^{\mathsf{T}} AC]_{rs}.$$

That is,

$$B = C^{\mathsf{T}} AC,$$

or, equivalently,

$$M(\varphi, \mathcal{B}_2) = M(\mathcal{B}_2, \mathcal{B}_1)^{\mathsf{T}} M(\varphi, \mathcal{B}_1) M(\mathcal{B}_2, \mathcal{B}_1)$$ (A.3)

which is the *formula of the change of basis* for bilinear maps.

In particular,

$$\det M(\varphi, \mathcal{B}_2) = \det M(\varphi, \mathcal{B}_1) \cdot (\det C)^2.$$

This enables us to define the *rank* and *discriminant* of a bilinear map. The *rank* of a bilinear map ϕ is the rank of the matrix associated to ϕ in any basis,

$$\operatorname{rank} \phi = \operatorname{rank} M(\phi, \mathcal{B}).$$

Recall that the rank of a matrix does not vary when we multiply this matrix, on the left or on the right, by an invertible matrix (see, for instance, [8], page 201). For this reason, and by (A.3), it makes sense to talk about the rank of ϕ.

The *discriminant* of ϕ is the discriminant of the matrix associated to ϕ in any basis,

$$\operatorname{discriminant} \phi = \operatorname{discriminant} M(\phi, \mathcal{B}).$$

Recall that *the discriminant of a matrix is its determinant, modulo squares.* Thus,

$$\operatorname{discriminant} M(\phi, \mathcal{B}_1) = \operatorname{discriminant} M(\phi, \mathcal{B}_2),$$

since

$$\frac{\det M(\varphi, \mathcal{B}_2)}{\det M(\varphi, \mathcal{B}_1)}$$

is a square. Thus it is meaningful to talk about the discriminant of ϕ, but not the determinant of ϕ.

Observe that two matrices have the same discriminant if and only if their determinants are different from zero and *the quotient of these determinants is a square.* Therefore, the discriminant strongly depends on the arithmetic of the underlying field.

For instance, on $\mathbb{C}$ all invertible matrices have the same discriminant, because every non-zero complex number has a square root; and on $\mathbb{R}$ two invertible matrices have the same discriminant if and only if their determinants have the same sign, because then their quotient is positive and therefore has a square root.

A.2.3 Symmetric Bilinear Maps

Definition A.3

A bilinear map φ on an $\mathbb{R}$-vector space E is *symmetric* if and only if, for all $u, v \in E$, we have $\varphi(u, v) = \varphi(v, u)$.

Observe that the standard scalar product is a symmetric bilinear map on $\mathbb{R}^n$.

Proposition A.4

A bilinear map φ on an $\mathbb{R}$-vector space E is symmetric if and only if its associated matrix with respect to a basis $\mathcal{B} = (e_1, \ldots, e_n)$ of E, $A = M(\varphi, \mathcal{B})$, is symmetric, that is $A = A^\mathsf{T}$.

Proof

It is clear that if φ is symmetric then its associated matrix A is symmetric, because $\varphi(e_i, e_j) = \varphi(e_j, e_i)$.

Conversely, let us assume A is symmetric.

First we observe that, if $c_1, c_2 \in M_{n \times 1}(\mathbb{R})$, the product of matrices $c_1^\mathsf{T} A c_2$ is a scalar (a 1×1 matrix), and hence we have

$$(c_1^\mathsf{T} A c_2)^\mathsf{T} = c_1^\mathsf{T} A c_2.$$

Thus, if $u, v \in E$ are the vectors with components c_1, c_2 in the basis $\mathcal{B}$, $c_1 = C(u, \mathcal{B})$ and $c_2 = C(u, \mathcal{B})$, we have

$$\varphi(u, v) = c_1^\mathsf{T} A c_2 = (c_1^\mathsf{T} A c_2)^\mathsf{T} = c_2^\mathsf{T} A^\mathsf{T} c_1 = c_2^\mathsf{T} A c_1 = \varphi(v, u).$$

$\square$

A.2.4 The Radical

We have already remarked in the introduction to this chapter that the concept of a bilinear map generalizes the concept of the standard scalar product on $\mathbb{R}^n$. However, with respect to this 'generalized scalar product', it is possible for a non-zero vector in the $\mathbb{R}$-vector space E to have modulus zero, that is, there may be a vector $u \in E$, $u \neq 0$, such that $\phi(u, u) = 0$. It is even possible for a non-zero vector $0 \neq u \in E$ to satisfy $\phi(u, v) = 0$ for all $v \in E$. For this reason, in order to study bilinear maps, it is useful to introduce the concept of the *radical*.

Definition A.5

The *radical* of a symmetric bilinear map ϕ of an $\mathbb{R}$-vector space E is the vector subspace of E defined by

$$\operatorname{rad}\phi = \{v \in E : \phi(v,w) = 0, \text{ for all } w \in E\}.$$

When $\operatorname{rad}\phi = \{\vec{0}\}$, we say that ϕ is *non-singular*, or *non-degenerate*.

Extending a basis of the radical to a basis of the whole space we obtain a basis $\mathcal{B}$ such that

$$M(\phi,\mathcal{B}) = \begin{pmatrix} 0 & 0 \\ 0 & A \end{pmatrix}$$

and a decomposition

$$E = \operatorname{rad}\phi \oplus F$$

such that ϕ restricted to F is non-singular. That is, the radical of $(F, \phi|F)$ is zero. Thus, the study of symmetric bilinear maps essentially reduces to the study of non-singular symmetric bilinear maps.

Note that the subspace F complementary to the radical is not at all unique (a basis of the radical can be extended to a basis of the whole space in many different ways). If we have

$$E = \operatorname{rad}\phi \oplus F',$$

and $\mathcal{B}'$ is a basis adapted to this decomposition (the first vectors are in $\operatorname{rad}\phi$ and the others in F'), then

$$M(\phi,\mathcal{B}') = \begin{pmatrix} 0 & 0 \\ 0 & A' \end{pmatrix}.$$

In this case, it is easy to see (change of basis) that there is an invertible matrix C such that

$$A' = C^{\mathsf{T}}AC.$$

Hence, discriminant $A' = $ discriminant A, and we call this common value the *discriminant of the non-singular part* of ϕ.

Finally, let us recall that the notation $E = F \perp H$ reads "E is equal to the orthogonal direct sum of F and H" and means $E = F \oplus H$ with $\phi(u,v) = 0$ for all $u \in F$ and $v \in H$. That is, E is the direct sum of two orthogonal subspaces.

In particular, the notation $\operatorname{rad}\phi \oplus F$ is equivalent to the notation $\operatorname{rad}\phi \perp F$, since the radical is orthogonal to any subspace.

A.3 Scalar Product

Definition A.6

We say that a bilinear map φ on an $\mathbb{R}$-vector space E is *positive definite* if for all $u \in E$ we have $\varphi(u, u) \geq 0$, and $\varphi(u, u) = 0$ if and only if $u = 0$.

Notice that the condition $\varphi(u, u) \geq 0$ is meaningless on arbitrary fields. Observe also that the standard scalar product is positive definite; and that if φ is positive definite, then $\operatorname{rad} \varphi = \{\vec{0}\}$.

Definition A.7 (Scalar product)

A *scalar product* on an $\mathbb{R}$-vector space E is a positive definite symmetric bilinear map φ on E.

Definition A.8

A *Euclidean* vector space is an $\mathbb{R}$-vector space together with a scalar product defined on it.

Thus, if φ is a scalar product on E, we say that (E, φ) is a Euclidean vector space. Sometimes, by abuse of notation, we say "the Euclidean vector space E" when it is clear which scalar product is being considered.

A scalar product on E enables us to define a distance between elements of E, angles, etc.

Concretely, if φ is a scalar product on E, we define the *modulus* or *norm* of a vector u of E by

$$|u| = \sqrt{\varphi(u, u)},$$

and the *cosine of the angle* between two non-zero vectors u and v of E by

$$\cos(u, v) = \frac{\varphi(u, v)}{|u| \cdot |v|}.$$

This definition makes sense because of the Cauchy-Schwarz inequality.

Proposition A.9 (The Cauchy-Schwarz Inequality)

Let (E, φ) be a Euclidean vector space. Then, for all $u, v \in E$, we have

$$\varphi(u, v)^2 \leq |u|^2 |v|^2.$$

Proof

Let $\lambda \in \mathbb{R}$. Then

$$0 \le \varphi(u + \lambda v, u + \lambda v) = \varphi(u, u) + 2\lambda\varphi(u, v) + \lambda^2\varphi(v, v).$$

Since the polynomial $\varphi(u, u) + 2\varphi(u, v)x + \varphi(v, v)x^2$ takes non-negative values for all $\lambda \in \mathbb{R}$, its discriminant must be negative or zero. Hence,

$$4\varphi(u, v)^2 - 4\varphi(u, u)\varphi(v, v) \le 0,$$

and this completes the proof.   □

Corollary A.10

Let (E, φ) be a Euclidean vector space. Then, for all $u, v \in E$, we have

$$|u + v| \le |u| + |v|.$$

Proof

Using the Cauchy-Schwarz inequality we have

$$
\begin{aligned}
|u + v|^2 &= \varphi(u + v, u + v) = |u|^2 + |v|^2 + 2\varphi(u, v) \\
&\le |u|^2 + |v|^2 + 2|u||v| \\
&= (|u| + |v|)^2,
\end{aligned}
$$

and this completes the proof.   □

Definition A.11

Let (E, φ) be a Euclidean vector space. We say that $u \in E$ is a *unit vector* if $|u| = 1$. Two vectors $u, v \in E$ are said to be *orthogonal* if $\varphi(u, v) = 0$. A basis $(e_1, \ldots, e_n)$ is called *orthogonal* if the e_i are orthogonal to each other, that is, $\varphi(e_i, e_j) = 0$, for all $i \ne j$. If, moreover, all the e_i are unit vectors, the basis is called *orthonormal*.

When a basis is orthonormal with respect to a scalar product φ we also say that this basis is φ-orthonormal. However, it is often the case that the scalar product is understood, in which case we simply say that the basis is orthonormal.

Therefore, the matrix of φ in an orthonormal basis $\mathcal{B} = (e_1, \ldots, e_n)$ is the identity matrix, because the component (i, j) of $M(\varphi, \mathcal{B})$ is

$$\varphi(e_i, e_j) = \delta_{ij}.$$

Recall that $\delta_{ij} = 1$, if $i = j$, and $\delta_{ij} = 0$, if $i \neq j$.

Observe that if $E = \mathbb{R}^n$, the canonical basis is orthonormal with respect to the standard scalar product.

Proposition A.12

Let $u_1, \ldots, u_k$ be non-zero vectors of a Euclidean vector space (E, φ). If they are orthogonal to each other, then they are linearly independent.

Proof

If

$$\lambda_1 u_1 + \cdots + \lambda_k u_k = 0,$$

then

$$0 = \varphi(u_j, 0) = \varphi\left(u_j, \sum_{i=1}^{i=k} \lambda_i u_i\right) = \lambda_j \varphi(u_j, u_j),$$

and hence $\lambda_j = 0$, for all $j = 1, \ldots, k$. $\square$

Theorem A.13 (Gram-Schmidt)

Let (E, φ) be a Euclidean vector space. Then there exists an orthonormal basis of E.

Proof

We begin with any basis $(u_1, \ldots, u_n)$ of E. Consider

$$E_1 = \langle u_1 \rangle,$$
$$E_2 = \langle u_1, u_2 \rangle,$$
$$\vdots$$
$$E_i = \langle u_1, \ldots, u_i \rangle,$$
$$\vdots$$
$$E_n = \langle u_1, \ldots, u_n \rangle.$$

The vector subspace E_1 admits an orthonormal basis. Simply take

$$e_1 = \frac{u_1}{\sqrt{\varphi(u_1, u_1)}}.$$

Let us assume, by induction, that E_k admits an orthonormal basis $(e_1, \ldots, e_k)$.

We construct an orthonormal basis of $E_{k+1} = \langle e_1, \ldots, e_k, u_{k+1} \rangle$ in the following way:

Let

$$e'_{k+1} = u_{k+1} - (\lambda_1 e_1 + \cdots + \lambda_k e_k),$$

so that e'_{k+1} is orthogonal to all e_i, $i = 1, \ldots, k$. For this to be the case, we must take

$$\lambda_i = \varphi(e_i, u_{k+1}), \quad \text{for } i = 1, \ldots, k.$$

Next we set

$$e_{k+1} = \frac{e'_{k+1}}{|e'_{k+1}|},$$

and we obtain an orthonormal basis $(e_1, \ldots, e_{k+1})$ of E_{k+1}. Thus, by induction, the proof is complete. □

This recursive method of constructing an orthonormal basis is known as the *Gram-Schmidt orthonormalization method*.

A.4 The Orthogonal Group

The set of all $n \times n$ real matrices such that $A^\mathsf{T} A = I_n$ form a group called the *orthogonal group*, denoted by $O(n)$. That is,

$$O(n) = \{A \in \mathcal{M}_{n \times n}(\mathbb{R}) : A^\mathsf{T} A = I_n\}.$$

Equivalently, A is an orthogonal matrix if its inverse coincides with its transpose. If $A \in O(n)$, we say that A is an *orthogonal matrix*.

These matrices appear in the study of the relationship between different orthonormal bases. We have seen, in the above Theorem A.13, that there is at least one basis of the Euclidean vector space (E, φ) orthonormal with respect to φ. However, this basis is not unique.

Formula (A.3) allows us to study the matrix of the change of basis between two orthonormal bases. Indeed, if $\mathcal{B}_1$ and $\mathcal{B}_2$ are two bases of E, we have

$$M(\varphi, \mathcal{B}_2) = M(\mathcal{B}_2, \mathcal{B}_1)^{\mathsf{T}} M(\varphi, \mathcal{B}_1) M(\mathcal{B}_2, \mathcal{B}_1).$$

If, moreover, $\mathcal{B}_1$ and $\mathcal{B}_2$ are orthonormal, we have

$$M(\varphi, \mathcal{B}_1) = M(\varphi, \mathcal{B}_2) = I_n,$$

and hence the above formula gives

$$I_n = M(\mathcal{B}_2, \mathcal{B}_1)^{\mathsf{T}} M(\mathcal{B}_2, \mathcal{B}_1).$$

That is, *the matrix of the change of basis between two orthonormal bases is an orthogonal matrix.*

A.4.1 Positive Definite Matrices

We have seen in Proposition A.4 that a bilinear map on E is symmetric if and only if its matrix, in any basis, is symmetric. It is very easy to see if a matrix is symmetric, but to distinguish, among symmetric matrices, those corresponding to scalar products is not immediate.

Definition A.14

A matrix $A \in \mathcal{M}_{n \times n}(\mathbb{R})$ is called *positive definite* if and only if

$$c^{\mathsf{T}} A c > 0$$

for all non-zero column matrices $c \in \mathcal{M}_{n \times 1}(\mathbb{R})$.

Proposition A.15

Let $A \in \mathcal{M}_{n \times n}(\mathbb{R})$. Then there exist a scalar product φ on E and a basis $\mathcal{B}$ of E such that $M(\varphi, \mathcal{B}) = A$ if and only if A is symmetric and positive definite.

Proof

See [8], page 447.   □

The most useful criterion to determine if a symmetric matrix is positive definite is the following.

Proposition A.16

A symmetric matrix $A = (a_{ij}) \in \mathcal{M}_{n \times n}(\mathbb{R})$ is positive definite if and only if

$$\begin{vmatrix} a_{11} & \cdots & a_{1i} \\ \vdots & & \vdots \\ a_{i1} & \cdots & a_{ii} \end{vmatrix} > 0,$$

for all $i = 1, \ldots, n$.

Proof

See [8], page 448.   □

A.5 Orthogonal Vector Subspaces

Let (E, φ) be a Euclidean vector space. Let F be a vector subspace of E. We define the *orthogonal* $F^\perp$ of F by

$$F^\perp = \{u \in E : \varphi(u, v) = 0, \text{ for all } v \in F\}.$$

Proposition A.17

Let F and H be vector subspaces of the Euclidean vector space E. Then we have:
  (i) $F^\perp$ is a vector subspace of E.
  (ii) If $F \subseteq H$, then $H^\perp \subseteq F^\perp$.
  (iii) $F \cap F^\perp = \{\vec{0}\}$.
  (iv) $F \subseteq (F^\perp)^\perp$.

Proof

The proof is left to the reader.   □

Proposition A.18

Let F be a vector subspace of a Euclidean vector space E. Then

$$E = F \perp F^\perp.$$

In particular,

$$\dim F^{\perp} = \dim E - \dim F.$$

Proof

We know that $F \cap F^{\perp} = \{\vec{0}\}$. Let us show that $E = F + F^{\perp}$. Take an orthonormal basis $(e_1, \ldots, e_k)$ of F. Using Steinitz's theorem (see [8], page 251) we complete these linearly independent vectors to a basis of E,

$$(e_1, \ldots, e_k, u_{k+1}, \ldots, u_n).$$

Next we apply the Gram-Schmidt orthonormalization method. This method transforms the vectors $u_{k+1}, \ldots, u_n$ to vectors $e_{k+1}, \ldots, e_n$, so that $(e_1, \ldots, e_k, e_{k+1}, \ldots, e_n)$ is an orthonormal basis of E with $e_{k+1}, \ldots, e_n \in F^{\perp}$. Hence, every element of E is the sum of an element of F and an element of $F^{\perp}$. □

Corollary A.19

Let F be a vector subspace of a Euclidean vector space E. Then

$$F^{\perp\perp} = F.$$

Proof

We know that $F \subseteq F^{\perp\perp}$ and also that

$$\dim F^{\perp\perp} = \dim E - \dim F^{\perp} = \dim E - (\dim E - \dim F) = \dim F.$$

Since F is of finite dimension, $F = F^{\perp\perp}$. □

A.6 Isometries

Definition A.20

Let (E, φ) be a Euclidean vector space. An *isometry* of E is a map $f : E \longrightarrow E$ preserving the scalar product, that is, such that

$$\varphi(u, v) = \varphi(f(u), f(v)), \quad \text{for all } u, v \in E.$$

Isometries are automatically linear:

Theorem A.21

Let (E, φ) be a Euclidean vector space. Let $f : E \to E$ be an isometry. Then f is linear.

Proof

A simple calculation shows that

$$\varphi(f(u+v) - f(u) - f(v), f(u+v) - f(u) - f(v)) = 0,$$

$$\varphi(f(\lambda u) - \lambda f(u), f(\lambda u) - \lambda f(u)) = 0.$$

See [8], page 453. □

Recall that $\lambda \in \mathbb{R}$ is an *eigenvalue* of an endomorphism $f : E \to E$ if there exists a non-zero vector $u \in E$ such that $f(u) = \lambda u$. In this case we say that u is an *eigenvector* of f with eigenvalue λ.

Theorem A.22

Let (E, φ) be a Euclidean vector space. Let $f : E \to E$ be an isometry. Then
  (i) For all $u \in E$, $|f(u)| = |u|$.
 (ii) For all $u, v \in E$, $\varphi(u, v) = 0$ if and only if $\varphi(f(u), f(v)) = 0$.
(iii) The map f is bijective.
(iv) If λ is an eigenvalue of f then $\lambda = 1$ or $\lambda = -1$.
 (v) Eigenvectors of f with different eigenvalues are orthogonal.

Proof

Parts (i) and (ii) are clear.
    (iii) If $f(u) = 0$, we have $0 = \varphi(f(u), f(u)) = \varphi(u, u)$, and hence $u = 0$. Thus, f is injective and, by the Isomorphism Theorem (see [8], page 284), f is bijective.
    (iv) Let $u \in E$ be an eigenvector of f with eigenvalue λ, that is, $f(u) = \lambda u$. Then we have

$$\varphi(u, u) = \varphi(f(u), f(u)) = \varphi(\lambda u, \lambda u) = \lambda^2 \varphi(u, u),$$

and hence $\lambda = \pm 1$.

(v) Let $u, v \in E$ be eigenvectors of f with different eigenvalues. By (iv), we can assume that $f(u) = u$ and $f(v) = -v$. Then

$$\varphi(u, v) = \varphi(f(u), f(v)) = \varphi(u, -v) = -\varphi(u, v),$$

and hence $\varphi(u, v) = 0$. $\square$

Proposition A.23

Let (E, φ) be a Euclidean vector space, f a map from E to E, and $(e_1, \ldots, e_n)$ a basis of E. Then f is an isometry if and only if it is linear and

$$\varphi(f(e_i), f(e_j)) = \varphi(e_i, e_j), \quad i, j = 1, \ldots, n.$$

Proof

If f preserves the scalar product, then, by Theorem A.21, it is linear; and it obviously preserves the scalar product between vectors of the basis.

Conversely, assume that f is linear and preserves the scalar product between vectors of the basis. Let $u = \sum_{i=1}^n u_i e_i$, $v = \sum_{j=1}^n v_j e_j$ be two vectors of E, with $u_i, v_j \in \mathbb{R}$. Then

$$\varphi(f(u), f(v)) = \sum_{i=1}^n \sum_{j=1}^n u_i u_j \varphi(f(e_i), f(e_j))$$

$$= \sum_{i=1}^n \sum_{j=1}^n u_i u_j \varphi(e_i, e_j) = \varphi(u, v).$$

$\square$

Proposition A.24

Let (E, φ) be a Euclidean vector space, and $f : E \longrightarrow E$ a linear map with matrix A in a basis $\mathcal{B}_1 = (e_1, \ldots, e_n)$ of E. Let G be the matrix of φ in this basis. Then f is an isometry if and only if $A^{\mathsf{T}} G A = G$.

Proof

In the usual notation, see pages 57 and 322, we have

$$A = M(f, \mathcal{B}_1, \mathcal{B}_1), \qquad G = M(\varphi, \mathcal{B}_1).$$

First we suppose that f is an isometry. Then, by Theorem A.22, f is bijective and $\mathcal{B}_2 = (f(e_1), \ldots, f(e_n))$ is a basis of E. By the formula of the change of basis (formula (A.3), page 323), we have

$$M(\varphi, \mathcal{B}_2) = M(\mathcal{B}_2, \mathcal{B}_1)^\mathsf{T} M(\varphi, \mathcal{B}_1) M(\mathcal{B}_2, \mathcal{B}_1);$$

but it is clear that

$$M(\mathcal{B}_2, \mathcal{B}_1) = M(f, \mathcal{B}_1, \mathcal{B}_1) = A,$$

and hence the formula of the change of basis implies

$$M(\varphi, \mathcal{B}_2) = A^\mathsf{T} G A.$$

But, since

$$\varphi(f(e_i), f(e_j)) = \varphi(e_i, e_j),$$

we have

$$M(\varphi, \mathcal{B}_2) = M(\varphi, \mathcal{B}_1) = G,$$

and hence

$$G = A^\mathsf{T} G A.$$

Conversely, the matricial equality $G = A^\mathsf{T} G A$ implies

$$\varphi(e_i, e_j) = \varphi(f(e_i), f(e_j)),$$

and, by Proposition A.23, f is an isometry.   □

Now we can prove that *the matrix of an isometry in an orthonormal basis is an orthogonal matrix.*

Corollary A.25

Let (E, φ) be a Euclidean vector space, $f : E \longrightarrow E$ a linear map, and $A = M(f, \mathcal{B})$ the matrix of f in an orthonormal basis $\mathcal{B}$ of E. Then f is an isometry if and only if $A^\mathsf{T} A = I_n$.

Proof

Apply Proposition A.24 with $G = I_n$.   □

Observe that this corollary is also true when A is the matrix of f in a basis $\mathcal{B} = (e_1, \dots, e_n)$ such that

$$\varphi(e_i, e_j) = a\delta_{ij}, \quad a \in \mathbb{R},$$

because in this case $G = M(\varphi, \mathcal{B}) = aI_n$ and $A^{\mathsf{T}}GA = G$ if and only if $A^{\mathsf{T}}A = I_n$.

A.7 Polynomials of Isometries

Let f be an isometry of a Euclidean vector space E. Observe that $(f - \mathrm{id})$ does not preserve, in general, the scalar product (the difference of isometries is not an isometry). More generally, if f is an isometry and $p(x)$ is a polynomial, the endomorphism $p(f)$ is not, in general, an isometry. However, endomorphisms obtained in this way have very interesting properties. For instance, if we denote the scalar product by $\varphi(u, v) = \langle u, v \rangle$, we have the following.

Proposition A.26

Let f be an isometry and $p(x) = a_0 + a_1 x + \cdots + a_n x^n$ be any polynomial. Then

$$\langle p(f)(u), v \rangle = \langle u, p(f^{-1})(v) \rangle, \quad \text{for all } u, v \in E.$$

Proof

$$\langle p(f)(u), v \rangle = \left\langle \sum_i a_i f^i(u), v \right\rangle = \sum_i a_i \langle f^i(u), v \rangle$$

$$= \sum_i a_i \langle u, f^{-i}(v) \rangle = \left\langle u, \sum_i a_i f^{-i}(v) \right\rangle$$

$$= \langle u, p(f^{-1})(v) \rangle.$$

$\square$

Corollary A.27

Let f be an isometry and $p(x)$ be any polynomial. Then

$$|p(f)(u)| = |p(f^{-1})(u)|, \quad \text{for all } u \in E.$$

Proof

$$\langle p(f)(u), p(f)(u)\rangle = \langle u, p(f^{-1})p(f)(u)\rangle$$
$$= \langle u, p(f)p(f^{-1})(u)\rangle$$
$$= \langle p(f^{-1})(u), p(f^{-1})(u)\rangle.$$

$\square$

Corollary A.28

Let f be an isometry and $p(x)$ be any polynomial. If, for some vector $u \in E$, we have $p(f)^2(u) = 0$, then $p(f)(u) = 0$.

Proof

Set $w = p(f)(u)$. Since $p(f)(w) = 0$, by the previous corollary we have $p(f^{-1})(w) = 0$, and hence

$$0 = \langle p(f^{-1})(w), u\rangle$$
$$= \langle w, p(f)(u)\rangle$$
$$= \langle w, w\rangle;$$

therefore $w = 0$, and the proof is complete.

$\square$

Evidently, this corollary also proves that if $p(f)^r(u) = 0$, $r \geq 2$, then $p(f)(u) = 0$.

Corollary A.29

Let f be an isometry. Let $p(x)$ be an irreducible monic polynomial divisor of the minimal polynomial $m_f(x)$ of f. Then

$$m_f(x) = p(x)q(x), \qquad \gcd(p(x), q(x)) = 1.$$

That is, the minimal polynomial $m_f(x)$ of an isometry f is equal to the product of all the irreducible factors of the characteristic polynomial $p_f(x)$ of f, with exponent 1.

Proof

Let us suppose $m_f(x) = p(x)^r q(x)$. Then, for all $v \in E$, $p(f)^r q(f)(v) = 0$, and, by Corollary A.28, $p(f)q(f)(v) = 0$. If $r > 1$, we would have a polynomial, $p(x)q(x)$, with degree less than the degree of the minimal polynomial, annihilating f. This is a contradiction, and hence we must have $r = 1$. This completes the proof.

Since the minimal polynomial has all the irreducible factors of the characteristic polynomial (see [8], page 350), this result says that the minimal polynomial $m_f(x)$ of an isometry f is equal to the product of all irreducible factors of the characteristic polynomial $p_f(x)$ of f, each factor with exponent 1. □

Proposition A.30

Let f be an isometry and let

$$m_f(x) = q_1(x) \cdots q_r(x)$$

be the factorization of the minimal polynomial of f in monic irreducible factors. Then

$$\ker q_i(f) \subset (\ker q_j(f))^{\perp}, \quad i \neq j,$$

$$\operatorname{Im} q_i(f) = (\ker q_i(f))^{\perp}.$$

Proof

Let us first show that $\ker q_i(f) \subset (\ker q_j(f))^{\perp}$, $i \neq j$.

For this we take $u \in \ker q_i(f)$ and we prove that $\langle u, v \rangle = 0$, for all $v \in \ker q_j(f)$, $i \neq j$.

Note that $q_i(f)(u) = 0$ and $q_j(f)(v) = 0$. Recall that, by Corollary A.27, we also have $q_j(f^{-1}(v)) = 0$. By Bézout's identity, there are polynomials $a(x), b(x)$ such that

$$q_i(x)a(x) + q_j(x)b(x) = 1.$$

In particular, $q_j(f)b(f)(u) = u$. Hence,

$$\langle u, v \rangle = \langle q_j(f)b(f)(u), v \rangle$$

$$= \langle b(f)(u), q_j(f^{-1})(v) \rangle$$

$$= 0.$$

Now we want to show that $\operatorname{Im} q_i(f) = (\ker q_i(f))^{\perp}$.

For this we take $w \in \operatorname{Im} q_i(f)$, that is, $w = q_i(f)(u)$, and we prove that $\langle w, v \rangle = 0$, for all $v \in \ker q_i(f)$. As before, we also have $q_i(f^{-1}(v)) = 0$. Hence,

$$\langle w, v \rangle = \langle q_i(f)(u), v \rangle = \langle u, q_i(f^{-1})(v) \rangle = 0.$$

This proves $\operatorname{Im} q_i(f) \subset (\ker q_i(f))^\perp$; but, by the Isomorphism Theorem (see [8], page 284), we have equality. This completes the proof. $\qquad\square$

Since we are working over the field $\mathbb{R}$, the irreducible factors of the characteristic polynomial of an isometry, and therefore also those of the minimal polynomial, have degree 1 or 2. In fact, the only factors that can appear in the factorization of the characteristic polynomial are $(x - 1), (x + 1)$ (the eigenvalues have modulus 1), $(x - a)^2 + b^2$, $b \neq 0$, and their powers.

But there is still one more constraint.

Proposition A.31

Suppose that the characteristic polynomial of an isometry f is divisible by $p(x) = (x - a)^2 + b^2$, $b \neq 0$. Then

$$a^2 + b^2 = 1.$$

Proof

We know that $\ker((f - a\,\mathrm{id})^2 + b^2\,\mathrm{id}) \neq \{\vec{0}\}$, see [8], page 363. Let $v \in \ker((f - a\,\mathrm{id})^2 + b^2\,\mathrm{id})$. The vectors v and $f(v)$ are linearly independent, since there are no eigenvectors in this kernel. Let $F = \langle v, f(v) \rangle$. Since $f^2(v)$ is a linear combination of v and $f(v)$, the vector subspace F is invariant under f. The matrix of $f_{|F}$ in the basis $\mathcal{B} = (v, f(v))$ is

$$M(f_{|F}, \mathcal{B}) = \begin{pmatrix} 0 & -(a^2 + b^2) \\ 1 & -2a \end{pmatrix}.$$

Since the determinant of an isometry is ± 1, we must have

$$a^2 + b^2 = 1,$$

and this completes the proof. $\qquad\square$

Theorem A.32

Let f be an isometry. We factorize the characteristic polynomial $p_f(x)$ of f into irreducible factors

$$p_f(x) = (x-1)^r(x+1)^s q_1(x)^{n_1} \cdots q_t(x)^{n_t},$$

with $q_i(x) = x^2 - 2a_i x + 1$, $0 \le a_i < 1$; $r, s, n_i \ge 0$, $i = 1,\ldots,t$. Then

$$E = \ker(f - \mathrm{id}) \oplus \ker(f + \mathrm{id}) \oplus \ker q_1(f) \oplus \cdots \oplus \ker q_t(f), \qquad (A.4)$$

and these subspaces are orthogonal to each other (or zero).

Proof

This is a consequence of Corollary A.29, Proposition A.30 and the decomposition theorem corresponding to the minimal polynomial, [8], page 364. □

Theorem A.33 (Canonical expression)

For each isometry f there exists an orthonormal basis $\mathcal{B}$ such that $M = M(f; \mathcal{B})$ is equal to

$$
\begin{pmatrix}
1 & & & & & & & & & & \\
 & \ddots & & & & & & & & & \\
 & & 1 & & & & & & & & \\
 & & & -1 & & & & & & & \\
 & & & & \ddots & & & & & & \\
 & & & & & -1 & & & & & \\
 & & & & & & \cos\alpha_1 & -\sin\alpha_1 & & & \\
 & & & & & & \sin\alpha_1 & \cos\alpha_1 & & & \\
 & & & & & & & & \ddots & & \\
 & & & & & & & & & \cos\alpha_m & -\sin\alpha_m \\
 & & & & & & & & & \sin\alpha_m & \cos\alpha_m
\end{pmatrix}
$$

Proof

We know that

$$E = \ker(f - \mathrm{id}) \oplus \ker(f + \mathrm{id}) \oplus \ker q_1(f) \oplus \cdots \oplus \ker q_t(f),$$

where $q_i(x) = x^2 - 2a_i x + 1$, $0 \le a_i < 1$; $r, s, n_i \ge 0$, $i = 1, \ldots, t$.

We have already seen, in the proof of Proposition A.31, that if $v \in \ker(f^2 - 2a_i f + \mathrm{id})$, then the subspace $F_{i1} = \langle v, f(v) \rangle$ of this kernel is invariant under f. In particular, the restriction $f_{|F_{i1}}$ of f to F_{i1} is an isometry on a vector space of dimension two. Since it does not admit the eigenvalue 1, there is an orthonormal basis $\mathcal{B}_{i1}$ of F_{i1} such that

$$M(f_{|F_{i1}}, \mathcal{B}_{i1}) = \begin{pmatrix} \cos \alpha_i & -\sin \alpha_i \\ \sin \alpha_i & \cos \alpha_i \end{pmatrix}, \quad 0 < \alpha \le \pi.$$

The minimal polynomial (or the characteristic polynomial) of $f_{|F_{i1}}$ is

$$x^2 - (2 \cos \alpha_i) x + 1,$$

and, since it must coincide with the irreducible $q_i(x)$, we have

$$a_i = \cos \alpha_i.$$

Thus we have the orthogonal direct sum decomposition

$$\ker q_i(f) = F_{i1} \oplus F_{i1}^{\perp}.$$

Next we take a vector $w \in F_{i1}^{\perp}$ and we repeat the process (i.e., we consider the subspace $F_{i2} = \langle w, f(w) \rangle$, etc.) until we obtain the orthogonal direct sum decomposition

$$\ker q_i(f) = F_{i1} \oplus \cdots \oplus F_{im_i}.$$

Note that on each F_{ij_i}, $j_i = 1, \ldots, m_i$, there is an orthonormal basis $\mathcal{B}_{ij_i}$ such that

$$M(f_{|F_{ij_i}}, \mathcal{B}_{ij_i}) = \begin{pmatrix} \cos \alpha_i & -\sin \alpha_i \\ \sin \alpha_i & \cos \alpha_i \end{pmatrix}, \quad 0 < \alpha_i \le \pi.$$

Finally, the orthonormal basis that we are looking for is given by $\mathcal{B} = (\mathcal{B}_1, \mathcal{B}_{-1}, \mathcal{B}_{11}, \ldots, \mathcal{B}_{tj_t})$ where $\mathcal{B}_1$ is an orthonormal basis of $\ker(f - \mathrm{id})$ and $\mathcal{B}_{-1}$ is an orthonormal basis of $\ker(f + \mathrm{id})$. The orthonormality of $\mathcal{B}$ follows from the orthogonal direct sum decomposition (A.4), page 340. This completes the proof. $\qquad \square$

A.8 The Orthogonal Groups $O(2)$ and $O(3)$

Notice that Theorem A.33 says that every orthogonal matrix A is conjugate, via an orthogonal matrix C, to the matrix M given in the same theorem, page 340. That is, $M = C^{\mathsf{T}} A C$.

In low dimensions this matrix M cannot have many different aspects. Concretely, in dimension two we have:

Theorem A.34

Let f be an isometry of a Euclidean vector space of dimension two. Then there is an orthonormal basis $\mathcal{B}$ of E such that

$$M(f,\mathcal{B}) = \begin{pmatrix} 1 & 0 \\ 0 & -1 \end{pmatrix} \quad \text{or} \quad M(f,\mathcal{B}) = \begin{pmatrix} \cos\alpha & -\sin\alpha \\ \sin\alpha & \cos\alpha \end{pmatrix},$$

with $0 \leq \alpha \leq \pi$.

Equivalently, for all matrices $A \in O(2)$ there is an orthogonal matrix C such that

$$C^T AC = \begin{pmatrix} 1 & 0 \\ 0 & -1 \end{pmatrix} \quad \text{or} \quad C^T AC = \begin{pmatrix} \cos\alpha & -\sin\alpha \\ \sin\alpha & \cos\alpha \end{pmatrix}.$$

And in dimension three:

Theorem A.35

Let f be an isometry of a Euclidean vector space of dimension 3. Then there exists an orthonormal basis $\mathcal{B}$ of E such that

$$M(f,\mathcal{B}) = \begin{pmatrix} \pm 1 & 0 & 0 \\ 0 & \cos\alpha & -\sin\alpha \\ 0 & \sin\alpha & \cos\alpha \end{pmatrix},$$

with $0 \leq \alpha \leq \pi$.

Equivalently, for all matrices $A \in O(3)$ there is an orthogonal matrix C such that

$$C^T AC = \begin{pmatrix} \pm 1 & 0 & 0 \\ 0 & \cos\alpha & -\sin\alpha \\ 0 & \sin\alpha & \cos\alpha \end{pmatrix}.$$

One can find a direct proof of these two theorems, which does not make use of Theorem A.31, in [8], Chapter 13.

A.9 Classification of Isometries

Definition A.36

Two isometries $\tilde{f}$ and $\tilde{g}$ of a Euclidean vector space E are called *isometrically similar* if and only if there exists an isometry $\tilde{h}$ of E such that $\tilde{f} = \tilde{h}^{-1} \circ \tilde{g} \circ \tilde{h}$.

Observe that if $\tilde{f}$ is *isometrically similar* to $\tilde{g}$, then $\tilde{f}$ is *similar as an endomorphism* to $\tilde{g}$ (Definition 3.1, page 91). Next we shall see that, curiously, the converse is also true.

Proposition A.37

Let $\tilde{f}$ and $\tilde{g}$ be two isometries of a Euclidean vector space E. Then:
(i) The matrices $M(\tilde{f}, \mathcal{B}_1)$ and $M(\tilde{f}, \mathcal{B}_2)$ of $\tilde{f}$ with respect to two orthonormal bases $\mathcal{B}_1$ and $\mathcal{B}_2$ of E are conjugated by an orthogonal matrix.
(ii) $\tilde{f}$ and $\tilde{g}$ are similar as isometries if and only if the matrices $M(\tilde{f}, \mathcal{B})$ and $M(\tilde{g}, \mathcal{B})$, where $\mathcal{B}$ is an orthonormal basis of E, are conjugated by an orthogonal matrix.
(iii) $\tilde{f}$ and $\tilde{g}$ are similar as isometries if and only if there are orthonormal bases $\mathcal{B}_1$ and $\mathcal{B}_2$ of E such that

$$M(\tilde{f}, \mathcal{B}_1) = M(\tilde{g}, \mathcal{B}).$$

Proof

The proof is as in Proposition 3.3, page 92, taking into account that the matrix of the change of basis between two orthonormal bases is an orthogonal matrix, and that the matrix of an isometry with respect to an orthonormal basis is an orthogonal matrix (Corollary A.25). □

From this result, and from Theorem A.33, page 340, we deduce the following important result.

Theorem A.38

Two isometries $\tilde{f}$ and $\tilde{g}$ are isometrically similar if and only if they have the same characteristic polynomial.

Proof

It is well known that similar endomorphisms have the same characteristic polynomial; see [8], page 332.

To prove the converse, we observe that Theorem A.33 tells us that if the characteristic polynomial of an isometry $\tilde{f}$ is

$$p_{\tilde{f}}(x) = (x-1)^r(x+1)^s(x^2 - 2a_1x + 1)^{n_1} \cdots (x^2 - 2a_tx + 1)^{n_t},$$

then there is an orthonormal basis $\mathcal{B}$ such that $M(\tilde{f}, \mathcal{B})$ is equal to the matrix M given in the same Theorem A.33, page 340.

But this matrix is determined by the numbers: r (the number of 1s), s (the number of -1s), n_i (the number of rotations by angle α_i) and a_i (the cosine of the angle of rotation, $a_i = \cos\alpha_i$).

Hence, if the isometries $\tilde{f}$ and $\tilde{g}$ have the same characteristic polynomial, there are (possibly different) orthonormal bases in which $\tilde{f}$ and $\tilde{g}$ have the same associated matrix. Hence, $\tilde{f}$ is isometrically similar to $\tilde{g}$, and this completes the proof.   □

EXERCISES

A.1. Let E be a Euclidean vector space. Prove that given two non-zero vectors $u, v \in E$, with the same norm, there is an isometry φ of E such that $\varphi(u) = v$.

A.2. Let E be a Euclidean vector space. Give a map $f : E \longrightarrow E$ preserving the norm but not the scalar product. Is there a linear map $f : E \longrightarrow E$ preserving the norm but not the scalar product?

A.3. Given the orthogonal matrix

$$A = \begin{pmatrix} \frac{1}{2} + \frac{\sqrt{2}}{4} & \frac{1}{2} & -\frac{1}{2} + \frac{\sqrt{2}}{4} \\ -\frac{1}{2} & \frac{\sqrt{2}}{2} & -\frac{1}{2} \\ -\frac{1}{2} + \frac{\sqrt{2}}{4} & \frac{1}{2} & \frac{1}{2} + \frac{\sqrt{2}}{4} \end{pmatrix},$$

find an orthogonal matrix C such that

$$C^{\mathsf{T}}AC = \begin{pmatrix} 1 & 0 & 0 \\ 0 & \frac{\sqrt{2}}{2} & -\frac{\sqrt{2}}{2} \\ 0 & \frac{\sqrt{2}}{2} & \frac{\sqrt{2}}{2} \end{pmatrix}.$$

A.4. Prove that the map $\varphi : \mathbb{R}^2 \times \mathbb{R}^2 \longrightarrow \mathbb{R}$ given by

$$\varphi((x,y),(z,t)) = 5xz - xt - yz + yt$$

is a scalar product. Find an orthonormal basis, the angle formed by the axes $x = 0$ and $y = 0$, and the norm of the vector $(1,1)$.

A.5. Find the expression, in the canonical basis of the affine space $\mathbb{R}^2$, of the scalar product on $\mathbb{R}^2$ defined by the condition that the vectors $(3,2)$ and $(2,3)$ form an orthonormal basis.

A.6. Let ϕ be the bilinear map of $\mathbb{R}^3$ given in the canonical basis by the matrix:

$$\begin{pmatrix} 1 & -1 & 1 \\ 0 & 1 & -1 \\ 0 & 0 & 1 \end{pmatrix}.$$

(a) Find the expression of $\phi((x_1, x_2, x_3), (y_1, y_2, y_3))$.

(b) Find the matrix of ϕ with respect to the basis $((1,0,0), (1,1,0), (1,1,1))$.

(c) Check the formula of the change of basis.

A.7. Let ϕ be the bilinear map of $\mathbb{R}^3$ given in the canonical basis by the matrix:

$$\begin{pmatrix} -1 & 0 & 1 \\ 0 & 2 & -1 \\ 1 & -1 & 3 \end{pmatrix}.$$

Calculate $\phi((1,0,1), (2,-1,0))$ and find the matrix of ϕ with respect to the basis $((1,1,0), (0,1,1), (0,0,1))$.

A.8. Let ϕ be the bilinear map of $\mathbb{R}^3$ given in the canonical basis by the matrix:

$$\begin{pmatrix} -1 & 1 & -1 \\ 1 & 2 & -2 \\ -1 & -2 & 3 \end{pmatrix}.$$

Find the matrix of ϕ with respect to the basis $((2,1,1), (-1,0,1), (2,-1,0))$.

A.9. Prove that the symmetric bilinear map ϕ of the vector space $\mathbb{R}^4$ given by

$$\phi((x,y,z,t), (x',y',z',t')) = xx' + yy' + zz' - tt'$$

is not positive definite. Prove that, in contrast, ϕ is positive definite when it is restricted to the vector subspace E of $\mathbb{R}^4$ defined by the equation

$$x + y + z - 2t = 0.$$

ϕ is called the Lorentz metric. See Exercise 5.19, page 173.

A.10. Study the orthogonal group $O(4)$.

Diagonalization of Bilinear Symmetric Maps

B.1 Introduction

Let E be a finite dimensional vector space over a field k of characteristic different from 2.

We shall see that given a symmetric bilinear map ϕ on E, there are bases in which the matrix of ϕ is diagonal. The elements on the diagonal of this matrix can be described with more precision with the knowledge of some properties of the arithmetic of the field k. We study the cases $k = \mathbb{R}$, $k = \mathbb{C}$, and k finite. We will see that the diagonal matrix associated to a symmetric bilinear map is, in some sense, unique. From this, we see how many symmetric bilinear maps there are on a given vector space, modulo a certain natural equivalence relation.

B.2 The Diagonalization Theorem

Definition B.1

A bilinear map ϕ on a k-vector space E is said to be *diagonalizable* if there is a basis $\mathcal{B}$ of E such that the matrix of ϕ with respect to this basis, $M(\phi, \mathcal{B})$, is a diagonal matrix.

We also say, in this case, that ϕ *diagonalizes* with respect to the basis $\mathcal{B}$.

A. Reventós Tarrida, *Affine Maps, Euclidean Motions and Quadrics*,
Springer Undergraduate Mathematics Series,
DOI 10.1007/978-0-85729-710-5, © Springer-Verlag London Limited 2011

Proposition B.2

Let ϕ be a bilinear map on a k-vector space E, and let $A = M(\phi, \mathcal{B})$ be the matrix of ϕ with respect to a basis $\mathcal{B}$ of E. Then ϕ is diagonalizable if and only if there is an invertible matrix C, with entries in k, such that the matrix

$$D = C^{\mathsf{T}} A C$$

is diagonal.

Proof

If ϕ diagonalizes with respect to a certain basis $\mathcal{B}'$, we take $C = M(\mathcal{B}', \mathcal{B})$, and then the result follows from the formula of the change of basis (A.3), page 323.

Conversely, if there exists an invertible matrix C, with entries in k, such that the matrix $D = C^{\mathsf{T}} A C$ is diagonal, we take $\mathcal{B}'$ as the only basis such that

$$C = M(\mathcal{B}', \mathcal{B}).$$

(That is, the columns of C are interpreted as the components of the vectors of the basis $\mathcal{B}'$ with respect to the basis $\mathcal{B}$.) The same formula (A.3) says that D is the matrix of ϕ in $\mathcal{B}'$, and hence, ϕ diagonalizes. □

If ϕ is a positive definite symmetric bilinear map on an $\mathbb{R}$-vector space E, there exists, by Proposition A.13, a basis orthonormal with respect to ϕ. The matrix of ϕ with respect to this basis is, therefore, the identity. In particular, *every positive definite symmetric bilinear map diagonalizes.*

If we remove the hypothesis that ϕ is positive definite, it is clear that orthonormal bases, that is, bases in which the matrix of ϕ is the identity, may not exist.

For instance, given

$$A = \begin{pmatrix} 0 & 1 \\ 1 & 0 \end{pmatrix}, \quad \text{or} \quad A = \begin{pmatrix} 1 & 0 \\ 0 & -1 \end{pmatrix},$$

it is easy to see that there is no invertible matrix C, with real coefficients, such that $C^{\mathsf{T}} A C = I_2$. It is sufficient to observe that we would have $\det A \cdot (\det C)^2 = 1$, and that the determinant of A is, in both cases, negative. Hence, if we fix a basis $\mathcal{B}$ of an $\mathbb{R}$-vector space E and we consider the symmetric bilinear map ϕ defined by $M(\phi, \mathcal{B}) = A$, where A is one of the two previous matrices, we obtain a symmetric bilinear map which does not admit an orthonormal basis. However, this does not mean that ϕ is not diagonalizable.

If we replace the $\mathbb{R}$-vector space E by an arbitrary k-vector space, the previous argument showing that orthonormal bases may not exist does not work. For instance, there is an invertible matrix C, with complex coefficients, such that $C^{\mathsf{T}} A C = I_2$, where A is one of the two previous matrices. It is sufficient to take

$$C = \begin{pmatrix} 1 & i \\ 1/2 & 1/2i \end{pmatrix}$$

in the first case, and

$$C = \begin{pmatrix} 1 & 0 \\ 0 & i \end{pmatrix}$$

in the second.

Hence, if ϕ is a bilinear symmetric map on a $\mathbb{C}$-vector space E with $M(\phi, \mathcal{B}) = A$, for some basis $\mathcal{B}$ of E, then there exists a basis $\mathcal{B}'$ of E such that $M(\phi, \mathcal{B}') = I_2$.

However, we are only interested in knowing whether a given symmetric bilinear map diagonalizes or not (independently of whether this diagonal matrix is the identity).

Theorem B.3 (Diagonalization)

Let ϕ be a symmetric bilinear map on a k-vector space E, with k a field of characteristic different from 2. Then there exists a basis of E with respect to which ϕ diagonalizes.

Proof

The proof is implicit in the following method, which is an adaptation of Gauss' method of solving linear systems by triangulation, which simultaneously diagonalizes ϕ and gives us explicitly the basis in which the diagonalization takes place.

> *Method to find the basis in which a symmetric bilinear map diagonalizes.*

To diagonalize a symmetric bilinear map we shall follow the following steps:
(a) First we write down the matrix of ϕ with respect to some basis $\mathcal{B} = (e_1, \ldots, e_n)$, $A = M(\phi, \mathcal{B})$, and, directly beneath this matrix, we write down the identity matrix.
(b) Next we diagonalize A using Gauss' method (see [8], page 196) with the precaution that every operation affecting rows must be repeated exactly

but affecting columns (for instance, if we substitute row 2 by the sum of rows 1 and 2, we must also substitute column 2 by the sum of columns 1 and 2). The added identity matrix is also affected by these changes on the columns.

(c) The basis in which ϕ diagonalizes is formed, once A has been diagonalized, by the columns of the matrix that has been modified from the original identity matrix. More precisely, each of these columns is formed by the components of the vectors of the new basis with respect to the initial one. See Example B.5, page 352.

The validity of the method is easy to justify. We simply observe that to change first the row i, F_i, by $F_i + \lambda F_j$, and then the column i, C_i, by $C_i + \lambda C_j$, corresponds to the change of basis

$$e'_k = e_k, \quad k \neq i,$$
$$e'_i = e_i + \lambda e_j.$$

In fact, the k-th component of the i-th row of the matrix of ϕ in this new basis is

$$\phi(e'_i, e'_k) = \phi(e_i, e_k) + \lambda\phi(e_j, e_k), \quad k \neq i, \tag{B.1}$$

$$\phi(e'_i, e'_i) = \phi(e_i, e_i) + 2\lambda\phi(e_i, e_j) + \lambda^2\phi(e_j, e_j). \tag{B.2}$$

By symmetry, this is also the k-th component of the column C_i. The elements of the matrix of ϕ in the initial basis, belonging neither to the row F_i nor to the column C_i, are not affected by this change of basis.

On the other hand, when we make the change

$$F'_i = F_i + \lambda F_j,$$

we obtain a new row that has k-th component

$$\phi(e_i, e_k) + \lambda\phi(e_j, e_k), \quad k = 1, \ldots, n,$$

expressions that, for $k \neq i$, coincide with (B.1).

However, the modification made on the row F_i affects all the columns of the matrix. Concretely, the term (i, i) of the column C_i, which was equal to $\phi(e_i, e_i)$, is now equal to $\phi(e_i, e_i) + \lambda\phi(e_j, e_i)$, and the term (i, j) of the column C_j, which was equal to $\phi(e_i, e_j)$, is now equal to $\phi(e_i, e_j) + \lambda\phi(e_j, e_j)$. Hence, when after the row transformation $F'_i = F_i + \lambda F_j$, we make the column transformation $C'_i = C_i + \lambda C_j$, we obtain that the k-th component of this new column C'_i is

$$\phi(e_k, e_i) + \lambda\phi(e_k, e_j), \quad k \neq i, \tag{B.3}$$

$$\phi(e_i, e_i) + \lambda\phi(e_j, e_i) + \lambda(\phi(e_i, e_j) + \lambda\phi(e_j, e_j)). \tag{B.4}$$

These two expressions coincide, respectively, with (B.1) and (B.2), and hence the method is justified. For another justification, see Observation B.6, page 353.

Let us see now how the performed changes affect the identity matrix which was placed below the initial matrix.

Operations on rows of A do not affect this identity matrix, but it is affected of course by operations on columns. Concretely, when we first perform the operation $C_i' = C_i + \lambda C_j$, we are replacing the column with a 1 in the i-th position (and zeros elsewhere) by a column with a 1 in the i-th position and a λ in the j-th position (and zeros elsewhere). Therefore, the elements of this column are the components, with respect to $\mathcal{B}$, of the vector $e_i' = e_i + \lambda e_j$, and this corresponds exactly to the change of basis performed ($e_i' = e_i + \lambda e_j$, $e_j' = e_j$, $j \neq i$). This is what happens in each of the successive steps of Gauss' method. Hence, the columns of the modified identity matrix are the components, with respect to $\mathcal{B}$, of the vectors of the new basis.

Finally, let us see that Gauss' method leads to diagonalization. If a_{11} is non-zero, then the operation

$$F_2' = F_2 - \frac{a_{12}}{a_{11}} F_1, \qquad C_2' = C_2 - \frac{a_{12}}{a_{11}} C_1$$

takes

$$A = \begin{pmatrix} a_{11} & a_{12} & \cdots \\ a_{12} & & \\ \vdots & & \end{pmatrix},$$

into

$$A' = \begin{pmatrix} a_{11} & 0 & \cdots \\ 0 & & \\ \vdots & & \end{pmatrix}.$$

That is, we have obtained a zero in the 2nd position of the first row and in the 2nd position of the first column.

Making the same transformation on the others rows and columns, that is,

$$F_i' = F_i - \frac{a_{1i}}{a_{11}} F_1, \qquad C_i' = C_i - \frac{a_{1i}}{a_{11}} C_1, \quad i = 2, \ldots, n,$$

we obtain a matrix in which the first row and the first column are formed by zeros, except for the term a_{11}. Repeating the process, now pivoting on the new term a_{22}, we reduce the dimension of the matrix we want to diagonalize, and in a finite number of steps we obtain the diagonalization.

If a_{11} is zero, we inspect the other terms on the diagonal. If a_{ii} is non-zero, $i \neq 1$, we permute the rows 1 and i and the columns 1 and i. This corresponds to the change of basis $e_1' = e_i$, $e_i' = e_1$, leaving invariant the remaining vectors of the basis. Thus the new term a_{11} is different from zero and we are in the previous case.

If all the elements on the diagonal are zero, but $a_{1i} \neq 0, i \neq 1$, the change $F_1' = F_1 + F_i$, $C_1' = C_1 + C_i$, which corresponds to the change of basis $e_1' = e_1 + e_i$, $e_j' = e_j, j \neq 1$, gives rise to a new $a_{11} \neq 0$, because the characteristic of the field is different from two (the new term a_{11} is equal to $2a_{1i}$), and we are in the previous case. This completes the proof.   □

These last two results, Proposition B.2 and Theorem B.3, allow us to solve the following matricial problem:

Corollary B.4

Given a symmetric matrix $A \in \mathcal{M}_{n \times n}(k)$, there exists an invertible matrix $C \in \mathcal{M}_{n \times n}(k)$ such that the matrix

$$D = C^{\mathsf{T}} A C,$$

is diagonal.

Proof

Let E be a k-vector space of dimension n and take any basis $\mathcal{B}$ of E. Let us consider then the bilinear map ϕ determined by the condition $A = M(\phi, \mathcal{B})$.

We know, by Theorem B.3, that ϕ diagonalizes; hence, by Proposition B.2, there exists an invertible $C \in \mathcal{M}_{n \times n}(k)$ such that the matrix

$$D = C^{\mathsf{T}} A C$$

is diagonal.   □

Gauss' method is essentially a mechanical way of applying the so called *method of completing the squares* discussed later (see Section B.5, page 367).

Example B.5

Let us consider the symmetric bilinear map ϕ of $\mathbb{R}^3$ given in the canonical basis by the matrix

$$A = \begin{pmatrix} 3 & -\frac{1}{2} & 0 \\ -\frac{1}{2} & 0 & 2 \\ 0 & 2 & -2 \end{pmatrix}.$$

Find a basis of $\mathbb{R}^3$ with respect to which ϕ diagonalizes.

Solution

The following scheme follows the previous steps

$$\begin{pmatrix} 3 & -\frac{1}{2} & 0 \\ -\frac{1}{2} & 0 & 2 \\ 0 & 2 & -2 \\ 1 & 0 & 0 \\ 0 & 1 & 0 \\ 0 & 0 & 1 \end{pmatrix} \xrightarrow{F_2 + \frac{1}{6}F_1} \begin{pmatrix} 3 & -\frac{1}{2} & 0 \\ 0 & -\frac{1}{12} & 2 \\ 0 & 2 & -2 \\ 1 & 0 & 0 \\ 0 & 1 & 0 \\ 0 & 0 & 1 \end{pmatrix} \xrightarrow{C_2 + \frac{1}{6}C_1} \begin{pmatrix} 3 & 0 & 0 \\ 0 & -\frac{1}{12} & 2 \\ 0 & 2 & -2 \\ 1 & \frac{1}{6} & 0 \\ 0 & 1 & 0 \\ 0 & 0 & 1 \end{pmatrix}$$

$$\xrightarrow{F_3 + 24F_2} \begin{pmatrix} 3 & 0 & 0 \\ 0 & -\frac{1}{2} & 2 \\ 0 & 0 & 46 \\ 1 & \frac{1}{6} & 0 \\ 0 & 1 & 0 \\ 0 & 0 & 1 \end{pmatrix} \xrightarrow{C_3 + 24C_2} \begin{pmatrix} 3 & 0 & 0 \\ 0 & -\frac{1}{12} & 0 \\ 0 & 0 & 46 \\ 1 & \frac{1}{6} & 4 \\ 0 & 1 & 24 \\ 0 & 0 & 1 \end{pmatrix}.$$

Hence the new basis is $\mathcal{B} = (u_1, u_2, u_3)$ with $u_1 = (1,0,0)$, $u_2 = (\frac{1}{6},1,0)$, $u_3 = (4,24,1)$, and in this basis

$$M(\phi, \mathcal{B}) = \begin{pmatrix} 3 & 0 & 0 \\ 0 & -\frac{1}{12} & 0 \\ 0 & 0 & 46 \end{pmatrix}.$$

□

Observation B.6

Another way to justify the above method is as follows. Recall that to make an *elementary transformation* on the rows of a matrix A corresponds to multiplying A on the left by an elementary matrix, and to make an elementary transformation on the columns of A corresponds to multiplying A on the right by an elementary matrix; see [8]. In order to perform the "same" elementary transformations on the rows as on the columns of A, if we multiply on the left by a matrix P we must multiply on the right by the matrix P^{T}. Recall also the properties of the block product of matrices, see [8], page 182.

With all of this in mind we review the above process of diagonalization given
in the proof of Theorem B.3, step by step.

(a) First we write

$$\left(\frac{A}{I}\right).$$

(b) Next we make an elementary transformation on the rows of A. This transformation does not affect the identity matrix placed below A. It corresponds
to forming the product

$$\left(\frac{P|O}{O|I}\right)\left(\frac{A}{I}\right) = \left(\frac{PA}{I}\right),$$

where P is the elementary matrix performing the elementary transformation on the rows of A, O is the zero matrix, and I the identity matrix.

(c) Next we perform, on the new matrix, the same elementary transformation
except on the columns. This corresponds to forming the product

$$\left(\frac{PA}{I}\right)P^{\mathsf{T}} = \left(\frac{PAP^{\mathsf{T}}}{P^{\mathsf{T}}}\right).$$

If, at this point, the matrix PAP^{T} is diagonal, we have finished. In fact, it is
enough to take $\mathcal{B}'$ as the basis determined by the formula

$$M(\mathcal{B}',\mathcal{B}) = P^{\mathsf{T}},$$

(that is, the columns of P^{T} are the components of the vectors of $\mathcal{B}'$ with respect
to $\mathcal{B}$; see page 322), and, by the formula of the change of basis, we will have

$$PAP^{\mathsf{T}} = M(\mathcal{B}',\mathcal{B})^{\mathsf{T}}M(\phi,\mathcal{B}')M(\mathcal{B}',\mathcal{B}) = M(\phi,\mathcal{B}'),$$

and $\mathcal{B}'$ is, therefore, the basis with respect to which ϕ diagonalizes.

Moreover, this calculation allows us to record this basis (i.e., the matrix P^{T})
in the position occupied, at the beginning of the calculation, by the identity
matrix.

If, on the contrary, once we have completed steps (a), (b) and (c), the matrix
PAP^{T} is still not diagonal, we continue the process repeating systematically
steps (b) and (c) until we reach the diagonalization.

B.3 Canonical Expression

Theorem B.3 says that there is a basis $\mathcal{B} = (e_1,\ldots,e_n)$ of E such that $\phi(e_i,e_j) =
a_{ij}\delta_{ij}$, with a_{ij} elements of the field k and δ_{ij} defined on page 328.

The elements appearing on the diagonal can be reduced to only a few ones by making use of some specific properties of the field k. We study the cases $k = \mathbb{C}$, $k = \mathbb{R}$ and k finite.

Theorem B.7 (Complex case)

Let ϕ be a symmetric bilinear map on a $\mathbb{C}$-vector space E. Then there exists a basis of E with respect to which ϕ diagonalizes, and the diagonal entries of this matrix are 1s and 0s.

Proof

First we find a basis with respect to which ϕ diagonalizes. Then we transform each non-zero element a_{ii} on the diagonal to $+1$. For this we make the following change of basis: if $a_{ii} = \phi(e_i, e_i) \neq 0$, we put $e_i' = \frac{1}{\sqrt{a_{ii}}} e_i$, and leave unchanged the other vectors of the basis. Note that we can make this change of basis because every non-zero element of $\mathbb{C}$ has a square root. Thus, reordering if necessary, we have obtained a new basis $\mathcal{B}$ in which the matrix of ϕ is

$$M(\phi, \mathcal{B}) = \begin{pmatrix} 1 & & & & & & \\ & \ddots & & & & & \\ & & 1 & & & & \\ & & & 0 & & & \\ & & & & \ddots & & \\ & & & & & 0 \end{pmatrix}.$$

$\square$

Theorem B.8 (Real case)

Let ϕ be a symmetric bilinear map on an $\mathbb{R}$-vector space E. Then there exists a basis of E with respect to which ϕ diagonalizes, and the diagonal entries of this matrix are ± 1 or 0.

Proof

First we find a basis with respect to which ϕ diagonalizes. Then we transform each non-zero element a_{ii} on the diagonal to $+1$ or -1. For this we make the following change of basis: if $\phi(e_i, e_i) = \pm a_i^2$, we put $e_i' = \frac{1}{|a_i|} e_i$ and leave unchanged the other vectors of the basis. Note that we can make this change of basis because every non-zero positive element of $\mathbb{R}$ has a square root. Thus, reordering if necessary, we have obtained a new basis $\mathcal{B}$ in which the matrix of ϕ is

$$
M(\phi, \mathcal{B}) =
\begin{pmatrix}
1 & & & & & & & & \\
& \ddots & & & & & & & \\
& & 1 & & & & & & \\
& & & -1 & & & & & \\
& & & & \ddots & & & & \\
& & & & & -1 & & & \\
& & & & & & 0 & & \\
& & & & & & & \ddots & \\
& & & & & & & & 0
\end{pmatrix}.
\tag{B.5}
$$

$\square$

On arbitrary fields we are faced with the problem of determining whether a given element has a square root or not. The complexity of the matrix depends on the arithmetic of the field. This is, in general, a difficult problem.

Finite fields have pleasant properties which allow us to progress further in this problem.

We shall use two of these properties:

Proposition B.9

Let k be a finite field of characteristic different from 2, and let $a, b \in k$ be non-squares. Then there exists a $c \in k$ such that $b = ac^2$.

Proof

See, for instance, [21], Theorem 8.3.1. $\square$

When we say that $a \in k$ is a *non-square* we mean that it has no square root, that is, there is no element $c \in k$ such that $a = c^2$.

Proposition B.10

Let k be a finite field of characteristic different from 2, and let $a \in k$. Then there exist $x, y \in k$ such that $a = x^2 + y^2$.

Proof

See, for instance, [21], Theorem 9.10.1.   □

From these two propositions we can already prove the following theorem (see, for instance, [21], Theorem 9.10.3).

Theorem B.11 (Finite case)

Let ϕ be a symmetric bilinear map on a k-vector space E, with k finite and of characteristic different from 2. Then there exists a basis of E with respect to which ϕ diagonalizes, and the diagonal is comprised of 0s, 1s and at most one non-square.

Proof

First we find a basis with respect to which ϕ diagonalizes. On the diagonal we shall have squares, non-squares and zeros.

By Proposition B.9 there exists a non-square $a \in k$ such that every non-square appearing on the diagonal is of the form ad_i^2 with $d_i \in k$. The change of basis $e_i' = \frac{1}{d_i}e_i$, leaving the other vectors unchanged, transforms all the non-squares on the diagonal into a.

The elements on the diagonal admitting square roots are of the form c_j^2 with $c_j \in k$. The change of basis $e_j' = \frac{1}{c_j}e_j$, leaving the other vectors unchanged, transforms all the squares on the diagonal into 1.

In this manner we have constructed a basis $\mathcal{B}'$ such that, reordering if necessary,

$$
M(\phi, \mathcal{B}') = \begin{pmatrix} a & & & & & & & & \\ & \ddots & & & & & & & \\ & & a & & & & & & \\ & & & 1 & & & & & \\ & & & & \ddots & & & & \\ & & & & & 1 & & & \\ & & & & & & 0 & & \\ & & & & & & & \ddots & \\ & & & & & & & & 0 \end{pmatrix}.
$$

Now, by Proposition B.10, there exist $x, y \in k$ such that $a = x^2 + y^2$.

Then we make the following change of basis: we transform the first two vectors of the basis, e_1, e_2, into the vectors

$$u_1 = \frac{x}{a} e_1 + \frac{y}{a} e_2,$$

$$u_2 = \frac{y}{a} e_1 - \frac{x}{a} e_2,$$

and leave the other vectors unchanged. Note that $\phi(u_1, u_1) = \phi(u_2, u_2) = 1$, $\phi(u_1, u_2) = 0$.

Repeating this process for the two following vectors e_3, e_4, and then for e_5, e_6, etc., we will obtain a new basis in which the matrix of ϕ is diagonal, with only ones and zeros on the diagonal (if the non-square a appeared on the diagonal an even number of times) or with only one a and ones and zeros on the diagonal (if the non-square a appeared on the diagonal an odd number of times).

Therefore, there exists a basis $\mathcal{B}$ such that the matrix $A = M(\phi, \mathcal{B})$ is equal to

$$A = \begin{pmatrix} 1 & & & & & \\ & \ddots & & & & \\ & & 1 & & & \\ & & & 0 & & \\ & & & & \ddots & \\ & & & & & 0 \end{pmatrix}, \quad \text{or} \quad A = \begin{pmatrix} a & & & & & \\ & 1 & & & & \\ & & \ddots & & & \\ & & & 1 & & \\ & & & & 0 & \\ & & & & & \ddots \\ & & & & & & 0 \end{pmatrix},$$

with a non-square. This completes the proof. $\qquad\square$

These results have the following matricial version (simply consider a given symmetric matrix as the matrix associated to a symmetric bilinear map):

Proposition B.12 (Complex case)

Given a complex symmetric matrix A, there exists a complex invertible matrix C such that the matrix

$$D = C^{\mathsf{T}} A C$$

is diagonal, and on the diagonal there are only 1s and 0s.

Proof

This is a consequence of B.7. $\qquad\square$

Proposition B.13 (Real case)

Given a real symmetric matrix A, there exists a real invertible matrix C such that the matrix

$$D = C^\mathsf{T} A C$$

is diagonal, and on the diagonal there are only -1s, 1s and 0s.

Proof

This is a consequence of B.8. □

Proposition B.14 (Finite case)

Given a symmetric matrix A with elements in a finite field k, of characteristic different from 2, there exists an invertible matrix C, also with elements in k, such that the matrix

$$D = C^\mathsf{T} A C$$

is diagonal, and the diagonal is comprised of only the elements $1 \in k$, $0 \in k$, and at most one non-square.

Proof

This is a consequence of B.11. □

We say, in the above three cases, that the matrix C diagonalizes A.

B.4 Uniqueness of the Canonical Expression

In this section we shall see that the canonical expression of a symmetric bilinear map, given in Section B.3, is unique up to the order of the elements on the diagonal, when $k = \mathbb{R}$ or $k = \mathbb{C}$; and unique up to the order, modulo squares, when k is finite.

B.4.1 The Complex Case

Theorem B.15

Let ϕ be a symmetric bilinear map on a $\mathbb{C}$-vector space E. Let us assume that ϕ diagonalizes with respect to certain bases $\mathcal{B}$ and $\mathcal{B}'$, with only 1s and 0s appearing on the respective diagonals. Then the number of 1s is the same in both matrices.

Proof

The number of 1s coincides with the rank of the matrix and the above two matrices are related by the formula of the change of basis

$$M(\phi, \mathcal{B}) = C^{\mathsf{T}} M(\phi, \mathcal{B}') C.$$

Since the rank of a matrix is invariant under left and right multiplication by invertible matrices, the number of 1s in both matrices must coincide.   □

Indeed, we can say that the above two matrices $M(\phi, \mathcal{B})$ and $M(\phi, \mathcal{B}')$ are equal up to the order of the elements on the diagonal. If we agree that, on the diagonal, we first put the 1s, followed by the zeros, then the above two matrices are equal. For this reason we say that this matrix is the canonical form or the *canonical expression* of ϕ.

B.4.2 The Real Case

Let ϕ be a symmetric bilinear map on an $\mathbb{R}$-vector space E. Let us assume that ϕ diagonalizes with respect to certain bases $\mathcal{B}$ and $\mathcal{B}'$, with only 1s, -1s and 0s appearing on the respective diagonals. Is the number of $+1$s the same in both matrices?

We will see that the answer is yes, but the rank argument used in the complex case is not sufficient here. That argument only tells us that the total number of ± 1s on the diagonal of $M(\phi, \mathcal{B})$ is the same as the total number of ± 1s on the diagonal of $M(\phi, \mathcal{B}')$, and we cannot conclude from this that the number of $+1$s in both diagonals coincide.

The correct argument comes from Sylvester's theorem.

Theorem B.16 (Sylvester's theorem)

Let E be an $\mathbb{R}$-vector space and let ϕ be a symmetric bilinear map on E. Then we can decompose E as a direct orthogonal sum of vector subspaces in such a way that

$$E = \operatorname{rad}\phi \perp E_+ \perp E_-,$$

with ϕ positive definite on E_+ and negative definite on E_-. The dimensions $\dim E_+$ and $\dim E_-$ do not depend on the decomposition.

Proof

The existence of the decomposition is clear, since it follows directly from the Diagonalization Theorem B.8.

Indeed, we know that there exists a basis $\mathcal{B} = (e_1, \ldots, e_n)$ with respect to which ϕ diagonalizes and such that

$$\phi(e_i, e_i) = 1, \quad i = 1, \ldots, r^+,$$

$$\phi(e_i, e_i) = -1, \quad i = r^+ + 1, \ldots, r^+ + r^-,$$

$$\phi(e_i, e_i) = 0, \quad i > r^+ + r^-.$$

Then simply take

$$E_+ = \langle e_1, \ldots, e_{r^+} \rangle,$$

$$E_- = \langle e_{r^+ + 1}, \ldots, e_{r^+ + r^-} \rangle.$$

Since

$$\operatorname{rad}\phi = \langle e_{r^+ + r^- + 1}, \ldots, e_n \rangle,$$

we have

$$E = \operatorname{rad}\phi \perp E_+ \perp E_-.$$

Observe that ϕ is negative definite on E_-. Indeed, the condition $\phi(e_i, e_i) < 0$, satisfied for all elements of the basis of E_-, implies that for each element $u = \sum u_i e_i \in E_-$ we also have $\phi(u, u) = -\sum(u_i)^2 \leq 0$.

Analogously we can see that ϕ is positive definite on E_+. This completes the first part of the proof.

Now we prove the independence of r^+ and r^- with respect to the decomposition.

This independence means that in every decomposition of E as an orthogonal direct sum of the radical of ϕ, a subspace on which ϕ is positive definite, and a subspace on which ϕ is negative definite, the dimensions of these last two subspaces are completely determined by ϕ.

Suppose that we have

$$E = \operatorname{rad}\phi \perp E_+ \perp E_- = \operatorname{rad}\phi \perp F_+ \perp F_-, \tag{B.6}$$

with ϕ positive definite on E_+ and on F_+ and negative definite on E_- and on F_-. We want to prove that $\dim F_+ = \dim E_+$. This also implies $\dim F_- = \dim E_-$.

Let us consider the linear map

$$F_+ \overset{i}{\hookrightarrow} E \overset{\mathrm{pr}}{\longrightarrow} E_+$$

obtained as the composition of the canonical inclusion i of F_+ into E with the projection pr of E onto E_+ determined by the direct sum decomposition

$$E = \operatorname{rad}\phi \perp E_+ \perp E_-,$$

that is,

$$\operatorname{pr}(e_0 + e_+ + e_-) = e_+, \quad e_0 \in \operatorname{rad}\phi, \; e_+ \in E_+, \; e_- \in E_-.$$

If we prove that this map $(\mathrm{pr} \circ i)$ is injective, we have finished, because injectivity implies

$$\dim F_+ \le \dim E_+,$$

and by symmetry, that is by exchanging the roles played by E_+ and F_+, we obtain the opposite inequality, and hence the equality of these dimensions.

To prove that $(\mathrm{pr} \circ i)$ is injective we take $e \in F_+$ such that $(\mathrm{pr} \circ i)(e) = 0$. We decompose e as

$$e = e_0 + e_+ + e_-, \quad e_0 \in \operatorname{rad}\phi, \; e_+ \in E_+, \; e_- \in E_-.$$

The image of e under $(\mathrm{pr} \circ i)$ is e_+ and hence $e_+ = 0$; in particular $e = e_0 + e_-$. But then

$$\phi(e,e) = \phi(e_-, e_-) \le 0,$$

since ϕ is negative definite on E_-. On the other hand, $\phi(e,e) \ge 0$, since ϕ is positive definite on F_+. In conclusion, $\phi(e,e) = 0$, and hence $e = 0$, again since ϕ is positive definite on F_+. Thus, $(\mathrm{pr} \circ i)$ is injective, and this completes the proof. $\square$

Notice that the subspaces E_+ and E_- are not invariant: the equality (B.6) does not imply $E_+ = F_+$. What is invariant is the dimension of E_+ and the dimension of E_-. We shall denote these dimensions by $r^+(\phi)$ and $r^-(\phi)$, respectively.

Theorem B.17

Let ϕ be a symmetric bilinear map on an $\mathbb{R}$-vector space E. Let $\mathcal{B}$ and $\mathcal{B}'$ be bases of E and let us suppose that the matrices $M(\phi, \mathcal{B})$ and $M(\phi, \mathcal{B}')$ are diagonal and that on the respective diagonals there are only ± 1s and 0s. Then these two matrices have the same number of 1s and the same number of -1s.

Proof

The number of 1s of $M(\phi, \mathcal{B})$ is equal to $r^+(\phi)$ (by the same argument as in the proof Sylvester's theorem), and this number, by Sylvester's theorem, does not depend on the basis.

  Since the rank is preserved, the number of -1s is also preserved.   □

We say that the above two matrices $M(\phi, \mathcal{B})$ and $M(\phi, \mathcal{B}')$ are equal up to the order of the elements on the diagonal. If we agree that, on the diagonal, we first put the 1s, then the -1s, and finally the zeros, the above two matrices are equal. For this reason we say that this matrix is the canonical form or the *canonical expression* of ϕ.

  Notice that

$$\boxed{r^+(\phi) = \text{the number of } 1\text{s in the canonical expression of } \phi.}$$

The *index* of ϕ is defined as

$$\boxed{\text{index }(\phi) = \min\{r^+(\phi), r^-(\phi)\}.}$$

  We call $r^+(\phi)$ the *positivity dimension* of ϕ, because it is the dimension of the largest vector subspace of E on which ϕ is positive definite; see Exercise B.5, page 380.

B.4.3 The Finite Case

Theorem B.18

Let ϕ be a symmetric bilinear map on a k-vector space E, with k finite and of characteristic different from 2. Let us suppose that ϕ diagonalizes with respect to some basis $\mathcal{B}$, and that the diagonal is comprised of 1s, 0s and at most one non-square $a \in k$.

Suppose that there is another basis $\mathcal{B}'$ with respect to which ϕ also diagonalizes, and that the diagonal is comprised of 1s, 0s and at most one non-square $b \in k$. Then the number of 1s in both matrices is the same, and b/a is a square.

Proof

The proof that both matrices have the same number of 1s on the diagonal follows from the fact that the matrices $A = M(\phi, \mathcal{B})$ and $A' = M(\phi, \mathcal{B}')$ have the same rank, and that the following can never occur, where a is non-square:

$$
A = \begin{pmatrix} 1 & & & & & & \\ & \ddots & & & & & \\ & & 1 & & & & \\ & & & 0 & & & \\ & & & & \ddots & & \\ & & & & & 0 & \end{pmatrix} \quad \text{and} \quad A' = \begin{pmatrix} a & & & & & & \\ & 1 & & & & & \\ & & \ddots & & & & \\ & & & 1 & & & \\ & & & & 0 & & \\ & & & & & \ddots & \\ & & & & & & 0 \end{pmatrix}.
$$

Indeed, these matrices are conjugate, in the sense that there is an invertible matrix P such that $A = P^\mathsf{T} A P$, and this conjugation induces (by the block product of matrices) a conjugation between the non-singular parts. Concretely, there is an invertible matrix C such that

$$
\begin{pmatrix} 1 & & \\ & \ddots & \\ & & 1 \end{pmatrix} = C^\mathsf{T} \begin{pmatrix} a & & \\ & 1 & \\ & & \ddots \\ & & & 1 \end{pmatrix} C,
$$

and this is impossible because of the invariance of the discriminant.

Finally, it is possible that

$$
A = \begin{pmatrix} a & & & & & & \\ & 1 & & & & & \\ & & \ddots & & & & \\ & & & 1 & & & \\ & & & & 0 & & \\ & & & & & \ddots & \\ & & & & & & 0 \end{pmatrix} \quad \text{and} \quad A' = \begin{pmatrix} b & & & & & & \\ & 1 & & & & & \\ & & \ddots & & & & \\ & & & 1 & & & \\ & & & & 0 & & \\ & & & & & \ddots & \\ & & & & & & 0 \end{pmatrix},
$$

with a and b non-squares, because, by Proposition B.9, there exists a $c \in k$ such that $b = ac^2$. Hence, this situation occurs when the relationship between the bases $\mathcal{B} = (e_1, \ldots, e_n)$ and $\mathcal{B}' = (e'_1, \ldots, e'_n)$ is given by $e'_1 = ce_1$, $e'_i = e_i$, $i = 2, \ldots, n$. □

In fact, we can say that the above two matrices $M(\phi, \mathcal{B})$ and $M(\phi, \mathcal{B}')$ are equal (modulo squares) up to the order of the elements on the diagonal. If we agree that on the diagonal we first put a non-square, then 1s, and finally zeros, the above two matrices are equal (modulo squares). For this reason we say that this matrix is the canonical form or the *canonical expression* of ϕ.

B.4.4 Matricial Version

Theorems B.15, B.17 and B.18 can be formulated matricially in the following way (simply consider a given symmetric matrix as the matrix associated to a bilinear map).

Proposition B.19

If the matrices $A \in \mathcal{M}_{n \times n}(\mathbb{C})$ and $D = C^{\mathrm{T}}AC$, with $C \in \mathcal{M}_{n \times n}(\mathbb{C})$ invertible, are diagonal with only 1s and 0s on the diagonal, then they have the same number of 1s.

Proof

This is a consequence of B.15.   □

Proposition B.20

If the matrices $A \in \mathcal{M}_{n \times n}(\mathbb{R})$ and $D = C^{\mathrm{T}}AC$, with $C \in \mathcal{M}_{n \times n}(\mathbb{R})$ invertible, are diagonal with only 1s, -1s and 0s on the diagonal, then they have the same number of 1s and the same number of -1s.

Proof

This is a consequence of B.17.   □

Observe that Sylvester's theorem allows us to define $r^+(A)$ for any real symmetric matrix A. Concretely, given a real symmetric matrix A there exists a real invertible matrix C such that the matrix $D = C^{\mathsf{T}}AC$ is diagonal with only ± 1s and 0s on the diagonal. We say that D is the canonical expression of A. We define

$$r^+(A) = \text{the number of 1s in the canonical expression of } A.$$

By the above proposition, this number is well defined, in the sense that it does not depend on the invertible matrix C yielding the canonical expression $C^{\mathsf{T}}AC$.

Analogously, we define

$$r^-(A) = \text{the number of } -1\text{s in the canonical expression of } A,$$

$$\text{index}(A) = \min\{r^+(A), r^-(A)\}.$$

Hence, we have the following.

Proposition B.21

Let A be a real symmetric matrix. Then for all real invertible matrices P we have

$$r^+(A) = r^+(P^{\mathsf{T}}AP),$$

$$r^-(A) = r^-(P^{\mathsf{T}}AP),$$

$$\text{index}(A) = \text{index}(\pm P^{\mathsf{T}}AP).$$

Proof

The matrices A and $P^{\mathsf{T}}AP$ have the same canonical expression, and the index does not change when we change the sign of a matrix. □

In particular, for every symmetric bilinear map ϕ of a real vector space, we have

$$r^+(\phi) = r^+(M(\phi, \mathcal{B})),$$

$$r^-(\phi) = r^-(M(\phi, \mathcal{B})),$$

$$\text{index}(\phi) = \text{index}(\pm M(\phi, \mathcal{B})),$$

for all bases $\mathcal{B}$ of E.

Proposition B.22

If the matrix $A \in \mathcal{M}_{n \times n}(k)$, with k finite and of characteristic different from 2, is diagonal with only 1s and 0s on the diagonal, and the matrix $B \in \mathcal{M}_{n \times n}(k)$ is diagonal with 1s, 0s and a single non-square on the diagonal, then there is no invertible matrix C such that $B = C^{\mathsf{T}} A C$.

Proof

This is a consequence of B.18.   □

B.5   The Method of Completing the Squares

In this section we shall give a second proof of the Diagonalization Theorem B.3 following the procedure known as the *method of completing the squares*.

   This method is based on the fact that symmetric matrices can be identified with homogeneous quadratic polynomials. Concretely, to each $n \times n$ symmetric matrix $A = (a_{ij})$, we associate the homogeneous quadratic polynomial

$$r(x) = x^{\mathsf{T}} A x = \sum_{i,j=1}^{n} a_{ij} x_i x_j$$

$$= \sum_{i<j}(a_{ij} + a_{ji})x_i x_j + \sum_{i} a_{ii} x_i x_j = \sum_{i \leq j} \bar{a}_{ij} x_i x_j,$$

with

$$\bar{a}_{ii} = a_{ii},$$

$$\bar{a}_{ij} = 2a_{ij}, \quad i < j,$$

and to each homogeneous quadratic polynomial

$$\sum_{i,j=1,i\leq j} x_i x_j \bar{a}_{ij}, \quad 1 \leq i \leq j \leq n,$$

we associate the symmetric matrix $A = (a_{ij})$ with

$$a_{ii} = \bar{a}_{ii},$$

$$a_{ij} = \bar{a}_{ij}/2, \quad i < j,$$

$$a_{ji} = \bar{a}_{ij}/2, \quad i < j.$$

For instance, to the symmetric matrix

$$\begin{pmatrix} a & b \\ b & c \end{pmatrix}$$

we associate the quadratic polynomial

$$\begin{pmatrix} x_1 & x_2 \end{pmatrix} \begin{pmatrix} a & b \\ b & c \end{pmatrix} \begin{pmatrix} x_1 \\ x_2 \end{pmatrix} = ax_1^2 + 2bx_1x_2 + cx_2^2,$$

and conversely (dividing the coefficient of x_1x_2 by 2 before placing it in the matrix).

This is, essentially, what we did in Section 8.2, page 225, however the quadratic polynomials considered there were not necessarily homogeneous.

Theorem B.23 (Diagonalization)

Let ϕ be a symmetric bilinear map on a k-vector space E with k a field of characteristic different from 2. Then there exists a basis of E with respect to which ϕ diagonalizes.

Proof

Method of completing the squares. Let $\mathcal{B}$ be a basis of E and set

$$A = M(\phi, \mathcal{B}).$$

By Theorem B.2, we must find an invertible matrix C such that $C^{\mathsf{T}}AC$ is diagonal.

Let $r(x) = x^{\mathsf{T}}Ax$ be the homogeneous quadratic polynomial associated to A. Concretely,

$$r(x) = \begin{pmatrix} x_1 & \cdots & x_n \end{pmatrix} A \begin{pmatrix} x_1 \\ \vdots \\ x_n \end{pmatrix} = \sum_{i,j=1}^{n} a_{ij}x_ix_j, \quad \text{where } A = (a_{ij}).$$

Let us try to make a change of variables in such a way that the polynomial written in these new variables contains only terms of the form $\bar{a}_{ii}\bar{x}_i\bar{x}_i$, that is, only square terms. To do this we shall follow the following procedure:

If $a_{11} \neq 0$ we take it as a common factor and, since $a_{1i} = a_{i1}$, we have

$$r(x) = a_{11}\left((x_1)^2 + 2x_1 \sum_{i=2}^{n} \frac{a_{1i}}{a_{11}}x_i \right) + \text{quadratic terms without } x_1$$

$$= a_{11}\left(\left(x_1 + \sum_{i=2} \frac{a_{1i}}{a_{11}}x_i\right)^2 - \left(\sum_{i=2} \frac{a_{1i}}{a_{11}}x_i\right)^2\right)$$

+ quadratic terms without x_1.

If we make now the change of variables

$$y_1 = x_1 + \sum_{i=2} \frac{a_{1i}}{a_{11}}x_i,$$

$$y_i = x_i, \quad i = 2,\ldots,n,$$

the previous expression is transformed into

$$r(x) = a_{11}(y_1)^2 + r(y_2,\ldots,y_n),$$

where $r(y_2,\ldots,y_n)$ is a quadratic polynomial in $y_2,\ldots,y_n$.

If $a_{11} = 0$, we inspect the elements a_{kk}, $k \neq 1$, in order to see if one of them is non-zero. In this case, we repeat the above steps but substituting a_{11} by a_{kk}. If, on the contrary, all $a_{ii} = 0$, $i = 1,\ldots,n$, we fix a pair $i \neq j$ such that $a_{ij} \neq 0$. Next we make the change of variables

$$y_i = x_i + x_j,$$

$$y_j = x_i - x_j,$$

$$y_k = x_k, \quad k \neq i,j,$$

so that

$$x_i x_j = \frac{1}{4}(y_i^2 - y_j^2).$$

Thus we have an expression of the polynomial in which some of the new a_{ii} are non-zero (the characteristic of k is different from 2), and we are, therefore, in the previous case.

In this way we have reduced the problem to the study of the homogeneous quadratic polynomial $r(y_2,\ldots,y_n)$, which has one variable less than the original polynomial. Iterating this process we arrive at a new system of variables $\bar{x}_1,\ldots,\bar{x}_n$, with respect to which $r(x)$ is expressed as a sum of squares.

Hence, in a finite number of steps, we will have rewritten the given homogeneous quadratic polynomial as

$$r(x) = a_1(\bar{x}_1)^2 + \cdots + a_r(\bar{x}_r)^2, \quad a_i \neq 0, \ i = 1,\ldots,r.$$

This finishes the *method of completing the squares*.

Next we shall use this method in order to find a basis of E with respect to which ϕ diagonalizes, or, equivalently, an invertible matrix C such that the matrix $C^\mathsf{T}AC$ is diagonal.

Let

$$
\begin{pmatrix} \bar{x}_1 \\ \vdots \\ \bar{x}_n \end{pmatrix} = B \begin{pmatrix} x_1 \\ \vdots \\ x_n \end{pmatrix}
$$

be the change of variables that has transformed the given homogeneous quadratic polynomial into a sum of squares. That is, the composition of all the changes of variables made during the application of the method of completing the squares.

We shall write the composition of all these changes of variables as

$$
\bar{x} = Bx.
$$

Equivalently,

$$
x = C\bar{x}, \quad \text{with } C = B^{-1}.
$$

This matrix C is the matrix that we are looking for.

Indeed, we have

$$
r(x) = x^\mathsf{T} A x = \bar{x}^\mathsf{T} C^\mathsf{T} A C \bar{x} = \bar{x}^\mathsf{T} D \bar{x},
$$

with $D = C^\mathsf{T} A C$. But the expression of $r(x)$ as a function of $\bar{x}$ contains only squares (that is, it contains only terms of the form $\bar{a}_{ii}\bar{x}_i\bar{x}_i$). Hence, given A, we have found a matrix C such that $C^\mathsf{T}AC$ is diagonal. This completes the proof. □

Now we repeat Example B.5 using the method of completing the squares.

Example B.24

Let ϕ be the symmetric bilinear map of $\mathbb{R}^3$ given, in the canonical basis, by the matrix

$$
A = \begin{pmatrix} 3 & -\tfrac{1}{2} & 0 \\ -\tfrac{1}{2} & 0 & 2 \\ 0 & 2 & -2 \end{pmatrix}.
$$

Find a basis of $\mathbb{R}^3$ with respect to which ϕ diagonalizes.

Solution

The homogeneous quadratic polynomial associated to A is

$$(x_1, x_2, x_3) \begin{pmatrix} 3 & -\frac{1}{2} & 0 \\ -\frac{1}{2} & 0 & 2 \\ 0 & 2 & -2 \end{pmatrix} \begin{pmatrix} x_1 \\ x_2 \\ x_3 \end{pmatrix} = 3x_1^2 - 2x_3^2 - x_1 x_2 + 4x_2 x_3.$$

Since the coefficient of x_1^2 is different from zero, we begin by "squaring" x_1.

$$3x_1^2 - 2x_3^2 - x_1 x_2 + 4x_2 x_3 = 3\left(x_1 - \frac{x_2}{6}\right)^2 - \frac{x_2^2}{12} - 2x_3^2 + 4x_2 x_3$$

$$= [\text{now squaring } x_2]$$

$$3\left(x_1 - \frac{x_2}{6}\right)^2 - \frac{1}{12}\left(x_2 - 24x_3\right)^2$$

$$+ 48x_3^2 - 2x_3^2$$

$$= 3\bar{x}_1^{\,2} - \frac{1}{12}\bar{x}_2^{\,2} + 46\bar{x}_3^{\,2}.$$

Hence, the change of variables is given by

$$\bar{x}_1 = x_1 - \frac{x_2}{6},$$

$$\bar{x}_2 = x_2 - 24x_3,$$

$$\bar{x}_3 = x_3.$$

Equivalently,

$$\begin{pmatrix} \bar{x}_1 \\ \bar{x}_2 \\ \bar{x}_3 \end{pmatrix} = \begin{pmatrix} 1 & -\frac{1}{6} & 0 \\ 0 & 1 & -24 \\ 0 & 0 & 1 \end{pmatrix} \begin{pmatrix} x_1 \\ x_2 \\ x_3 \end{pmatrix},$$

or,

$$\begin{pmatrix} x_1 \\ x_2 \\ x_3 \end{pmatrix} = \begin{pmatrix} 1 & \frac{1}{6} & 4 \\ 0 & 1 & 24 \\ 0 & 0 & 1 \end{pmatrix} \begin{pmatrix} \bar{x}_1 \\ \bar{x}_2 \\ \bar{x}_3 \end{pmatrix}.$$

The columns of the matrix of the change of coordinates are the elements of the basis with respect to which ϕ diagonalizes. Hence this basis is (u_1, u_2, u_3) where $u_1 = (1, 0, 0)$, $u_2 = (\frac{1}{6}, 1, 0)$, $u_3 = (4, 24, 1)$. Therefore we have

$$\begin{pmatrix} 1 & \frac{1}{6} & 4 \\ 0 & 1 & 24 \\ 0 & 0 & 1 \end{pmatrix}^{\mathsf{T}} \begin{pmatrix} 3 & -\frac{1}{2} & 0 \\ -\frac{1}{2} & 0 & 2 \\ 0 & 2 & -2 \end{pmatrix} \begin{pmatrix} 1 & \frac{1}{6} & 4 \\ 0 & 1 & 24 \\ 0 & 0 & 1 \end{pmatrix} = \begin{pmatrix} 3 & 0 & 0 \\ 0 & -\frac{1}{12} & 0 \\ 0 & 0 & 46 \end{pmatrix}.$$

□

Observe that in this example the bilinear map ϕ is given in the canonical basis of $\mathbb{R}^3$. This basis is orthonormal with respect to the standard scalar product. However, the basis with respect to which ϕ diagonalizes is not orthonormal. Is it possible to find an orthonormal basis with respect to which ϕ diagonalizes? We study this question in Appendix C.

Observation B.25 (Quadratic forms)

If a symmetric bilinear map $\phi : E \times E \longrightarrow k$ is applied only to pairs of equal vectors (u, u), we obtain a *quadratic form*. That is, a *quadratic form q on E* is the restriction of a symmetric bilinear map on E to the diagonal of $E \times E$.

Concretely, the *quadratic form q associated to a symmetric bilinear map ϕ* is the map

$$q : E \longrightarrow k$$

given by

$$q(v) = \phi(v, v).$$

Observe that $q(\lambda v) = \lambda^2 q(v)$, but not every map $q : E \longrightarrow k$ with this property is a quadratic form. See Exercise B.8, page 380.

If we know $q(v)$, for all $v \in E$, then ϕ is completely determined, because

$$\phi(u, v) = \frac{1}{2}(q(u + v) - q(u) - q(v)).$$

(Recall that we are assuming that the characteristic of k is different from 2.) That is, if two symmetric bilinear maps coincide on the diagonal of $E \times E$, then they are equal. Therefore we have a bijection between symmetric bilinear maps and quadratic forms.

When we write a quadratic form in coordinates, we obtain a homogeneous quadratic polynomial.

Concretely, if we fix a basis $(e_1, \ldots, e_n)$ of E and write a vector $x \in E$ in coordinates, $x = \sum_i x_i e_i$, we obtain

$$q(x) = \phi\left(\sum_i x_i e_i, \sum_j x_j e_j\right) = \sum_{ij} x_i x_j a_{ij} = x^\mathsf{T} a x,$$

where $a_{ij} = \phi(e_i, e_j)$, $a = (a_{ij})$ and $x^\mathsf{T} = (x^1, \ldots, x^n)$. Conversely, given a homogeneous quadratic polynomial and a basis, the above formula defines a quadratic form.

Thus, once we fix a basis of E, we have a bijection between *quadratic forms, homogeneous quadratic polynomials and symmetric matrices.*

B.6   Classification of Symmetric Bilinear Maps

The problem that we consider in this section is the following: given two symmetric bilinear maps ϕ and ϕ' on the same k-vector space E, are there bases $\mathcal{B}$ and $\mathcal{B}'$ with respect to which ϕ and ϕ' have the same matrix, that is, such that $M(\phi, \mathcal{B}) = M(\phi', \mathcal{B}')$?

When this occurs we say that ϕ and ϕ' are *equivalent*.

Definition B.26

Two symmetric bilinear maps ϕ and ϕ' on a k-vector space E are said to be *equivalent* if their associated matrices, in possibly different bases, coincide.

Proposition B.27

Let E be a k-vector space and let ϕ and ϕ' be symmetric bilinear maps on E. Let $\mathcal{B}$ and $\mathcal{B}'$ be bases of E (not necessarily different). Then ϕ and ϕ' are equivalent if and only if there exists an invertible matrix C such that

$$M(\phi, \mathcal{B}) = C^{\mathsf{T}} M(\phi', \mathcal{B}') C.$$

Proof

If ϕ and ϕ' are equivalent, then there exist bases $\mathcal{B}_1$ and $\mathcal{B}_2$ such that

$$M(\phi, \mathcal{B}_1) = M(\phi', \mathcal{B}_2).$$

By the formula of the change of basis, we have

$$M(\mathcal{B}_1, \mathcal{B})^{\mathsf{T}} M(\phi, \mathcal{B}) M(\mathcal{B}_1, \mathcal{B}) = M(\mathcal{B}_2, \mathcal{B}')^{\mathsf{T}} M(\phi', \mathcal{B}') M(\mathcal{B}_2, \mathcal{B}').$$

Since the matrix of a change of basis is invertible, we have

$$M(\phi, \mathcal{B}) = C^{\mathsf{T}} M(\phi', \mathcal{B}') C, \quad \text{with } C = M(\mathcal{B}_2, \mathcal{B}') M(\mathcal{B}_1, \mathcal{B})^{-1}.$$

Observe that if $\mathcal{B} = \mathcal{B}'$, then $C = M(\mathcal{B}_2, \mathcal{B}_1)$, that is, C is the matrix of the change of basis between the two bases in which ϕ and ϕ' have respectively the same matrix.

Conversely, if there exists an invertible matrix C such that

$$M(\phi, \mathcal{B}) = C^{\mathsf{T}} M(\phi', \mathcal{B}') C,$$

then the basis $\mathcal{B}_2$ determined by the condition

$$C = M(\mathcal{B}_2, \mathcal{B}')$$

satisfies

$$M(\phi, \mathcal{B}) = M(\phi', \mathcal{B}_2),$$

and hence ϕ and ϕ' are equivalent. $\square$

In particular,

$$\det M(\phi, \mathcal{B}) = \det M(\phi', \mathcal{B}') \cdot (\det C)^2.$$

Thus, *if two symmetric bilinear maps are equivalent, they have the same* rank *and the same* discriminant.

However, two symmetric bilinear maps can have the same rank and discriminant without being equivalent. For instance, the symmetric bilinear maps of the vector space $\mathbb{R}^2$ given in the canonical basis by the matrices

$$\begin{pmatrix} 1 & 0 \\ 0 & 1 \end{pmatrix} \quad \text{and} \quad \begin{pmatrix} -2 & 0 \\ 0 & -1 \end{pmatrix}$$

have the same rank and discriminant, but they are not equivalent.

The fact that the discriminant plays an important role in the classification of symmetric bilinear maps, and the fact that, on arbitrary fields, it is difficult to determine which elements have square roots, are the reasons why the classification problem on arbitrary fields is rather involved. We shall study here only the simplest cases: $k = \mathbb{R}$, $k = \mathbb{C}$ and k finite.[1]

Proposition B.28

Two symmetric bilinear maps ϕ and ϕ' on a k-vector space E are equivalent if and only if there exists an isomorphism $h : E \longrightarrow E$ such that

$$\phi(u, v) = \phi'(h(u), h(v)), \quad \text{for all } u, v \in E.$$

Proof

If there exists an isomorphism h with this property we have, by formula (A.2),

$$M(\phi, \mathcal{B}) = M(h, \mathcal{B})^{\mathsf{T}} M(\phi', \mathcal{B}) M(h, \mathcal{B}),$$

---

[1] The classification for $k = \mathbb{Q}$ can be found in [6].

for any basis $\mathcal{B}$ of E. Hence, by Proposition B.27, ϕ and ϕ' are equivalent. Conversely, if there exists an invertible matrix C such that

$$M(\phi, \mathcal{B}) = C^{\mathsf{T}} M(\phi', \mathcal{B}) C,$$

we define h by the condition

$$M(h, \mathcal{B}) = C.$$

Since C is invertible, h is an isomorphism and we have

$$M(\phi, \mathcal{B}) = M(h, \mathcal{B})^{\mathsf{T}} M(\phi', \mathcal{B}) M(h, \mathcal{B}).$$

This implies, again by formula (A.2), that

$$\phi(u, v) = \phi'(h(u), h(v)), \quad \text{for all } u, v \in E,$$

and this completes the proof.   □

Theorem B.29 (Classification on $\mathbb{C}$)

Two symmetric bilinear maps on a $\mathbb{C}$-vector space are equivalent if and only if they have the same number of 1s in their canonical expression. That is, the rank classifies.

Proof

We know, by Theorem B.7, that there exist bases $\mathcal{B}$ and $\mathcal{B}'$ such that

$$M(\phi, \mathcal{B}) = \begin{pmatrix} I_r & O \\ O & O \end{pmatrix},$$

$$M(\phi', \mathcal{B}') = \begin{pmatrix} I_s & O \\ O & O \end{pmatrix},$$

where I_r and I_s are the $r \times r$ and $s \times s$ identity matrices, respectively.

Let us suppose that ϕ is equivalent to ϕ'. Then, by Proposition B.27, there exists an invertible matrix C such that

$$\begin{pmatrix} I_r & O \\ O & O \end{pmatrix} = C^{\mathsf{T}} \begin{pmatrix} I_s & O \\ O & O \end{pmatrix} C.$$

But this implies $r = s$, since, as we have already remarked, the rank of a matrix is invariant under left and right multiplication by invertible matrices.

The converse is evident, since there are bases (the bases $\mathcal{B}$ and $\mathcal{B}'$ with respect to which ϕ and ϕ' have the canonical expression) such that $M(\phi,\mathcal{B}) = M(\phi',\mathcal{B}')$. $\qquad\square$

This theorem can also be stated in the form *two symmetric bilinear maps on a $\mathbb{C}$-vector space are equivalent if and only if they have the same rank, or if and only if they have the same canonical expression.*

Theorem B.30 (Classification on $\mathbb{R}$)

Two symmetric bilinear maps on an $\mathbb{R}$-vector space are equivalent if and only if they have the same number of 1s and -1s in their canonical expression. That is, the rank and the number of 1s classify.

Proof

This is an immediate consequence of Sylvester's theorem. Indeed, let us first suppose that ϕ is equivalent to ϕ'. Let $\mathcal{C}$ be a basis with respect to which ϕ has the canonical expression, with $r^+(\phi)$ 1s and $r^-(\phi)$ -1s on the diagonal. Let $\mathcal{C}'$ be a basis with respect to which ϕ' has the canonical expression, with $r^+(\phi')$ 1s and $r^-(\phi')$ -1s on the diagonal.

By Proposition B.27 there exists an invertible matrix C such that

$$M(\phi,\mathcal{C}) = C^\mathsf{T} M(\phi',\mathcal{C}')C.$$

But this implies, by Proposition B.20, $r^+(\phi) = r^+(\phi')$.

The converse is evident, since there are bases (the bases $\mathcal{B}$ and $\mathcal{B}'$ with respect to which ϕ and ϕ' have the canonical expression) such that $M(\phi,\mathcal{B}) = M(\phi',\mathcal{B}')$. $\qquad\square$

This theorem can also be stated in the form *two symmetric bilinear maps on an $\mathbb{R}$-vector space are equivalent if and only if they have the same rank and the same positivity dimension, or if and only if they have the same canonical expression.*

Theorem B.31 (Classification when k is finite)

Let k be a finite field. Two symmetric bilinear maps on a k-vector space are equivalent if and only if they have the same rank and the same number of 1s in their canonical expression. That is, the rank and the discriminant of the non-singular part classify.

Proof

If ϕ and ϕ' are equivalent, the canonical expressions given by Theorem B.11 are conjugate. But this implies, by the block product of matrices, that the matrices corresponding to the non-singular parts are also conjugate. Hence, they have the same rank, and, either they both have only 1s on the diagonal, or they both have only one non-square and 1s on the diagonal (see the proof of Theorem (B.18), page 363). Therefore, in both cases they have the same number of 1s.

Conversely, if ϕ and ϕ' have the same rank and the same number of 1s in their canonical expressions, either there are only 1s and 0s on the diagonal, in which case the matrices of ϕ and ϕ' are equal and we have finished, or the first element on the diagonal of both matrices is a non-square. But given two non-squares a and b, there exists a $c \in k$ such that $b = ac^2$ (Proposition B.9, page 356). Hence, these matrices are conjugated by the matrix

$$\begin{pmatrix} c & & & \\ & 1 & & \\ & & \ddots & \\ & & & 1 \end{pmatrix},$$

and so ϕ is equivalent to ϕ'. □

This theorem can also be stated in the form *two symmetric bilinear maps on a k-vector space, with k finite, are equivalent if and only if they have the same canonical expression modulo squares.*

B.7 Projective Classification

Observe that, in the real case, the matrices

$$\begin{pmatrix} 1 & 0 \\ 0 & 1 \end{pmatrix} \quad \text{and} \quad \begin{pmatrix} -1 & 0 \\ 0 & -1 \end{pmatrix}$$

are not equivalent.

However, if we consider these matrices as quadratic polynomials they have the same zeros. For this reason it is convenient to introduce the concept of *projectively equivalent* symmetric bilinear maps.

Definition B.32

Two symmetric bilinear maps ϕ and ϕ' on an $\mathbb{R}$-vector space E are said to be *projectively equivalent* if their associated matrices, in possibly different bases, coincide up to the sign.

Proposition B.33

Let E be an $\mathbb{R}$-vector space and let ϕ and ϕ' be symmetric bilinear maps on E. Let $\mathcal{B}$ be a basis of E. Then ϕ and ϕ' are projectively equivalent if and only if there exists an invertible matrix C such that

$$M(\phi, \mathcal{B}) = \pm C^{\mathsf{T}} M(\phi', \mathcal{B}) C.$$

Proof

This is a consequence of Proposition B.27.   □

Vector subspaces of dimension 2 admitting a basis with respect to which

$$\phi = \begin{pmatrix} 1 & 0 \\ 0 & -1 \end{pmatrix}$$

are called *hyperbolic planes*. They are characterized as the non-singular vector subspaces of dimension two with an isotropic vector. Recall that *non-singular* means that there is no non-zero vector ϕ-orthogonal to all vectors, see Definition A.5, page 325. An *isotropic* vector is a vector u such that $\phi(u, u) = 0$.

Observe that the *number of hyperbolic planes*, which appear by grouping together pairs $\{1, -1\}$ in the canonical expression of ϕ, *coincides with the index of ϕ*.

Proposition B.34

Let E be an $\mathbb{R}$-vector space and let ϕ and ϕ' be symmetric bilinear maps on E. Then ϕ and ϕ' are projectively equivalent if and only if they have the same rank and the same index.

Proof

As a consequence of Propositions B.21 and B.33, if ϕ and ϕ' are projectively equivalent they have the same rank and the same index.

Conversely, if they have the same rank and the same index, they will have, up to the sign, the same canonical expression (in appropriate bases). In fact, if $\rho = \text{rank}(\phi) = \text{rank}(\phi')$ and $i = \text{index}(\phi) = \text{index}(\phi')$, it is clear that the $\rho - 2i$ remaining elements of the diagonal, once we have removed the pairs $\{1, -1\}$ in the respective canonical expressions, are all equal to 1 or all equal to -1.

Again by Proposition B.33, ϕ and ϕ' are projectively equivalent. □

Theorem C.14, page 390, provides a quick method of calculating the index.

EXERCISES

B.1. Consider the bilinear map $\phi : \mathbb{R}^3 \times \mathbb{R}^3 \to \mathbb{R}^3$ defined by

$$\phi((x_1, x_2, x_3), (y_1, y_2, y_3))$$

$$= x_1 y_1 - x_1 y_3 + x_2 y_2 - x_3 y_1 + x_3 y_2 + 2x_3 y_3 + x_2 y_3.$$

(a) Find the matrix of ϕ with respect to the canonical basis of $\mathbb{R}^3$.
(b) Consider the change of coordinates of $\mathbb{R}^3$ given by

$$\begin{cases} x' = 2x_1 + 2x_2 + 3x_3, \\ x'_2 = -x_1 - x_3, \\ x'_3 = -x_2 - x_3. \end{cases}$$

Find, with respect to these new coordinates, the analytical expression of the quadratic form associated to ϕ.

B.2. Find the symmetric matrix associated to each one of the following quadratic polynomials:

$Q_1(x) = x_1^2 + x_2^2,$ $Q_5(x) = 2x_1^2 - 4x_1 x_2 + 2x_2^2 - 2x_2 x_3,$

$Q_2(x) = x_1^2 + 2x_1 x_2 + x_2^2,$ $Q_6(x) = x_1^2 + x_2^2 + x_3^2 - 4x_1 x_3,$

$Q_3(x) = x_1^2 + 2x_1 x_2 + 2x_2^2,$ $Q_7(x) = 2x_1 x_2 - 2x_1 x_3,$

$Q_4(x) = x_1^2 + 6x_1 x_2 + 8x_2^2,$ $Q_8(x) = 2x_1^2 - x_2 x_3.$

B.3. Find the canonical expressions of the symmetric bilinear maps defined, with respect to the canonical bases of $\mathbb{R}^2$ or $\mathbb{R}^3$, by the eight symmetric matrices found in the previous exercise. Find their indices. Specify in each case the change of basis used to find the canonical expression.

B.4. Express the quadratic polynomial $q(x) = 2x_1^2 - x_2^2 + 2x_1x_3 - 4x_2x_3$ in terms of (x_1', x_2', x_3') if

$$\begin{cases} x_1 = x_1' + x_2' - 2x_3', \\ x_2 = -2x_1' + x_2', \\ x_3 = x_2' + x_3'. \end{cases}$$

B.5. Let ϕ be a symmetric bilinear map of a real vector space E. Prove that

$$r^+(\phi) = \max\{\dim F : F \text{ vector subspace of } E \\ \text{and } \phi|F \text{ positive definite}\}$$

and that

$$r^-(\phi) = \max\{\dim F : F \text{ vector subspace of } E \\ \text{and } \phi|F \text{ negative definite}\}.$$

B.6. Classify the symmetric bilinear map ϕ of $\mathbb{R}^4$ that gives rise to the quadratic form

$$q(x) = x_1^2 + 3x_2^2 + 4x_1x_2 + 2x_1x_3 + 2x_1x_4 + 2x_2x_4 - 2x_3x_4,$$

and find a basis $\mathcal{B}$ with respect to which $M(\phi, \mathcal{B})$ is the canonical expression of ϕ.

B.7. Classify the symmetric bilinear map ϕ of $\mathbb{R}^3$ that gives rise to the quadratic form

$$q(x) = 2x_1^2 + 2x_2^2 + 2x_3^2 + 2x_1x_2 + 2x_1x_3 + 2x_2x_3$$

and find a basis $\mathcal{B}$ with respect to which $M(\phi, \mathcal{B})$ is the canonical expression of ϕ.

B.8. Find a map $q : \mathbb{R}^2 \longrightarrow \mathbb{R}^2$ satisfying $q(\lambda v) = \lambda^2 q(v)$ for all $v \in \mathbb{R}^2$ and for all $\lambda \in \mathbb{R}$, but that is not a quadratic form.

$$C$$

Orthogonal Diagonalization

C.1 Introduction

We shall see that, given two symmetric bilinear maps on a finite dimensional $\mathbb{R}$-vector space E, one of them positive definite, there is a basis of E with respect to which the respective associated matrices are simultaneously diagonal. For this reason this result is called the *Simultaneous Diagonalization Theorem*, although it is also known as the *Spectral Theorem*.

C.2 Associated Endomorphism

Definition C.1

Let ϕ and g be symmetric bilinear maps of a real vector space E, with g positive definite. The endomorphism *associated* to ϕ and g is the unique endomorphism f of E such that

$$\phi(u,v) = g(f(u),v), \quad u,v \in E.$$

Observe that g is a scalar product.

Notice that the formula $\phi(u,v) = g(f(u),v)$ determines f, because an endomorphism is determined by its value on a basis. If we take, for instance, $\mathcal{B} = (e_1,\ldots,e_n)$, an orthonormal basis with respect to g, the above formula implies

$$f(e_i) = \sum_j \phi(e_i, e_j) e_j,$$

and hence f is completely determined.

This expression of $f(e_i)$ tells us that

$$M(f, \mathcal{B}) = M(\phi, \mathcal{B}).$$

The reader is invited to pay particular attention to the previous equation: the first matrix is associated to a linear map while the second matrix is associated to a bilinear map!

Therefore, roughly speaking, the associated endomorphism is obtained by thinking of the matrix of ϕ with respect to an orthonormal basis as if it was the associated matrix of a linear map.

If $\mathcal{B}$ is an arbitrary basis, the condition $\phi(u, v) = g(f(u), v)$ implies

$$M(\phi, \mathcal{B}) = M(f, \mathcal{B})^{\mathsf{T}} M(g, \mathcal{B}).$$

Thus,

$$M(f, \mathcal{B})^{\mathsf{T}} = M(\phi, \mathcal{B}) M(g, \mathcal{B})^{-1},$$

and hence, transposing, and using the fact that the matrices of ϕ and g are symmetric, we have

$$\boxed{M(f, \mathcal{B}) = M(g, \mathcal{B})^{-1} M(\phi, \mathcal{B}).}$$

If $\mathcal{B}$ is not g-orthonormal, $M(f, \mathcal{B})$ need not be symmetric. This formula is very useful because it enables us to calculate the endomorphism associated to ϕ and g very quickly.

C.3 Self-adjoint Endomorphisms

We have just seen that if we have two symmetric bilinear maps ϕ and g on an $\mathbb{R}$-vector space E, with g positive definite, then there is an endomorphism f of E (the associated endomorphism) satisfying, because of the symmetry of ϕ,

$$\boxed{g(f(u), v) = g(u, f(v))}$$

for all $u, v \in E$.

Endomorphisms satisfying this property are said to be *self-adjoint* with respect to g.

Definition C.2 (Self-adjoint endomorphism)

Let f be an endomorphism of a Euclidean vector space (E, g). Then f is *self-adjoint with respect to g* if and only if $g(f(u), v) = g(u, f(v))$, for all $u, v \in E$.

When the scalar product g is understood we simply say that f is self-adjoint.

Notice that if ϕ is a symmetric bilinear map on (E, g), *the endomorphism associated to ϕ and g is self-adjoint with respect to g*.

Analogously, given a self-adjoint endomorphism f of a Euclidean vector space (E, g), we can define a symmetric bilinear map ϕ on E by the formula

$$\phi(u, v) = g(f(u), v).$$

Then f is the endomorphism associated to ϕ and g. Note that ϕ is symmetric if and only if f is self-adjoint.

Proposition C.3

Let f be a self-adjoint endomorphism of a Euclidean vector space (E, g). Then eigenvectors of f with different eigenvalues are orthogonal.

Proof

Suppose that $f(u) = \lambda u$, $f(v) = \mu v$, with $u \neq 0$, $v \neq 0$, $\lambda \neq \mu$.
Then

$$\lambda g(u, v) = g(\lambda u, v) = g(f(u), v)$$
$$= g(u, f(v)) = g(u, \mu v) = \mu g(u, v).$$

Hence, $g(u, v) = 0$, and this completes the proof. $\qquad\square$

Proposition C.4

Let f be an endomorphism of a Euclidean vector space (E, g). Then the matrix of f in an orthonormal basis is symmetric if and only if f is self-adjoint.

Proof

Let $\mathcal{B} = (e_1, \ldots, e_n)$ be an orthonormal basis. Put

$$f(e_i) = \sum_k a_{ki} e_k.$$

Then

$$g(f(e_i), e_j) = g\left(\sum_k a_{ki}e_k, e_j\right) = a_{ji},$$

$$g(e_i, f(e_j)) = g\left(e_i, \sum_k a_{kj}e_k\right) = a_{ij}.$$

Hence, if f is self-adjoint, we have $a_{ij} = a_{ji}$, that is, the matrix of f in $\mathcal{B}$, $M(f, \mathcal{B})$, is symmetric.

Conversely, if this matrix is symmetric, that is $a_{ij} = a_{ji}$, we have

$$g(f(e_i), e_j) = g(e_i, f(e_j)), \quad i, j = 1, \ldots, n,$$

and it follows easily from this that

$$g(f(u), v) = g(u, f(v)), \quad \text{for all } u, v \in E,$$

and this completes the proof.   □

C.4 Orthogonal Diagonalization of Symmetric Matrices

We begin with a simple observation.

Proposition C.5

Let A be a 2×2 real symmetric matrix. Then the eigenvalues of A are real numbers.

Proof

Let

$$A = \begin{pmatrix} a & b \\ b & c \end{pmatrix}.$$

The characteristic polynomial is

$$p_A(x) = x^2 - (a+c)x + (ac - b^2).$$

The discriminant is

$$\Delta = (a+c)^2 - 4(ac - b^2) = (a-c)^2 + 4b^2 \geq 0,$$

and hence there are two different real roots or a double real root. □

This result is true in arbitrary dimensions, but it is not easy to generalize the above proof. We must pass to the complex numbers.

Theorem C.6

Let A be an $n \times n$ real symmetric matrix. Then the eigenvalues of A are real numbers.

Proof

We consider A as a linear map from $\mathbb{C}^n$ to $\mathbb{C}^n$, concretely

$$A(u) = Au, \quad u \in \mathbb{C}^n,$$

where Au means the matrix A multiplied by the matrix of n rows and one column formed by the components of u,

$$u = \begin{pmatrix} u_1 \\ \vdots \\ u_n \end{pmatrix}.$$

The characteristic polynomial of A has n complex roots, counting multiplicities. Let $\lambda \in \mathbb{C}$ be one such root. Since the roots of the characteristic polynomial are the eigenvalues of A, there exists a $u \in \mathbb{C}^n$ such that

$$A(u) = \lambda u.$$

Multiplying the matricial equality $Au = \lambda u$ by $\bar{u}^\mathsf{T}$, we get

$$\bar{u}^\mathsf{T} A u = \lambda \bar{u}^\mathsf{T} u = \lambda \|u\|^2. \tag{C.1}$$

Taking the conjugate of (C.1) we obtain

$$u^\mathsf{T} A \bar{u} = \bar{\lambda} \|u\|^2, \tag{C.2}$$

since A and $\|u\|$ are real.

Transposing (C.1) we obtain

$$u^\mathsf{T} A\bar{u} = \lambda \|u\|^2, \tag{C.3}$$

since $A = \bar{A}^\mathsf{T}$ and $\lambda\|u\|^2$ is a real number.

Equating (C.2) and (C.3) we get $\lambda = \bar{\lambda}$, that is $\lambda \in \mathbb{R}$, and this completes the proof.   □

In particular, we have proved that the characteristic polynomial of A factorizes as a product of linear polynomials

$$p_A(x) = (x - \lambda_1)\ldots(x - \lambda_n), \quad \lambda_i \in \mathbb{R}, \ i = 1,\ldots,n,$$

where these λ_i are not necessarily distinct.

Observe that we have also proved that the eigenvector associated to the eigenvalue $\lambda \in \mathbb{R}$ is real, since its components are solutions of a homogeneous linear system with real coefficients.

C.5 The Spectral Theorem

Proposition C.7

Let f be a self-adjoint endomorphism of a Euclidean vector space (E, g) of dimension n. Then the characteristic polynomial of f factorizes over $\mathbb{R}$ as a product of linear factors. That is,

$$p_f(x) = (x - \lambda_1)\ldots(x - \lambda_n), \quad \lambda_i \in \mathbb{R}, \ i = 1,\ldots,n.$$

Proof

The matrix A of f in an orthonormal basis is symmetric, as we have seen in Proposition C.4. The characteristic polynomial of f is the characteristic polynomial of A, and, by Proposition C.6, the n eigenvalues (counting multiplicities) of an $n \times n$ real symmetric matrix are real numbers. This completes the proof.   □

Note that the λ_i can be equal to each other. Hence, we cannot directly deduce from this proposition that a self-adjoint endomorphism diagonalizes. Note also that this proposition tells us that the eigenvalues of a self-adjoint endomorphism are real numbers.

Theorem C.8 (Spectral Theorem)

Let (E, g) be a Euclidean vector space, and let ϕ be a symmetric bilinear map on E. Then there exists an orthonormal basis with respect to which ϕ diagonalizes. This basis is formed by the normalized eigenvectors of the associated endomorphism.

Proof

Let f be the endomorphism of E associated to ϕ and g, that is,

$$g(f(u), v) = \phi(u, v).$$

By Proposition C.7, there exist a $\lambda \in \mathbb{R}$ and an $e_1 \in E$ such that $f(e_1) = \lambda e_1$. This vector e_1 can be taken to be a unit vector, $\|e_1\| = 1$. Let

$$F = \langle e_1 \rangle^{\perp} = \{v \in E : g(e_1, v) = 0\}.$$

Since g is positive definite we have $\dim F = n - 1$.

Next we use induction on the dimension n of E.

If $n = 1$, the matrix of ϕ in any basis is a 1×1 matrix and hence is diagonal. Therefore, it is sufficient to take as orthonormal basis a vector $e \in E$ such that $g(e, e) = 1$. The vector e is, of course, an eigenvector of the associated endomorphism.

Assume that the result is true for $n - 1$. In particular, we can apply the result to F, since ϕ and g are automatically symmetric bilinear maps on F, and g is positive definite on F. Hence, there exists a g-orthonormal basis $(e_2, \ldots, e_n)$ of F with respect to which ϕ diagonalizes.

The basis $(e_1, e_2, \ldots, e_n)$ is clearly g-orthonormal and, since

$$\phi(e_1, e_j) = g(f(e_1), e_j) = g(\lambda e_1, e_j) = 0, \quad j = 2, \ldots, n,$$

the matrix of ϕ with respect to this basis is diagonal.

Moreover,

$$\phi(e_i, e_j) = a_{ij} \delta_{ij}$$

implies

$$g(f e_i, e_j) = a_{ij} \delta_{ij},$$

that is,

$$f(e_i) = a_{ii} e_i,$$

and hence *the vectors of the g-orthonormal basis $\mathcal{B}$ with respect to which ϕ diagonalizes are, automatically, eigenvectors of f, and the terms on the diagonal of $M(\phi, \mathcal{B})$ are the eigenvalues of f.* □

We have already remarked that this theorem is also known as the *Simultaneous Diagonalization Theorem.*

Corollary C.9

Let f be a self-adjoint endomorphism of a Euclidean vector space (E, g). Then f diagonalizes.

Proof

Simply apply the Spectral Theorem to the scalar product g and the symmetric bilinear map ϕ defined by

$$\phi(u, v) = g(f(u), v).$$

□

As a consequence of the Diagonalization Theorem for endomorphisms (see for instance [8], page 337), we have the following result on the dimension of the space of eigenvectors, of a given eigenvalue, of a self-adjoint endomorphism.

Proposition C.10

Let

$$p_f(x) = \Pi_i(x - \lambda_i)^{d_i}$$

be the factorization into linear factors of the characteristic polynomial of a self-adjoint endomorphism f. Then

$$\dim \ker(f - \lambda_i \operatorname{id}) = d_i.$$

Thus, in practice, to find the basis given by the Spectral Theorem (that is, the basis with respect to which the simultaneous diagonalization of ϕ and g takes place) we will first find the eigenvalues of the associated endomorphism, and then for each eigenvalue λ we will find a g-orthonormal basis of the vector subspace $\ker(f - \lambda \operatorname{id})$. The union of these bases is the desired basis.

The Spectral Theorem has the following matricial version.

Theorem C.11 (Matricial version of the Spectral Theorem)

Let A be a real symmetric matrix. Then there exists an orthogonal matrix C such that $D = C^\mathsf{T} A C$ is diagonal. The elements on the diagonal are the eigenvalues of A.

Proof

Let (E, g) be a Euclidean vector space, and let $\mathcal{B}$ be an orthonormal basis. We define a symmetric bilinear map ϕ on E by the formula

$$M(\phi, \mathcal{B}) = A.$$

By the Spectral Theorem, there exists an orthonormal basis $\mathcal{B}'$ with respect to which ϕ diagonalizes. Therefore, we have

$$D = M(\phi, \mathcal{B}') = C^\mathsf{T} M(\phi, \mathcal{B}) C = C^\mathsf{T} A C,$$

where $C = M(\mathcal{B}', \mathcal{B})$ is the matrix of the change of basis between orthonormal bases and is, therefore, orthogonal.

Moreover, since $C^\mathsf{T} = C^{-1}$, A and D have the same characteristic polynomial and, hence, the same eigenvalues. $\qquad\square$

Observe that a real symmetric matrix can be considered as a linear map from $\mathbb{R}^n$ to $\mathbb{R}^n$ which is self-adjoint with respect to the standard scalar product on $\mathbb{R}^n$. The previous Corollary C.9 tells us that all eigenvalues of A are real numbers, as we already knew from Proposition C.6.

C.6 Quick Calculation of the Positivity Dimension

We recall, without proof, the following result on the number of positive roots of a polynomial (see, for instance, [25], Theorem 11.6.1).

Theorem C.12 (Descartes' theorem)

The number of positive roots r^+ of a polynomial with real coefficients, counted according to their multiplicity, is less than or equal to the number v of sign differences between consecutive non-zero coefficients of the polynomial (i.e., $r^+ \leq v$).

If all roots of the polynomial are real, then $r^+ = v$.

Corollary C.13

The number of positive eigenvalues of a real symmetric matrix is equal to the number of sign differences between consecutive non-zero coefficients of its characteristic polynomial.

Proof

This is a consequence of Descartes' theorem and Proposition C.6.   □

Theorem C.14

Let ϕ be a symmetric bilinear map on an $\mathbb{R}$-vector space E of dimension n. Let $\mathcal{B}$ a basis of E. Then

$$r^+(\phi) = v,$$

where v is the number of sign differences between consecutive non-zero coefficients of the characteristic polynomial of the matrix $A = M(\phi, \mathcal{B})$.

Proof

By Proposition B.21, $r^+(\phi) = r^+(A)$, where $r^+(\phi)$ is the number of 1s in the canonical expression of ϕ. By Proposition C.11, there exists an orthogonal matrix C such that the matrix

$$D = C^T A C = C^{-1} A C$$

is diagonal, and the elements on the diagonal are the eigenvalues of A. If we put $D = (d_{ij})$, we have $d_{ii} = \pm a_i^2$, $i = 1, \ldots, \mathrm{rank}(\phi)$, for some $a_i \in \mathbb{R}$. Then the matrix $Q = P^T D P$, where $P = (p_{ij})$ is an invertible diagonal matrix with

$$p_{ii} = \frac{1}{|a_i|}, \quad i = 1, \ldots, \mathrm{rank}(\phi),$$

$$p_{jj} = 1, \quad j = \mathrm{rank}(\phi) + 1, \ldots, n,$$

is the canonical expression of A. Again by Proposition B.21, $r^+(A) = r^+(Q)$, and by Corollary C.13,

$$r^+(\phi) = r^+(A) = r^+(Q)$$

$$= \text{the number of positive eigenvalues of } A$$

$$= v,$$

and this completes the proof.   □

In particular, we have

$$\text{index}(\phi) = \min\{v, \text{rank}(\phi) - v\}.$$

Example C.15

Find the canonical expression of the symmetric bilinear map ϕ on $\mathbb{R}^2$ given with respect to the canonical basis by

$$M(\phi, \mathcal{C}) = \begin{pmatrix} 5 & 2 \\ 2 & 1 \end{pmatrix}.$$

Solution

The characteristic polynomial is $x^2 - 6x + 1$. The signs of the coefficients are $+, -, +$, and therefore there are two changes of sign. Hence, $r^+ = 2$ and $i = 0$. This means that there exists a basis $\mathcal{B}$ such that

$$M(\phi, \mathcal{B}) = \begin{pmatrix} 1 & 0 \\ 0 & 1 \end{pmatrix}.$$

We can find this basis by using either Gauss' method or the method of completing the squares.

For instance, by using Gauss' method we would proceed as follows: We put

$$\begin{pmatrix} 5 & 2 \\ 2 & 1 \\ \hline 1 & 0 \\ 0 & 1 \end{pmatrix}.$$

We make the transformation $F_2 \rightarrow F_2 - \frac{2}{5}F_1$ and we get

$$\begin{pmatrix} 5 & 2 \\ 0 & \frac{1}{5} \\ \hline 1 & 0 \\ 0 & 1 \end{pmatrix}.$$

This forces us to make the transformation $C_2 \to C_2 - \frac{2}{5}C_1$ and we get

$$
\begin{pmatrix}
5 & 0 \\
0 & \frac{1}{5} \\
\hline
1 & -\frac{2}{5} \\
0 & 1
\end{pmatrix}.
$$

Hence, the matrix of ϕ in the basis $u = \frac{1}{\sqrt{5}}(1,0), v = \frac{1}{\sqrt{5}}(-2,5)$ is the identity. That is,

$$
C^{\mathsf{T}} AC = \begin{pmatrix} 1 & 0 \\ 0 & 1 \end{pmatrix},
$$

where

$$
C = \begin{pmatrix} \frac{1}{\sqrt{5}} & -\frac{2}{\sqrt{5}} \\ 0 & \sqrt{5} \end{pmatrix}.
$$

$\square$

Observe that this matrix C is not orthogonal. In fact, it is clear that there does not exist an orthogonal matrix C such that

$$
C^{\mathsf{T}} \begin{pmatrix} 5 & 2 \\ 2 & 1 \end{pmatrix} C = \begin{pmatrix} 1 & 0 \\ 0 & 1 \end{pmatrix}.
$$

Nevertheless, we have:

Example C.16

Given the matrix

$$
A = \begin{pmatrix} 5 & 2 \\ 2 & 1 \end{pmatrix},
$$

find a real orthogonal matrix C such that the matrix

$$
D = C^{\mathsf{T}} AC
$$

is diagonal.

Solution

To find C we only need to calculate the eigenvectors and the eigenvalues of A (see Proposition C.11).

It is easy to see that $(1, \sqrt{2} - 1)$ is an eigenvector with eigenvalue $3 + 2\sqrt{2}$ and that $(1, -\sqrt{2} - 1)$ is an eigenvector with eigenvalue $3 - 2\sqrt{2}$.

Since the basis must be orthonormal with respect to the standard scalar product on $\mathbb{R}^2$, we take as eigenvectors

$$u = \frac{1}{a}(1, \sqrt{2} - 1), \quad \text{with } a = \sqrt{4 - 2\sqrt{2}},$$

$$v = \frac{1}{b}(1, -\sqrt{2} - 1), \quad \text{with } b = \sqrt{4 + 2\sqrt{2}}.$$

Hence,

$$C^{\mathsf{T}} AC = \begin{pmatrix} 3 + 2\sqrt{2} & 0 \\ 0 & 3 - 2\sqrt{2} \end{pmatrix},$$

with

$$C = \begin{pmatrix} \frac{1}{a} & \frac{1}{b} \\ \frac{\sqrt{2}-1}{a} & \frac{-\sqrt{2}-1}{b} \end{pmatrix}.$$

$\square$

Observe that this calculation can be used to find the canonical expression asked for in Example C.15. It is enough to change the above orthonormal basis (u, v) by

$$u' = \frac{u}{\sqrt{3 + 2\sqrt{2}}},$$

$$v' = \frac{v}{\sqrt{3 - 2\sqrt{2}}}.$$

Then we have

$$P^{\mathsf{T}} AP = I_2,$$

with

$$P = \begin{pmatrix} 1/b & 1/a \\ (\sqrt{2} - 1)/b & (-\sqrt{2} - 1)/a \end{pmatrix},$$

but this matrix is not orthogonal.

EXERCISES

C.1. Let ϕ and g be bilinear maps on $\mathbb{R}^2$ given, with respect to the canonical basis $\mathcal{C}$, by

$$M(\phi,\mathcal{C}) = \begin{pmatrix} 1 & 2 \\ 2 & 3 \end{pmatrix}, \qquad M(g,\mathcal{C}) = \begin{pmatrix} 1 & 2 \\ 2 & 8 \end{pmatrix}.$$

(a) Prove that ϕ and g are symmetric and that g is positive definite.

(b) Find the endomorphism $f : \mathbb{R}^2 \longrightarrow \mathbb{R}^2$ associated to ϕ and g.

(c) Find a g-orthonormal basis with respect to which ϕ diagonalizes.

C.2. Let ϕ and g be bilinear maps on $\mathbb{R}^2$ given, in the canonical basis $\mathcal{C}$, by

$$M(\phi,\mathcal{C}) = \begin{pmatrix} 2 & 4 \\ 4 & 19 \end{pmatrix}, \qquad M(g,\mathcal{C}) = \begin{pmatrix} 1 & 2 \\ 2 & 5 \end{pmatrix}.$$

(a) Prove that ϕ and g are symmetric and that g is positive definite.

(b) Find the endomorphism $f : \mathbb{R}^2 \longrightarrow \mathbb{R}^2$ associated to ϕ and g.

(c) Find a g-orthonormal basis with respect to which ϕ diagonalizes.

C.3. Given the symmetric matrix

$$A = \begin{pmatrix} 3/2 & 0 & -\sqrt{3}/2 \\ 0 & 2 & 0 \\ -\sqrt{3}/2 & 0 & 5/2 \end{pmatrix},$$

find an orthogonal matrix P such that the matrix $D = P^\mathsf{T} A P$ is diagonal.

C.4. Let ϕ_1 and ϕ_2 be bilinear maps on $\mathbb{R}^3$ given, in the canonical basis $\mathcal{C}$, by

$$M(\phi_1,\mathcal{C}) = \begin{pmatrix} 1 & 2 & 0 \\ 2 & 4 & \sqrt{6} \\ 0 & \sqrt{6} & 1 \end{pmatrix}, \qquad M(\phi_2,\mathcal{C}) = \begin{pmatrix} 1 & 2 & 0 \\ 2 & 5 & 0 \\ 0 & 0 & 1 \end{pmatrix}.$$

(a) Classify ϕ_1 and ϕ_2, that is, find their canonical expressions.

(b) Find a basis $\mathcal{B}$ such that the matrices $M(\phi_1,\mathcal{B})$ and $M(\phi_2,\mathcal{B})$ are diagonal.

C.5. Find, in each case, an orthogonal matrix diagonalizing the given matrix.

(a) $\begin{pmatrix} 3 & 4 \\ 4 & -3 \end{pmatrix}.$

(b) $\begin{pmatrix} 2 & 1 \\ 1 & 2 \end{pmatrix}.$

(c) $\begin{pmatrix} 1 & -1 & -1 \\ -1 & 1 & -1 \\ -1 & -1 & 1 \end{pmatrix}.$

(d) $\begin{pmatrix} 3 & 2 & 2 \\ 2 & 2 & 0 \\ 2 & 0 & 4 \end{pmatrix}.$

(e) $\begin{pmatrix} -1 & 2 & 2 \\ 2 & -1 & 2 \\ 2 & 2 & 1 \end{pmatrix}.$

(f) $\begin{pmatrix} 1 & -1 & 0 \\ -1 & 2 & -1 \\ 0 & -1 & 1 \end{pmatrix}.$

(g) $\begin{pmatrix} 1 & -1 \\ -1 & 1 \end{pmatrix}.$

(h) $\begin{pmatrix} 1 & -1 & 0 & 0 \\ -1 & 0 & 0 & 0 \\ 0 & 0 & 0 & 0 \\ 0 & 0 & 0 & 2 \end{pmatrix}.$

(i) $\begin{pmatrix} 1 & 0 & 1 & 0 \\ 0 & 0 & 2 & 0 \\ 1 & 2 & 0 & 0 \\ 0 & 0 & 0 & 2 \end{pmatrix}.$

C.6. Using an orthogonal change of variables, write the quadratic form

$$q(x) = 5x_1^2 + 5x_2^2 + 8x_3^2 - 8x_1x_2 + 4x_1x_3 + 4x_2x_3$$

as a linear combination of squares, and classify it.

D
Polynomials with the Same Zeros

D.1 Introduction

We shall study here a simplified version of Hilbert's zeros theorem, which establishes the relationship between two quadratic polynomials with the same zeros.

D.2 The Nullstellensatz

Theorem D.1 (Nullstellensatz)

Let $r(x), s(x)$ be polynomials in n variables on an algebraically closed field k. If $s(x)$ vanishes on the zeros of $r(x)$, then there exist an $m \in \mathbb{N}$ and a polynomial $q(x)$ such that $s^m(x) = r(x) \cdot q(x)$.

Since $x = (x_1, \ldots, x_n)$, the zeros of the polynomials are in k^n.

This theorem is a simplified version of Hilbert's *Nullstellensatz* (zeros theorem). Observe that if $r(x)$ and $s(x)$ have the same degree and $r(x)$ is irreducible (i.e., it is not equal to the product of polynomials of smaller degree), then $s(x) = \lambda r(x)$, with $0 \neq \lambda \in k$.

If we restrict to quadratic polynomials, those appearing in the study of quadrics, we can be a little more precise.

A. Reventós Tarrida, *Affine Maps, Euclidean Motions and Quadrics*,
Springer Undergraduate Mathematics Series,
DOI 10.1007/978-0-85729-710-5, © Springer-Verlag London Limited 2011

Theorem D.2 (Quadratic Nullstellensatz)

Let $r(x), s(x)$ be quadratic polynomials in n variables on an algebraically closed field k. If $r(x)$ and $s(x)$ have the same zeros, then there exists a $\lambda \in k$, $\lambda \neq 0$, such that $s(x) = \lambda r(x)$.

Proof

Since $s(x)$ vanishes on the zeros of $r(x)$, we have $s^m(x) = r(x) \cdot q(x)$, that is $r(x)$ divides $s^m(x)$. If $r(x)$ is irreducible, it must divide $s(x)$, and since $r(x)$ and $s(x)$ have the same degree, we must have $s(x) = \lambda r(x)$ and we have finished. If $r(x)$ factorizes as a product of two different polynomials of degree 1, each of them must divide $s(x)$, and hence we also have $s(x) = \lambda r(x)$. Finally, if $r(x) = r_1(x)^2$, $r_1(x)$ must divide $s(x)$, and hence $s^2(x) = r(x) \cdot q(x)$.

Since we are also assuming that $r(x)$ vanishes on the zeros of $s(x)$, the same argument proves that $r(x) = \mu s(x)$ when $s(x)$ is irreducible or a product of two different polynomials of degree 1, or $r^2(x) = s(x) \cdot p(x)$ when $s(x) = s_1(x)^2$.

From this we deduce that $s_1(x)$ divides $r_1(x)$, and hence, also in this case, $s(x) = \lambda r(x)$. $\qquad\square$

The hypothesis on the degree is essential. For instance, the polynomials $r(x) = x_1 x_2^2$ and $s(x) = x_1^2 x_2$ have the same zeros but one is not a multiple of the other.

We are interested in a result similar to that of Theorem D.2, but in the real case. Since $\mathbb{R}$ is not algebraically closed, we need a new version of this theorem.

We begin with the following lemma.

Lemma D.3

Let $r(x)$ be a *homogeneous* quadratic polynomial, with n variables and real coefficients, such that $r(x) = 0$ for all points $x \in \mathbb{R}^n$ with first coordinate $x_1 \neq 0$. Then $r(x)$ is the zero polynomial.

Proof

If $n = 1$, $r(x) = cx^2$ and the lemma is clear.

If $n > 1$, the result is an immediate consequence of the following expression for $r(x)$:

$$r(x_1, \ldots, x_n) = cx_1^2 + x_1 r_1(x_2, \ldots, x_n) + r_2(x_2, \ldots, x_n),$$

with r_i homogeneous polynomials of degree i, $i = 1, 2$. Then it is clear that $r(1, 0, \ldots, 0) = 0$ implies $c = 0$, and $r(x)$ should be, for each $(x_2, \ldots, x_n)$ fixed, a polynomial of degree 1 in x_1 with infinitely many roots. Since this is impossible, we must have $r_1 = r_2 = 0$, and hence $r(x) = 0$. □

Note that the condition $x_1 \neq 0$ for the points where $r(x) = 0$ can be replaced by the hypothesis $r(x) = 0$ on all points $(x_1, \ldots, x_n)$ that do not belong to a given hyperplane of $\mathbb{R}^n$, since, changing the affine frame, this hyperplane may be written as $x_1 = 0$.

Note that the theorem is true for arbitrary (i.e., not necessarily quadratic) homogeneous polynomials, and that the condition of vanishing outside a hyperplane can be replaced by the condition of vanishing on a non-empty open set.

Theorem D.4 (Real quadratic Nullstellensatz)

Let $r(x)$, $s(x)$ be quadratic polynomials over $\mathbb{R}$, with n variables. If the set of zeros of $s(x)$ coincides with the set of zeros of $r(x)$, and there is at least one regular zero with respect to $r(x)$, then there exists a $\lambda \in \mathbb{R}$, $\lambda \neq 0$, such that $s(x) = \lambda r(x)$.

Proof

Let $r(x) = x^\mathsf{T} A x + B x + C$ and $s(x) = x^\mathsf{T} A' x + B' x + C'$. Let $p = (p_1, \ldots, p_n)$ be a regular zero of $r(x)$. This means, see Definition 8.15, page 248, that $r(p) = 0$ and that $\operatorname{grad} r(p) = 2p^\mathsf{T} A + B \neq 0$. First we prove that $\operatorname{grad} s(p)$ is also nonzero.

For each vector $v = (v_1, \ldots, v_n) \in \mathbb{R}^n$, we consider the intersection of the straight line $p + tv$ of $\mathbb{R}^n$, $t \in \mathbb{R}$, with the set of zeros of $r(x)$.

We have

$$
\begin{aligned}
0 &= (p + tv)^\mathsf{T} A (p + tv) + B(p + tv) + C \\
&= t^2 v^\mathsf{T} A v + 2p^\mathsf{T} A tv + p^\mathsf{T} A p + B p + B tv + C \\
&= t^2 v^\mathsf{T} A v + 2p^\mathsf{T} A tv + B tv \\
&= t^2 v^\mathsf{T} A v + t L v,
\end{aligned}
$$

with $L = 2p^\mathsf{T} A + B = \operatorname{grad} r(p) \neq 0$.

Analogously, cutting the same straight line with the set of zeros of $s(x)$ we have

$$
t^2 v^\mathsf{T} A' v + t L' v = 0, \tag{D.1}
$$

with $L' = 2p^{\mathsf{T}}A' + B' = \operatorname{grad} s(p)$. But any root $t \in \mathbb{R}$ of this equation must also be root of the equation

$$t^2 v^{\mathsf{T}} A v + t L v = 0, \tag{D.2}$$

and conversely (since the sets of zeros coincide).

Since $L \neq 0$, there exists a vector $v \in \mathbb{R}^n$ such that $L v \neq 0$ and such that $v^{\mathsf{T}} A v \neq 0$. Otherwise we would have, by Lemma D.3, $A = 0$, and $r(x)$ would not be a quadratic polynomial.

If $L' = 0$, (D.1) and (D.2) corresponding to this direction v are

$$\begin{cases} t^2 v^{\mathsf{T}} A' v = 0, \\ t^2 v^{\mathsf{T}} A v + t L v = 0, \end{cases}$$

and they must have the same solutions in t. But this is impossible, and hence $L' \neq 0$.

Since the gradient of $r(x)$ at the point p is not zero, we can apply the Implicit Function Theorem to the equation $r(x) = 0$. This theorem says that we can (locally) solve for one of the variables $x_1, \ldots, x_n$ in terms of the others (see Observation D.5). Hence, there are curves $\gamma_i(t)$, $i = 2, \ldots, n$ a $\mathbb{R}^n$ such that

$$r(\gamma_i(t)) = 0,$$

$$\gamma_i(0) = p,$$

$$\dot\gamma_i(0) = e_i,$$

where $e_2, \ldots, e_n$ are $n-1$ linearly independent vectors, orthogonal to the vector $L = \operatorname{grad} r(p)$. Orthogonality follows from the chain rule applied to $r(\gamma_i(t)) = 0$.

Concretely, if we have solved for x_1, $x_1 = x_1(x_2, \ldots, x_n)$, these curves are given by

$$\gamma_i(t) = (x_1(p_2, \ldots, p_i + t, \ldots, p_n), p_2, \ldots, p_i + t, \ldots, p_n) \quad i = 2, \ldots, n.$$

Since the zeros of $r(x)$ and $s(x)$ coincide, we have $s(\gamma_i(t)) = 0$, and, by the chain rule,

$$\operatorname{grad} s(p) \cdot \dot\gamma_i(0) = 0.$$

Thus, the vector subspace $\langle e_2, \ldots, e_n \rangle$ is orthogonal to L and L', and hence there exists a $\lambda \in \mathbb{R}$, $\lambda \neq 0$, such that

$$L' = \lambda L.$$

Let v be a vector that does not belong to the hyperplane Π defined by

$Lx = 0$, that is, such that $Lv \neq 0$. We will also have $L'v \neq 0$. The equations

$$\begin{cases} t^2 v^{\mathsf{T}} A'v + t\lambda Lv = 0, \\ t^2 v^{\mathsf{T}} Av + tLv = 0, \end{cases}$$

have the same solutions, and hence

$$v^{\mathsf{T}} A'v = \lambda v^{\mathsf{T}} Av, \quad \text{for all } v \notin \Pi$$

(note that $v^{\mathsf{T}} A'v = 0$ if and only if $v^{\mathsf{T}} Av = 0$).

By Lemma D.3, we have

$$v^{\mathsf{T}} A'v - \lambda v^{\mathsf{T}} Av = 0, \quad \text{for all } v \in \mathbb{R}^n,$$

and hence $A' = \lambda A$.

Now we easily deduce that $B' = \lambda B$ and $C' = \lambda C$, and hence $s(x) = \lambda r(x)$. This completes the proof.   □

For instance, $s(x) = x_1^2 + x_2^2$ is not a multiple of $r(x) = x_1^2 + 5x_2^2$, although $s(x)$ vanishes on the zeros of $r(x)$ (the point $(0,0)$). But $r(x)$ has no regular zeros, because the gradient of $r(x)$, $\operatorname{grad} r(x) = (2x_1, 10x_2)$, vanishes at $(0,0)$.

However, all zeros of $r(x) = x_1^2 + x_2^2 - 1$ are regular, since $\operatorname{grad} r(x) = (2x_1, 2x_2)$ only vanishes at the point $(0,0)$, which is not a zero of $r(x)$. Hence, if a quadratic polynomial vanishes on the circle $x_1^2 + x_2^2 = 1$, it is a multiple of $r(x)$.

Observation D.5

In the proof of Theorem D.4 we have used the Implicit Function Theorem. In fact, we can avoid the use of this strong result by proving directly that we can solve for one variable in terms of the others in the equation $r(x) = 0$. Concretely, since $\operatorname{grad} r(p) \neq 0$, we may assume, for instance, that $\frac{\partial r(p)}{\partial x_1} \neq 0$. Then, we can solve for x_1 in terms of the other variables:

$$x_1 = x_1(x_2, \ldots, x_n).$$

In fact, $r(x)$ can be considered as a polynomial of second degree in x_1, in which the coefficients depend on the other variables, as we have seen in Lemma D.3.

To solve for x_1 from the equation of second degree

$$r(x) = cx_1^2 + x_1 r_1 + r_2 = 0,$$

we only need that the discriminant $\Delta = r_1^2 - 4cr_2$ is positive or zero.

Since $r_1 = r_1(x_2, \ldots, x_n)$ and $r_2 = r_2(x_2, \ldots, x_n)$ are polynomials in the variables $x_2, \ldots, x_n$, the discriminant Δ is a function of the variables $x_2, \ldots, x_n$.

But we know that $\Delta \geq 0$ at the point p, since $p = (p_1, \ldots, p_n)$ is a solution of $r(x) = 0$. Moreover, the condition on the gradient tells us that $\Delta \neq 0$ at p, since otherwise p_1 would be a double root of the polynomial $cp_1^2 + p_1 r_1(p_2, \ldots, p_n) + r_2(p_2, \ldots, p_n)$, and hence we would have

$$0 = cp_1^2 + p_1 r_1 + r_2 = c\left(p_1 + \frac{r_1}{2c}\right)^2 = \frac{1}{4c}\left(\frac{\partial r(p)}{\partial x_1}\right)^2 \neq 0.$$

Thus, $\Delta > 0$ in p, and hence $\Delta > 0$ in a neighborhood of p. Therefore, in this neighborhood we can write $x_1 = x_1(x_2, \ldots, x_n)$ with $p_1 = x_1(p_2, \ldots, p_n)$.

If $c = 0$, the argument is essentially the same, since

$$\frac{\partial r(p)}{\partial x_1} = r_1(p_2, \ldots, p_n) \neq 0,$$

and we can solve for $x_1 = x_1(x_2, \ldots, x_n)$ from $x_1 r_1 + r_2 = 0$.

EXERCISES

D.1. Find polynomials $r(x)$ and $s(x)$ satisfying the hypothesis of Theorem D.1, and such that $s^m(x) = r(x) \cdot q(x)$ for different values of $m \in \mathbb{N}$.

D.2. Find quadratic polynomials $r(x)$ and $s(x)$ over $\mathbb{R}$ such that the set of zeros of $s(x)$ coincides with the set of zeros of $r(x)$, this set is not reduced to a point, and $r(x)$ is not a multiple of $s(x)$.

D.3. Prove Lemma D.3 for homogeneous polynomials, not necessarily quadratic, vanishing on a non-empty open set.

D.4. Study the general version of the Nullstellensatz. Use, for instance, [10].

Bibliography

[1] Artin, E.: Algèbre geométrique. Gauthier-Villars, Paris (1972)

[2] Audin, M.: Geometry. Universitext. Springer, Berlin (2003). Original French edition: Géométrie. Ed. Belin, Paris (1972)

[3] Ayres, F.: Projective Geometry. McGraw-Hill, New York (1967)

[4] Berger, M.: Geometry I, II. Universitext. Springer, Berlin (2003). Original French edition: Géométrie, vols. 1–5. CEDIC and F. Nathan, Paris (1997)

[5] Beutelspacher, A., Rosenbaum, U.: Projective Geometry. Cambridge University Press, Cambridge (1998)

[6] Cassels, J.W.S.: Rational Quadratic Forms. Monographs of the London Math. Soc., vol. XVI (1978)

[7] Castellet, M., Llerena, I.: Àlgebra lineal i geometria. Collection Manuals of the Autonomous University of Barcelona, vol. 1 (1988)

[8] Cedó, F., Reventós, A.: Geometria plana i àlgebra lineal. Collection Manuals of the Autonomous University of Barcelona, vol. 39 (2004)

[9] Costa, A., Lafuente, J.: Geometrías lineales y grupos de transformaciones. Collection Cuadernos of the UNED (1991)

[10] Cox, D.A., Little, J.B., O'Shea, D.: Ideals, Varieties, and Algorithms. Springer, Berlin (2007)

[11] Coxeter, H.S.M.: Projective Geometry. Springer, Berlin (1987)

[12] Dou, A.: Evolució dels fonaments de la matemàtica i relacions amb la física. Inaugural Lecture of the Course 1987–1988, Autonomous University of Barcelona

[13] Euclid: The Thirteen Books of Euclid's Elements. Dover, New York (1956). Translation and commentary by Sir Thomas L. Heath

[14] Fishback, W.T.: Projective and Euclidean Geometry. Wiley, New York (1969)

A. Reventós Tarrida, *Affine Maps, Euclidean Motions and Quadrics*, 403
Springer Undergraduate Mathematics Series,
DOI 10.1007/978-0-85729-710-5, © Springer-Verlag London Limited 2011

[15] Hartshorne, R.: Foundations of Projective Geometry. Benjamin, New York (1967)

[16] Hausner, M.: A Vector Space Approach to Geometry. Dover, New York (1998). Originally published in Prentice-Hall Mathematics Series, Prentice-Hall, New York (1965)

[17] Hernández, E.: Álgebra y geometría. Addison-Wesley, Autonomous University of Madrid, Reading (1994)

[18] Heyting, A.: Axiomatic Projective Geometry, 1st edn. North-Holland, Amsterdam (1963); 2nd edn. (1980)

[19] Klein, F.: Elementary Mathematics from an Advanced Standpoint: Geometry. Dover, New York (2004)

[20] Lang, S.: Algebra, 3rd edn. Addison-Wesley, Reading (1971)

[21] Nart, E.: Grups abelians finitament generats i formes quadràtiques. Publications of the Autonomous University of Barcelona, Bellaterra (1995)

[22] Prasolov, V.V., Tikhomirov, V.M.: Geometry. Translations of Mathematical Monographs, vol. 200. American Mathematical Society, Providence (2001)

[23] Puig Adam, P.: Curso de geometria métrica, 13th edn. Euler, Madrid (1986). Two volumes

[24] Reventós Tarrida, A.: Geometria axiomàtica. Publications of the Institut d'Estudis Catalans, IEC, CVI (1993)

[25] Reventós Tarrida, A.: Geometría projectiva. Collection Materials of the Autonomous University of Barcelona, vol. 85 (2000)

[26] Reventós Tarrida, A.: Un nou món creat del no-res. Bull. Catalan Soc. Math. **19**, 47–83 (2004). Also: Conferència de Sant Albert, Faculty of Sciences of the Autonomous University of Barcelona, 17 November 2004

[27] Roe, J.: Elementary Geometry. Oxford University Press, London (1993)

[28] Roman, S.: Advanced Linear Algebra. Graduate Texts in Mathematics. Springer, Berlin (2008)

[29] Samuel, P.: Projective Geometry. Springer, Berlin (1988)

[30] Santaló, L.: Geometría proyectiva. Eudeba, Buenos Aires (1966)

[31] Scherk, P., Lingenberg, R.: Rudiments of Plane Affine Geometry. University of Toronto Press, Toronto (1975)

[32] Sidler, J.C.: Géométrie Projective. InterEditions, Paris (1993)

[33] Snapper, E., Troyer, R.J.: Metric Affine Geometry. Academic Press, San Diego (1971)

[34] Szmielew, W.: From Affine to Euclidean Geometry: An Axiomatic Approach. Reidel, Dordrecht (1983)

[35] Teixidor, J.: Geometría analítica. Publications of the University of Barcelona, Barcelona (1967)

[36] Wylie, C.R. Jr.: Introduction to Projective Geometry. McGraw-Hill, New York (1970)

[37] Xambó, S.: Geometria. Publications of the Polytechnical University of Catalunya, Barcelona (2000)

[38] Xambó, S.: Álgebra lineal y geometrías lineales. Eunibar, Barcelona (1994)

Author Index

A. Reventós Tarrida, *Affine Maps, Euclidean Motions and Quadrics*,
Springer Undergraduate Mathematics Series,
DOI 10.1007/978-0-85729-710-5, © Springer-Verlag London Limited 2011

Subject Index

A. Reventós Tarrida, *Affine Maps, Euclidean Motions and Quadrics*,
Springer Undergraduate Mathematics Series,
DOI 10.1007/978-0-85729-710-5, © Springer-Verlag London Limited 2011